Negative Neighbourhood Reputation and Place Attachment

The concept of territorial stigma, as developed in large part by the urban sociologist Loïc Wacquant, contends that certain groups of people are devalued, discredited and tainted by the reputation of the place where they reside.

This book argues that this theory is more relevant and comprehensive than others that have been used to frame and understand ostracised neighbourhoods and their populations (for example, segregation and the racialisation of place) and allows for an inclusive interpretation of the many spatial facets of marginalisation processes. Advancing conceptual understanding of how territorial stigmatisation and its components unfold materially as well as symbolically, this book presents a wide range of case studies from the Global South and Global North, including an examination of recent policy measures that have been applied to deal with the consequences of territorial stigmatisation. It introduces readers to territorial stigmatisation's strategic deployment but also illustrates, in a number of regional contexts, the attachments that residents at times develop for the stigmatised places in which they live and the potential counter-forces that are developed against territorial stigmatisation by a variety of different groups.

Paul Kirkness is a Lecturer in Human Geography at the University of Erlangen-Nürnberg, Germany.

Andreas Tijé-Dra is a Human Geographer at the University of Erlangen-Nürnberg, Germany.

T0174860

Global Urban Studies
Series Editor: Laura Reese

Providing cutting-edge interdisciplinary research on spatial, political, cultural and economic processes and issues in urban areas across the US and the world, volumes in this series examine the global processes that impact and unite urban areas. The organising theme of the book series is the reality that behaviour within and between cities and urban regions must be understood in a larger domestic and international context. An explicitly comparative approach to understanding urban issues and problems allows scholars and students to consider and analyse new ways in which urban areas across different societies and within the same society interact with each other and address a common set of challenges or issues. Books in the series cover topics which are common to urban areas globally, yet illustrate the similarities and differences in conditions, approaches and solutions across the world, such as environment/brownfields, sustainability, health, economic development, culture, governance and national security. In short, the Global Urban Studies book series takes an interdisciplinary approach to emergent urban issues using a global or comparative perspective.

Published:

Governing Urban Regions Through Collaboration
A View from North America
Joël Thibert

From Local Action to Global Networks: Housing the Urban Poor
Edited by Peter Herrle, Astrid Ley and Josefine Fokdal

Cities at Risk
Planning for and Recovering from Natural Disasters
Edited by Pierre Filion, Gary Sands and Mark Skidmore

Negative Neighbourhood Reputation and Place Attachment
The Production and Contestation of Territorial Stigma
Edited by Paul Kirkness and Andreas Tijé-Dra

Negative Neighbourhood Reputation and Place Attachment

The Production and Contestation of Territorial Stigma

Edited by
Paul Kirkness and Andreas Tijé-Dra

Routledge
Taylor & Francis Group

LONDON AND NEW YORK

First published 2017 by Routledge

2 Park Square, Milton Park, Abingdon, Oxon OX14 4RN

605 Third Avenue, New York, NY 10017

Routledge is an imprint of the Taylor & Francis Group, an informa business

First issued in paperback 2021

Publisher's Note

The publisher has gone to great lengths to ensure the quality of this reprint but points out that some imperfections in the original copies may be apparent.

British Library Cataloguing in Publication Data
A catalogue record for this book is available from the British Library

Library of Congress Cataloging in Publication Data
A catalog record for this book has been requested

ISBN: 978-1-4724-7552-7 (hbk)
ISBN: 978-0-367-21882-9 (pbk)

Typeset in Times New Roman
by Swales & Willis Ltd, Exeter, Devon, UK

Contents

vi *Contents*

Illustrations

Figures

Tables

Box

Contributors

Adefemi Adekunle is a researcher whose focus is based around scrutinising young people on their own terms. He is Lecturer at Newman University in the 'Working with Children and Young People' team in Birmingham. As a PhD student at UCL, he worked at the Runnymede Trust, researching the intersection of race, space, youth and identity. His present research interests are based around his work and experiences as a detached youth worker.

Kathy Arthurson is Director of Neighborhoods, Housing and Health at Flinders Research Unit, Flinders University of South Australia. Her past experiences as a senior policy analyst in a range of positions including public health, housing and urban policy are reflected in the nature of her research, which is applied research grounded in broader concepts concerning social inclusion, inequality, and social justice. She is currently completing an Australian Research Council funded Future Fellowship exploring the links between urban planning policy and health.

Everaldo Batista da Costa is Professor at the University of Brasília (Brazil) and holds a PhD in Geography (University of São Paulo, Brazil). He is currently doing postdoctoral research at the University of São Paulo and the National Autonomous University of Mexico. He is particularly interested in urban geography and explores urbanisation and patrimonialisation processes in Latin American cities.

Lee Crookes is a University Teacher in the Department of Urban Studies and Planning at the University of Sheffield. Lee's PhD conceptualised the implementation of the British Labour government's Housing Market Renewal programme as a form of state-led 'gentrification by bulldozer' and, focusing on three urban areas in the north of England, his research examined the impacts of the programme on those households whose homes were targeted for demolition. Having witnessed the harm that planning can do to people and places, Lee has been working on a collaborative research project with the Town and Country Planning Association that seeks to reconnect planning to the lives of ordinary people through the reinvention of the ideas and practice of social town planning. Working closely with marginalised communities, Lee is also committed to engaged scholarship and community-led research.

Michael Darcy is Associate Professor in Social Sciences and a member of the Urban Research Program at Western Sydney University. His research focuses on housing and social justice with a particular emphasis on deeply engaged and activist methodologies.

Klaus Geiselhart PhD is lecturer at the Institute of Geography, University of Erlangen-Nuremberg, where he teaches social geography, urban studies, geographies of health and theory of sciences. His research interest is on stigmatisation and discrimination, public health, traditional medicine and epistemology of medicine. Since 2004, he has been conducting research projects on livelihoods and health-related issues in southern Africa, with a special focus on Botswana.

Christoph Haferburg teaches Human Geography at the Institute of Geography at the University of Erlangen, Germany and is Visiting Associate Professor in the School of Architecture and Planning at the University of the Witwatersrand, South Africa. With a background in urban, political and social geography, his research interests include all aspects of urban development, especially in South and Southern Africa. Informed by social theory, governance and conflict studies, his recent publications discuss applications of theories of practice as well as concepts of cities on a global scale, particularly regarding the societal and spatial aspects of the urban condition.

Marie Huchzermeyer teaches in the School of Architecture and Planning at the University of the Witwatersrand in Johannesburg, where she convenes a Masters degree in housing. She is the current director of the Centre for Urbanism and Built Environment Studies (CUBES). Her research spans urban and housing policy, informality and rights, and she has worked comparatively within Africa and across several continents, examining in particular the formation and treatment of informal settlements and of private rental housing.

Shadia Husseini de Araújo is Professor at the University of Brasília (Brazil). She holds a PhD in Geography (University of Münster, Germany). Her current research interests lie particularly in post- and de-colonial thought, Muslim identities, the global halal market, and diverse economies. More recently, she has also been working on gentrification processes and residential displacement in Brazilian cities.

Hamish Kallin teaches Human Geography at the University of Edinburgh. His work focuses on shifting formations of capitalism and the constant remaking of the Scottish landscape. In particular, his work looks at the intersection between state power, gentrification and urban regeneration.

Paul Kirkness holds a PhD in Geography (The University of Edinburgh). He teaches Urban Geography at the Institute of Geography, Friedrich-Alexander-Universität Erlangen-Nürnberg. He is particularly interested in issues regarding both the production and the contestation of neighbourhood stigma and has also been working on the consequences of urban policies that advocate for the demolition of stigmatised urban areas. He was the network facilitator for Advanced Urban Marginality network and is now correspondence editor for the journal *CITY: analysis of urban trends, culture, theory, policy, action.*

Gaja Maestri is a PhD candidate in Human Geography at Durham University (UK), where she has also worked as teaching assistant in the Department of Geography. She holds an MSc in Political Sociology from the London School of Economics and Political Sciences (2011) and a BA and MA in Sociology from the University of Trento (Italy). Her PhD research, which was funded by the UK Economic and Social Research Council, focused on the role of urban policies in the production of spatial segregation and on strategies of citizenship claiming and political mobilisation for the access to housing.

Pascale Nédélec is Assistant Professor at the Ecole Normale Supérieure (ENS) in Paris. Alumna of the ENS in Lyon, she holds a PhD in Geography (University of Lyon, France). At the crossroads between social and cultural geography, her research focuses on innovative social and political appropriation dynamics and mobilisations in relation with the production of the city; as well as discourse analysis and urban representations, especially normative discourses on the urban experience and construction processes of urban imaginaries. Particularly interested in American cities, her dissertation demonstrated that Las Vegas is fundamentally hybrid; torn between exceptionalism – as a gaming destination – and urban banality.

Lucas Pohl is a PhD student at the Department of Human Geography in Frankfurt am Main, Germany. His PhD project focuses on abandonment and ruination of skyscrapers, so-called 'Ghostscrapers', but he has also worked on urban issues such as homelessness, right-wing extremism and social movements. In general, he is interested in spatial theory, urban politics and philosophy, based on post-Marxist, post-structural and psychoanalytical thought.

Dallas Rogers is a research member of the Institute for Culture and Society at Western Sydney University. His *Global Real Estate Project* investigates the relationships between globalising urban space, discourse and technology networks, and relational examinations of housing poverty and wealth. He publishes on urban and housing issues in academic and industry journals, has appeared in domestic and international media, participated in a parliamentary briefing and contributed to the United Nations Special Rapporteur report on global investment and speculation in land and real estate. Dallas is interested in using digital media techniques to communicate research findings to non-academic audiences. His 2015 radio documentary *Searching for the Mousetribe, in the Confucian City* – a story about the urban poor in Beijing who call underground air-raid bunkers their homes – was part of the 2015 Festival of Urbanism. Please connect with Dallas on social media to share your research or to find out more about the *Global Real Estate Project:* https://dallasrogers.live

Tom Slater is Reader in Urban Geography at the University of Edinburgh. He has research interests in the institutional arrangements producing and reinforcing urban inequalities, and in the ways in which marginalised urban dwellers organise against injustices visited upon them. He has written extensively on gentrification (notably the co-authored books, *Gentrification*, 2008 and *The Gentrification Reader*, 2010), displacement from urban space, territorial stigmatisation, welfare reform, and social movements. Since 2010 he has delivered lectures in 18 different countries on these issues. His work has been translated into eight different languages and circulates widely to inform struggles for urban social justice. For more information, see www.geos.ed.ac.uk/homes/tslater

Andreas Tijé-Dra is a PhD candidate in Geography, at the Institute of Geography, Friedrich-Alexander-Universität Erlangen-Nürnberg. His research areas are centred on Cultural Geography and Social and Political Geography and he is the author of several papers (in German) on discourse analysis, the discursive production of space and the French *banlieues*.

Yunpeng Zhang is a postdoctoral researcher at the University of Leuven, Belgium. He is a member of the REFCOM (Real Estate/Finance Complex) research group, funded by the European Research Council. He has a sustained interest in gentrification, residential displacement, neoliberalisation and popular resistances in contemporary China. His current project explores and compares the mechanisms and processes of financialised urban space production in Shanghai and Hong Kong.

Acknowledgements

The editors of this volume would like to thank everyone that we have been in contact with for the production of this book. Thanks to Katy Crossan, for approaching us at the start, and then those who worked to help us finalise the production: Sophie Iddamalgoda, Priscilla Corbett and Faye Leerink. We would also like to thank each of the wonderful authors who have produced the chapters without which there would be no book. Their work – as well as their thoughts and ideas – has been exceptional. We both would really like to thank Jim Kirkness, who read each chapter attentively, making useful linguistic suggestions for readability and accessibility. Andreas and I are both heavily indebted to him.

Paul Kirkness would like to thank a few extra people:

I would particularly like to thank Tom Slater for all his help over the past few years. Working with him, first for my PhD, then in Amsterdam and for the Advanced Urban Marginality network has been an absolute pleasure and the source of much learning. Lynn Staeheli has also been of invaluable help in the course of my studies and I would never have completed my thesis without her, meaning that this book would never have seen the light of day. Dan Swanton and Bob Catterall, as well as Georg Glasze, Pierre Foucault and Brendan Strady are all to be thanked for their patience and help over the past couple of years, providing me with the time, resources and intellectual challenges that permitted Andreas and myself to move the project of this book forward.

Finally, I would also like to thank those who put up with the extra hours of work during the editing process, Matilda and Freya, my adorable daughters (who probably didn't notice) and Susi, who certainly did! Thank you so much for the patience, the understanding, the help and the unconditional support.

Chapter 11: This chapter draws on material in and excerpts from 'Televised territorial stigma: how social housing tenants experience the fictional media representation of estates in Australia', in *Environment and Planning A 2014*, Volume 46, pp. 1334–1350, © 2015. Permission from SAGE Publications, RightsLink licence No. 3752710992335.

1 Introduction

Exploring urban tainted spaces

Paul Kirkness and Andreas Tijé-Dra

Identifying 'rough' neighbourhoods on Google Maps has become something of a sport for those keen to procrastinate. Several websites will point curious 'virtual tourists' to some of the urban areas in which 'intrepid' Google car drivers have once dared to venture. Identifying a police stand-off or a supposed gang member pointing a gun at the street-car camera are seen as signs that a neighbourhood is a tough one and an otherwise 'no-go area'. A prostitute and her client are identified and used as justification for an area's general vice and depravity. These visuals are spread across a number of websites[1] with titles that explicitly mark urban areas out as dangerous, immoral and inhospitable places.

The tabloid press, in countries such as the United Kingdom, is keen to reiterate the dangers involved in venturing beyond certain urban boundaries and to identify the 'worst neighbourhoods' in a city or a country (Kearns *et al.*, 2013). GIS techniques are put to use in creating 'crime maps' that often have potent material consequences for those living within designated high-crime neighbourhoods (Ratcliffe, 2002). More often than not, the boundaries that are constructed are internalised by those living outside the urban area described as exhibiting traits that are beyond the 'norm'. A telling example of this internalisation is given by the sociologist Azouz Begag when describing a bus journey during which he ventured into a stigmatised *banlieue* to the north of Paris. At some point the driver pointed his finger to the street and Begag describes the scene: 'he was showing me an invisible frontier which he was drawing with his finger, one that only he seemed to be able to see, as if the entrance to a new city' (Begag, 2002: 266). Begag insists that the fracture between the core and its periphery has grown larger in several contexts, to the extent where it is now almost visible, and the fears and fantasies have taken shape on maps. But this in no way applies only to France, as has been demonstrated by a number of researchers in other contexts (see e.g. Arthurson, 2004; Beach and Sernhede, 2011; Cohen, 2013; Fortuijn and van Kempen, 1998; Slater and Anderson, 2012).

Of course, the study of stigma is at the heart of several studies of social psychology, but it was also highlighted as a sociological issue in the seminal work of Erving Goffman (1963). There he argued that individuals become disqualified from society because of 'abominations of the body' (such as physical disabilities), 'blemishes of the individual character' (addiction or unemployment, for instance) and 'the tribal stigma of race, nation and religion', which can be passed down from generation to generation. It is Loïc Wacquant (2007) who adds territorial stigma to this list, arguing that it too, leads to individuals being 'discredited' and 'disqualified'. Certainly, the examples with which we have begun the introduction to this edited book point to the existence and persistence of stigmatised places within our cities. These have significant material and structural consequences for those residing in these areas, and this has most notably been demonstrated in a number of works by Wacquant (e.g. 2007, 2008). These range from the impossibility of finding a job as a consequence of address discrimination, to the impossibility of obtaining mortgage credits, to racialisation (Kirkness, 2013; Slater and Anderson, 2012) and limits access to health services (Pearce, 2012). Fire brigades and ambulances often tread cautiously when entering such stigmatised neighbourhoods.[2] There are also more mundane – but no less significant – consequences, such as the impossibility of ordering a taxi or having food delivered to one's home (see Slater, in press, p. 9). These may well be less significant than the hindrance that one's address can have in terms of obtaining employment, but they have the potential to become additional reasons to develop shame and frustration.

The examples that we have listed provide elements that feed the fascination that stigmatised neighbourhoods exert on those who know little or nothing about them. The media, political discourse and everyday conversations may well participate in forging representations about what these neighbourhoods are like in the collective imaginary. As a result of these stigmatic images, several areas are framed as 'no-go areas', limiting the interactions between residents and non-residents within the neighbourhood. Such ignorance makes research into neighbourhood stigma all the more vital and this book is one attempt to move forwards in this direction.

The fact that territorial stigmatisation has become recognised as, not only a serious structural problem, but also as an object of social scientific inquiry is in large part due to Wacquant's important work on the topic. As is now well known, his research has focused on comparative research between a stigmatised housing estate in the French suburban town of La Courneuve and a heavily defamed ghetto in the city of Chicago. Both these neighbourhoods, 'la cité des 4000' and Chicago's so-called 'South Side', have achieved 'the status of national eponym for all the evils and dangers now believed to afflict the dualized city' (Wacquant, 2007, p. 67). As the editors of this volume on the topic of territorial stigma, we are deeply indebted to Wacquant for opening up an avenue that has allowed the authors of this book, as well as many others throughout the world, to focus on the enduring problem and the structural consequences of negative place reputations. All the chapters in this volume relate in several ways to the work that Wacquant has conducted. It is worth noting, however, that much of the ethnographic research on which Wacquant's work is based was conducted several years ago. Much of the data on which his principal works are founded were gathered in the 1990s (see Wacquant, 1992). As such, the chapters in this book hope to constitute a development in the research on territorial stigma, as well as to update the ways in which it is endured and contested around the world.

Wacquant has not devoted much writing to the contestations of territorial stigmatisation. This has, however, emerged in a number of recent pieces of academic inquiry, pointing to the fact that this is an important part of the process of territorial stigmatisation. These show that it is resisted, rejected, appropriated or merely ignored. Slater and Anderson (2012) found that residents of St Paul's, in Bristol, were keen to deflect the stigmatising imagery that has been attached to the neighbourhood for decades. They found that little or none of the 'common knowledge' pertaining to St Paul's corresponded to the lived experiences of residents. In Aalborg East, a deprived neighbourhood in the north of Denmark, Jensen and Christensen (2012) found that, at worst, residents were ambivalent about the urban area and that they did not seem to internalise the negative reputations. They were fully aware of the problems that one had to negotiate as a result of being a resident of the neighbourhood, but they found several ways to cope with this and some described their sentiments of attachment to the place. Garbin and Millington (2012) returned to the neighbourhood of 'les 4000' that had been the subject of Wacquant's research in the 1990s. What they found is that contestation lived side by side with feelings of fear and the internalisation of neighbourhood stigma. Kirkness (2014) describes the mundane ways in which residents of two of Nîmes' so-called 'sensitive urban zones' have learnt to cope with stigma and move beyond the consequences that it imposes on their everyday lives. And there are plenty of other examples of research pointing to the fact that submissive strategies of internalisation of stigma are not the only response to negative place reputations (see Purdy, 2003; McKenzie, 2012). This edition provides both a synopsis as well as a further step towards the synthesis of current approaches heading in this direction.

Contestation is one thing, and learning to cope with life in stigmatised neighbourhoods is another. Wacquant, Slater and Pereira (2014: p.7) develop a useful list of the ways in which stigma can be coped with, dividing these strategies between those that are somewhat submissive and those that can be perceived as recalcitrant or resistive in some sense. In the former, the authors classify such tactics as blaming one's neighbour for the poor reputation of the neighbourhood, or withdrawing into the private space of the home. Resistive strategies involve defending the neighbourhood and its residents but the authors also note that they might involve the hyperbolic claims of belonging to a certain space. Lisa McKenzie (2012), in her study of St Ann's, in Liverpool, noted that residents at times used the term 'ghetto' to describe where they lived. When they did so, it was with a form of pride and belonging, in a sense a way to shine a positive light on the neighbourhood. She also notes the very strong feelings of 'being and belonging' to the area among the residents. Some of Garbin and Millington's respondents argued that they had a 'right to be other' (2012: p. 2075).

Although resistance to stigmatic representations does exist in a number of neighbourhoods around the globe, the fact that some people are capable of feeling place attachments to areas that are deemed to be threatening by anybody outside the neighbourhood is an important step towards the negation of the power of stigma.

The importance of the fact that stigma is not experienced in the same way or constituted in similar ways across regions and neighbourhoods is given a perfect illustration in Pascale Nédelec's chapter, the first of this volume. She shows that stigma is not reserved for the socially and economically deprived neighbourhoods of our cities, but that it can be activated for a host of different reasons. Her study of territorial stigmatisation in Las Vegas, otherwise known as 'Sin City', is an intriguing account of the way in which, even in relatively wealthy urban areas, territorial stigma can hinder place attachment. This is very unlike the ways in which some attachments are developed by residents of poorer neighbourhoods, according to a number of accounts. In such places, residents have had to make do with less, which can be seen as something that they can ultimately be proud of. Importantly, in such contexts, social networks are often very useful, thus adding feelings of attachment to community and place (see McKenzie, 2015).

Lucas Pohl's chapter focuses on the city of Zwickau, in which the far-right terror group known as the *Nationalsozialistischer Untergrund* (National Socialist Underground) had taken up residence. In the final days before the group was discovered, the house where they were said to have lived was set on fire and two out of the three were found shot in a burning mobile home. The town's reputation has become deeply intertwined with the events, creating a certain type of stigma that the city has attempted to cope with, articulating a number of tropes in order to do so. This has everything to do with image policies countering the logic of uncertain reputations.

Territorial stigma can also become intertwined with ethnic categorising. This is not unheard of in the context of the US ghetto, where separating the racialisation that afflicts African- Americans and the stigmatisation that is imposed onto 'notorious ghettos' is all but impossible. Gaja Maestri's chapter shows that the stigma of belonging to Roma communities, both in France and in Italy, is imposed onto the land where these communities reside. Thus, whether the Roma live in so-called illegitimate settlements, settlements that are established by municipalities, or whether they end up living in squats, Roma stigma is ultimately territorialised.

Christoph Haferburg and Marie Huchzermeyer have written about one of the unfortunate (and in some countries, illegal) actions taken by real estate promoters and developers: redlining. Their chapter focuses on the neighbourhood of Brixton, in Johannesburg, South Africa, in which redlining was pursued as a result of territorial stigmatisation. As the authors explain, Brixton suffered from stigma for so long that loans and insurance were denied to those residing there. When these were not denied, the cost of borrowing was otherwise selectively raised as a result of the area's reputation. The authors focus on the ways in which some residents have been able to contest the practice of redlining through collective action.

The first four chapters of this book thus look at the symbolic and material practices of coping with stigma. In the second main section of the book, we look at the demonstrations of place attachment that prevail in the face of 'common sense' understandings. One of these, which is often put forward by urban policymakers, is that residents would be better off were they to live elsewhere and away from stigmatised neighbourhoods. The two chapters by Lee Crookes and

Hamish Kallin offer accounts of urban renewal programmes that have had the result of reproducing territorial stigmatisation with the somewhat complicit participation of the urban real estate market. They thus focus their attention on the links between state-led (or state-sanctioned) gentrification and territorial stigmatisation. Crookes focuses on the extensive housing clearance programme in northern England, known as Housing Market Renewal (HMR) and Kallin has conducted extensive qualitative research in Craigmillar, in Edinburgh. In his chapter, he coins the fascinating concept of the 'reputational gap', thus furthering Neil Smith's famous work on the 'rent gap' (Smith, 1987).

In the seventh chapter, we show that there is at times a strong sense of belonging in stigmatised neighbourhoods in France and that these are ignored by paternalistic urban policies. When it is articulated within the lyrics of a rapper, place attachment to a stigmatised neighbourhood is considered to be an artist's own personal experience, and one that clashes with the descriptions of dilapidation and concrete jungles that can be found in broader discourses. We have examined rap music from artists that reside in Marseille and the purpose of the chapter has been to confront their vivid descriptions of belonging – both to the neighbourhood and the community residing there – to ethnographic fieldwork conducted in housing estates, with 'ordinary' residents in the town of Nîmes.

Our point is to highlight the everyday appropriations of territorialised space and to confront these to recent policy making decisions that are being taken in the French context, much of which have tended to promote the demolition of social housing neighbourhoods. In China, place attachments are also ignored when confronted with the high-speed megaprojects that cities and state provinces establish. Yunpeng Zhang explores these in the context of Shanghai's Great Expo in the Chapter 8. The 'domicides' that have been occurring as a result of Shanghai's Expo concerned housing that has been discredited by mainstream discourses, which has further legitimised the eradication of entire neighbourhoods. Sometimes for a number of generations, these homes had been made and remade, and their residents had become used to imagining the future of these structures before they were often forcefully ripped away from them. Similarly, Brazil's *favelas* have seen what might have been perceived as temporary housing become increasingly built up and formalised. Concrete, bricks and stone have, in many areas, replaced the corrugated tin roofs and the temporary walls, paving the way to the establishment of lasting neighbourhoods. The Varjão, a stigmatised Brasília neighbourhood, is one such urban area, which is discussed at length in Shadia Husseini de Araújo and Everaldo Batista da Costa's chapter. Among other things, their contribution describes the continuous resident struggles that have partly led to the legalisation of settlements, but it also highlights the different outcomes of participative urban policies and urban politics.

The book's final main section focuses on a number of methodological reflections revolving around contemporary geographical research that has examined the stigmatisation of urban space. These are divided between chapters that deal more explicitly with conceptual concerns (Chapters 13 and 14) as opposed to the chapters that delve explicitly into methods and practical research (Chapters 11 and 12). In the latter group, Dallas Rogers, Kathy Arthurson and Michael Darcy have brilliantly succeeded in redressing the boundary between 'researcher' and 'the researched', in a housing estate located near Sydney. They have, in a sense, turned a number of residents from the estate into social researchers who attempt to understand their neighbourhood's reputation. This technique is one that Adefemi Adekunle has also undertaken, when he enrolled a number of 'young' residents of the neighbourhood that he has studied: Islington, within the city of London. His background in social work has enabled him to develop strong relationships with the youths that are at the heart of his concerns. His work uncovers what stigma means to the young people who reside within stigmatised neighbourhoods and, in some way, it also highlights the ways in which youths also construct categories of threatening spaces. It provides an insightful analysis into sociology, as well as to the social psychology of growing up in a potentially stigmatised neighbourhood. Being both young, and originating from places with bad reputations, has meant that these youths are doubly stigmatised. Adekunle's work has uncovered the emotional geographies of these research participants: their fears, their concerns, their needs and desires.

Adekunle does not refer to territorial stigmatisation at length, and nor does he need to. The spatial categories that the youths with which he works use to construct the city are often not as stable as those applying to so-called ghettos or *banlieues*. The labels and the ways in which the city is experienced is mobile and transitory, and it operates differently according to the time of day. For Klaus Geiselhart, territorial stigmatisation, as defined by Loïc Wacquant, is in and of itself problematic. He astutely differentiates between the terms 'stigmatisation' and 'discrimination' to argue that what we are talking about when we discuss territorial stigma is in fact a form of 'territory-ism', operating in similar ways to racism or sexism.

Finally, we finish the section with an interview with Tom Slater, who is arguably one of the foremost thinkers on issues of territorial stigmatisation – as well as urban geography more generally. His responses to our questions are powerful in that they move beyond strict academic rules and have allowed Slater to speak openly about a host of issues linked to his own research on territorial stigmatisation and its links to other major issues in urban studies. His critical thinking about the role of the university is enlightening in many ways. He reminds us that the role of university is to encourage critical thinking and quotes Stuart Hall when he says that, 'The university is a critical institution or it is nothing.'

Territorial stigmatisation is clearly a concept that is still in need of much scrutiny and this book moves us in that general direction. We hope that the chapters in the volume will allow for critical discussions to progress theoretically and methodologically. There is still much to be learnt about how places and their residents are devalued. We hope that this book will draw attention to the fact that, if stigma is indeed produced throughout the globe with dramatic consequences, it is at times contested and negotiated, opening up a small window of hope that suggests that there are ways to tackle stigma and that we can collectively learn from these.

Notes

1 See http://www.neighborhoodscout.com/neighborhoods/crime-rates/25-most-dangerous-neighborhoods/ or http://www.dailymail.co.uk/news/article-2422963/Bad-Streetview-Wrong-tracks-revealed-Googles-roaming-camera-cars.html
2 In the city of Nîmes, a French city in southern France, as a result of a robbery emergency doctors refused to visit a number of neighbourhoods after nightfall, with potentially very serious consequences for residents in need of medical help (Boudes, 2012).

References

Arthurson, K. (2004), 'From stigma to demolition: Australian debates about housing and social exclusion', *Journal of Housing and the Built Environment*, 19: 255–270.

Beach, D. and Sernhede, O. (2011), 'From learning to labour to learning for marginality: School segregation and marginalization in Swedish suburbs', *British Journal of Sociology of Education*, 32(2): 257–274.

Begag, A. (2002), 'Frontières géographiques et barrières sociales dans les quartiers de banlieue', *Annales de Géographie*, 111(625): 265–284.

Boudes, R. (2012), 'Un docteur de SOS Médecin dépouillé de nuit à Valdegour', *Midi Libre*, Available online at: http://www.midilibre.fr/2012/10/24/un-docteur-de-sos-medecins-depouille-de-nuit-a-valdegour,583321.php (accessed 30 May 2016).

Cohen, N. (2013), 'Territorial stigma formation in the Israeli city of Bat Yam, 1950–1983: Planning, people and practice', *Journal of Historical Geography*, 39(1): 113–124.

Fortuijn, J. D. and van Kempen, E. (1998), 'Territoriale stigmatisering in Nederlandse Steden', *Geografie*, 7:32–34.

Garbin, D. and Millington, G. (2012), 'Territorial stigma and the politics of resistance in a Parisian *banlieue*: La Courneuve and beyond', *Urban Studies*, 49(10): 2067–2083.

Goffman, E. (1963), *Stigma: Notes on the management of spoiled identity*, London: Penguin Books.

Jensen, S. Q. and Christensen, A.-D. (2012), 'Territorial stigmatization and local belonging: A study of the Danish neighbourhood Aalborg East', *City: analysis of urban trends, culture, theory, policy, action*, 16(1–2): 74–92.

Kearns, A., Kearns, O. and Lawson, L. (2013), 'Notorious places: Image, reputation, stigma. The role of newspapers in area reputations for social housing estates', *Housing Studies*, 28(4): 579–598.

Kirkness, P. (2013), 'Territorial Stigmatisation of French Housing Estates: From Internalisation to Coping with Stigma', unpublished PhD thesis, University of Edinburgh.

Kirkness, P. (2014), 'The *cités* strike back: Restive responses to territorial taint in the French banlieues', *Environment and Planning A*, 46(6): 1281–1296.

McKenzie, L. (2012), 'A narrative from the inside, studying St Ann's in Nottingham: Belonging, continuity and change', *The Sociological Review*, 60: 457–475.

McKenzie, L. (2015), *Getting By: Estates, Class and Culture in Austerity Britain*, Bristol: Policy Press.

Pearce, J. (2012), 'The "blemish of place": Stigma, geography and health inequalities', *Social Science and Medicine*, 75(11): 1921–1924.

Purdy, S. (2003), '"Ripped off" by the system: Housing policy, poverty, and territorial stigmatization in Regent Park Housing Project, 1951–1991', *Labour/Le Travail*, 52: 45–108.

Ratcliffe, J. H. (2002), 'Damned if you don't, damned if you do: Crime mapping and its implications for the real world', *Policing and Society*, 12(3): 211–225.

Slater, T. and Anderson, N. (2012), 'The reputational ghetto: Territorial stigmatization in St Paul's, Bristol', *Transactions of the Institute of British Geographers*, 37(4): 530–546.

Slater, T. (in press), 'Territorial stigmatization: Symbolic defamation and the contemporary metropolis', in J. Hannigan and G. Richards (eds) *The Handbook of New Urban Studies*, London: Sage.

Smith, N. (1987), 'Gentrification and the rent gap', *Annals of the Association of American Geographers*, 77(3): 462–465.

Wacquant, L. (1992), 'Banlieues françaises et ghetto noir américain: de l'amalgame à la comparaison', *French Politics and Society*, 10(4): 81–103.

Wacquant, L. (2007), 'Territorial stigmatization in the age of advanced marginality', *Thesis Eleven* 91(1): 66–77.

Wacquant, L. (2008), *Urban Outcasts: A Comparative Sociology of Urban Marginality*, Cambridge: Polity Press.

Wacquant, L., Slater, T. and Pereira, V. B. (2014), 'Territorial stigmatisation in action', *Environment and Planning A*, 46(6): 1–11.

2 The stigmatisation of Las Vegas and its inhabitants

The other side of the coin

Pascale Nédélec

Introduction

Three buddies wake up from a bachelor party in Las Vegas, with no memory of the previous night and the bachelor missing. They make their way around the city in order to find their friend before his wedding.[1] This is the synopsis of the American blockbuster, *The Hangover*, released in 2009, whose huge popular success anchored even deeper in the public opinion, not only in the United States but also in the rest of the world, what is to be expected when visiting – that is, partying in – Las Vegas.

The city of Las Vegas is indeed a world-renowned and iconic entertainment centre that conjures up images of some of the world's largest hotel-casinos and their exuberant architecture, a free-wheeling, anything goes atmosphere, as well as intense partying. Undoubtedly, the rise of Las Vegas as the gaming capital of the world is an emblematic success that has brought economic wealth and demographic growth to the entire urban area. In less than half a century, tourist visitation was multiplied by six (1970[2]–2014), reaching a record number of 41.12 million visitors in 2014. That same year, tourism supported 43 per cent of jobs in Southern Nevada (LVCVA, 2015). The population of the urbanised area of Las Vegas grew proportionally with the rise of tourism, having been the fastest-growing metropolis in the United States between 1990 and 2005,[3] with a total of 2,069,681 million people calling Las Vegas home in 2014 (Census Bureau, 2015). In the public opinion, Las Vegas is thus associated with the tourist enclaves of the Strip and Fremont Street that concentrate the mega hotel-casino resorts and other tourist attractions. However, these 'iconic' parts of the city only represent a fraction of the entire urban area (Nédélec, 2012). The spatial disconnection between the overbearing weight of those enclaves as compared to the vast urbanised area foreshadows the influence of tourism on local inhabitants. Despite the very popular saying in the United States, what happens in Las Vegas *does not* stay in Las Vegas. The marketing of a city that became a national – if only symbolic – outlet for all earthly pleasures and forms of entertainment has encouraged the rise of stereotypes that portray Vergas as a loose and sinful destination in which gamblers, mobsters and prostitutes are (supposedly) the norm.

This chapter's argument rests upon the following question: what are the implications of Las Vegas' aggressive self-marketing as 'Sin City' for the local population?

Could the received wisdom about this wild tourist destination have direct and profound consequences on the sense of belonging and sense of community for local residents, and therefore on place attachment more generally? These initial questions lead to the main hypothesis of the chapter: Las Vegas' economic success has resulted in a paradoxical situation; that is, the stigmatisation of the city and its residents. The negative reputation of Las Vegas is studied here through the lens of stigma, in line with the seminal work of Goffman (1963). This chapter intends to highlight the historical and social construction of what I would call 'Las Vegas stigma', intrinsically intertwined with the popularisation of gambling and the rise of tourism. Stigma cannot only be understood as a static condition, but as the result of a social construct within a given society and historical period. The process of stigmatisation, interweaving that of the city and of its population, is deemed especially interesting in that it theorises the 'contamination' from the city's popular representations to its population. Of note, the study is not limited to specific neighbourhoods but relates to the entire Las Vegas urban area.[4] What makes Las Vegas' case so original and important to study is that stigmatisation does not overlap with the relegation of socially and economically challenged neighbourhoods – even though the urban area is not deprived of such spaces – but that it stems from the incredible success of tourism, hence the idea of 'the other side of the coin' that was suggested in the title of this chapter. By focusing on Las Vegas, my goal is to offer another perspective on spatial and place attachment, contrasting with the dominating studies on social or racial discrimination, segregated neighbourhoods and margins of society (Wacquant, 2007; Slater *et al.*, 2014).

The chapter is based on an in-depth study of the urban representations surrounding Las Vegas, and a fuller characterisation of the identity construction of Las Vegans (Nédélec, 2013). In particular, it relies on semi-structured interviews with Las Vegans from all walks of life, analysed in combination with large-scale statistical surveys (Harwood and Freeman, 2004; Futrell *et al.*, 2010). Qualitative and quantitative data are used to support the general demonstration.

Starting with a presentation of the scientific literature on stigma and stigmatisation, the first part of this chapter identifies the genesis of Las Vegas stigma. I shall then discuss the process that enables the stigmatisation of so-called Sin City to affect ordinary Las Vegans. Finally, this chapter draws out the implications of the stigmatisation process for the identity construction of the local population, showing what it is like for Las Vegans to live with stigma and how they strategically (or tactically) learn to cope with it.

The genesis of Las Vegas stigma

Stigmatisation and urban environments

What do we mean when we state that a neighbourhood has a 'bad reputation', something that this volume wishes to address in light of its title? Looking at definitions from the *Oxford English Dictionary* gives a working base. 'Reputation' is there defined as 'A widespread belief that someone or something has a particular characteristic.' When thinking about negative reputation, the term stigma comes to mind as a possible catalyst; a stigma being defined as 'A mark of disgrace associated with a particular circumstance, quality, or person.' Therefore, stigma leads to 'stigmatising', which is itself defined as to 'Describe or regard as worthy of disgrace or great disapproval', and is a synonym for 'disparage', 'vilify', 'pour scorn on' and 'discredit' (OED, 2015). The work of the sociologist Erving Goffman (1963) explains and opens up articulations of negative reputation, stigma and stig-matisation, revealing with precision the causal links between state and process.

Following Goffman (1963), an original stigma is defined as a mark of abnor-mality, a social construct that is constantly reorganising itself under the influence of the evolutions of a society's cultural customs and social mores, supporting stigmatisation processes. Even though Goffman's analyses deal essentially with physical stigma, and all sorts of disabilities, they are transposable to cities insofar as stigma is defined as a 'discrediting differentness', expressing a deviation from the 'normative expectations' erected by a given society. In the case of cities, col-lective representations fastened on place and a set of certain urban characteristics gradually evolve into a blemish or 'stained identity' discrediting a specific locale.

The 'discrediting differentness' that is at the source of stigma stems from spe-cific historical events, as illustrated by the works of historian Johnathan Foster (2009, 2012). Historical events and representations of those events stand at the core of the stigmatisation process in that they are appropriated by the media, the opinion setters and then the public opinion as character settings for an entire city. One historical 'anecdote' is gradually transformed into the main commonly known aspect of an urban area, dominating everything else. For example, the 16th Street Baptist Church bombing in 1963, an act of white supremacist terrorism, led to the stigmatisation of Birmingham, Alabama, as a racist city (ibid.). Stereotypes fuel this process, since they determine the analysis and perception of our sur-rounding environment. As stated by Lippmann ([1922] 2012):

> The subtlest and most pervasive of all influences are those which create and maintain the repertory of stereotypes. We are told about the world before we see it. We imagine most things before we experience them. And those pre-conceptions, unless education has made us acutely aware, govern deeply the whole process of perception.
>
> (Lippmann, 2012: p. 49)

According to Lippmann, in order to deal with the increasingly growing quan-tity of available information in our daily life, generalisations, and their extended stereotypes, tend to reduce singularities via a shortcut process, thus allowing a reduction of the quantity of necessary information needed to reach a conclusion.

From these 'subtl[e] influences' stem our understanding of a city's stigma: a par-ticular circumstance at a particular historical period is erected as the essential and inescapable characteristic of a place. Since the word stigma has negative overtones, we focus here on the selection of depreciatory or simplistic characteristics, echoing the 'disgrace' mentioned in dictionaries. The driving forces of this very selective process of information hierarchisation, which has to do with power relations, are composed of the media, opinion setters and public officials, relayed by public opinion in general. This selected particular circumstance sets up the way in which people perceive a place and what they then come to expect from it, building restrictive and at times negative expectations. As a result, a stigmatised city enters a vicious circle where the 'normative expectations' theorised by Goffman are erected as the norm for a place, even if based on a depreciative and partial understanding of the city's history.

Therefore, the importance of the stigmatisation process coincides with the strength of stereotypes and clichés resulting from stigmas. Whereas there has thankfully been a welcome evolution of mentalities and a strong political will to challenge racial stereotypes in many societies, spatial stereotypes still pervade contemporary perceptions of the world. The history of American intellectualism, studied by philosophers Morton and Lucia White who 'examine[d] the intellectual roots of anti-urbanism and ambivalence toward urban life in America' (White and White, 1962, p. 16), has shown a strong tradition of denigrating urban environments as places of moral depravation. This led to cities being considered at the heart of all contemporary evils, in opposition to the countryside, where one could live in symbiosis with nature and its redeeming properties.

Sociologists have been especially keen to demonstrate this process in urban environments, giving birth to the notion of 'territorial stigmatisation', popularised by Wacquant (2007; Slater *et al.*, 2014) and combining Goffman's (1963) take on stigma and Bourdieu's (1991) theories of symbolic power and 'site effects' (1993).

In the case of Las Vegas, urban stigmatisation does not rely upon one generic spatial category, as illustrated by the ghetto or the '*banlieues*' in the French context (see Kirkness, 2014). Nor does it relate to the stigma that has attached itself to the industrial or the Soviet city. In fact, it is anchored in the history of this particular place: a place that went from isolated desert outpost to 'Gambling Mecca' and the self-proclaimed 'Entertainment Capital of the World'.

The rise of Las Vegas as a gambling destination

The process of Las Vegas territorial stigmatisation cannot be understood without taking into consideration the driving forces that explain the rise of this tourist destination. Going back to the origins of Las Vegas allows us to identify the specific historical events that have shaped its stigmatisation. It is a very specific period in Las Vegas' history that has determined the perception of the city in the American and the world public opinion, setting the roots for a moral condemnation that has constantly been reinforced to this day.

Established along the Old Spanish Trail, set up by Spanish explorers[5] to connect settlements in New Mexico to the Californian Coast (Moehring and Green, 2005), the site that was to become Las Vegas was initially chosen by Mormon missionaries between 1855–1857, to convert the indigenous Paiute tribes settled in the area. The place became known as Las Vegas, meaning 'the Meadows' in Spanish, and it was a providential stop-over en route to the Pacific side of the continent, in large part because of its artesian wells, providing vital water in the especially harsh environment of the Mojave Desert. After the Mormon outpost's failure, the site ceased to grow until promoters of transcontinental railroads realised the potential offered by its strategic location. Therefore, the official foundation of Las Vegas in 1905, and its incorporation as a city, on 16 March 1911, stem from the completion of the San Pedro, Los Angeles and Salt Lake City railroad in January 1905 and the subsequent construction of a train station (Moehring and Green, 2005). The railroad ensured the economic prosperity of the small town until the 1930s. Yet, Las Vegas' historical past as a railroad town, or even its water wealth, is not the first aspect of the city that comes to mind when it is being discussed today.

The rise of Las Vegas as a tourist destination started in the 1930s, as a direct consequence of the official legalisation of gambling in Nevada in 1931 (Roske, 1990; Moody, 1994), which made this state the only one in the United States where money games were legal. Due to this political decision, for almost 50 years Las Vegas, and to a lesser extent Reno, enjoyed a strict monopoly on legal gambling. The privilege ended in 1978 with the successive legalisation of gambling, firstly in New Jersey, and then in almost every single state in the United States.[6] Notably based on the generalisation of casinos within Indian territories, gambling gradually lost its subversive nature and became a commonplace leisure activity (Findlay, 1986; Raento, 2003; Schwartz, 2006).

If gambling is no longer a privilege of Nevada, Las Vegas has become associated with customs and habits that, if not legally forbidden, are at least morally condemned elsewhere in America. Las Vegas' permissiveness does stand out in the United States, with its quick weddings and divorces and the freedom to drink alcohol on the streets in tourist neighbourhoods (Moehring and Green, 2005). The stereotype according to which one can still do in Vegas what is considered reprehensible elsewhere dominates all representations: it is for example commonly assumed that prostitution is legal in Las Vegas (Brents *et al.*, 2010). In fact, according to the legislation of Nevada,[7] prostitution can only be legal in counties of less than 400,000 inhabitants, which excludes Clark County where Las Vegas stands.

In Vegas, it is imagined that one can escape the social and moral norms that are in place elsewhere. It is assumed that one can party, drink excessive quantities of alcohol, forget all social manners and respectability, and do this without consequences – in theory. It is also commonly believed that one can engage easily in sexual encounters. The 'Las Vegas experience', as it is imagined, is perfectly embodied in the above-mentioned film, *The Hangover* (2009).

To understand the transition from historical characteristics to stereotypes about the 'Las Vegas experience', one has to question the underlying driving force behind the spread of clichés about the city: the local tourism industry.

The paradoxical success of the Las Vegas stigma

When tourism first emerged in the 1930s and 1940s, the main actors of the local tourism industry had marketed Las Vegas as a 'Frontier Town', with the first casinos of the Strip replicating dude ranches and enacting scenes from the Wild West. Already, alcohol and money games were at the heart of everything (Gragg, 2013).

With the banalisation of gambling nationwide, the local tourism promoters feared that Las Vegas had lost its appeal and clientele, thus weakening the first economic sector of the city. Tourist marketing thus changed focus and began to emphasise the overall atmosphere of liberty and licentiousness in order to attract people long-ing for some respite from their puritan and corseted homes. In the 2000s, at the instigation of the Las Vegas Convention and Visitors Authority, Las Vegas was mostly branded around the notion of 'adult freedom'. Embodied by the emblem-atic slogans 'Sin City' or 'what happens in Vegas, stays in Vegas' (Friess, 2004), the local tourism industry has heavily played into the wild card of Las Vegas being a free-wheeling, anything goes kind of place.

With such successful ad campaigns broadcast nationally, the Las Vegas tour-ism industry fashioned and crystallised what I call 'tourism imaginaries', defined as 'a collection of images and representations initially produced and used within the tourism sphere and then spread to the rest of society', leading to the progres-sive association of values with specific places (Nédélec, in press). What used to be mere branding materials are gradually appropriated, and most importantly, internalised by the media, opinion setters and ultimately by the public opinion, as social constructs and valued representations. They shape a larger 'social imagi-nary' (Zukin *et al.*, 1998) that lies at the heart of the stigmatisation of Las Vegas. The tourism imaginaries that made the success of Las Vegas as a tourist desti-nation, became an unavoidable and even expected substrate for any analysis or commentary about the city, as a local journalist explains humorously:

> Though it's not against the law to gamble here or stay up past a decent hour or indulge yourself in ways you'd never dream of back home, it's also not against the law to refrain from these earthly pursuits. But then what would our visitors have to report to the boys at the bank or the women in the bridge club [. . .]? The folks in Abilene want the unfiltered dirt, and it's the returning tourists' job to dish it out.
>
> (Sheehan, 2008)

Hence, a vicious circle is created: to sell the destination to tourists, Las Vegas is mar-keted as a heaven for earthly pleasures, a place to escape from society's 'normative expectations' of good and respectable behaviour. This branding strategy constantly entertains the idea that Las Vegas is a place without morals, a representation that is reinforced through popular media and artistic productions (Eumann, 2005; Gragg, 2013) which affect the public opinion. Those peculiar tourism imaginaries have merged with the overall imaginaries of Las Vegas as the city of sin. Moral condem-nation of Las Vegas, both in the public opinion and the intellectual sphere (Nédélec, in press), appears to be a side effect of this circular spread of imaginaries. In the end, the perpetual evocation of Las Vegas stigma, as a perverted and sinful place, explains the 'stained identity' coined by Goffman (1963) and the negative reputa-tions that feed into the Las Vegas stigmatisation process.

Las Vegas negative stigma may have deeper material consequences than urban policies or urban projects: its perceptions and representations have direct impacts on identity construction and consolidation – or the lack thereof – among the local Las Vegan population.

From the stigmatisation of the city to the stigmatisation of its inhabitants

Confusing tourism marketing and daily life in Las Vegas

Since Las Vegas is known for indulgent behavior, how can it be deemed suitable to accommodate honest people? Surely no law-abiding, honest citizen would establish residence in such a place? This line of thought supports and explains how the stigma of a city led to the stigmatisation of its inhabitants. Goffman (1963) theorises three types of stigma. The third form of stigmatisation that he identifies is one that he calls 'tribal stigma'.[8] These stigmas are associated with an individual and 'can be transmitted through lineages and equally contaminate all members of a family' (Goffman, 1963: p. 14). There is some logical link here to territorial stigmatisation, where 'the family' might be expanded and come to designate a local population of a neighbourhood or that of an entire city. The most interesting idea here is the notion of 'contamination' and the idea of the propagation of stigmatisation among a social group.

The interweaving of city stigmatisation and that of its inhabitants is anchored in the confusion between tourism imaginaries and the reality of the everyday lives of Las Vegans. The most explicit example of the amalgamation between the tourism industry and the local population is illustrated by what a local journalist coined the 'Silly Vegas Questions' (Reza, 2013), designing the kind of questions that everyone who grew up in Las Vegas had to answer:

> 'Do you live in a hotel? Is Wayne Newton your neighbor? Which casino does your mom work in? Is she a showgirl? A hooker? A stripper? Is your dad in the Mafia? Is he a gambler? Do you eat in casinos? Where do you play outside?' When I was a little Las Vegan, any trip outside Nevada would subject me to this litany of Silly Vegas Questions, and many more like them.
>
> (Reza, 2013)

Both in the local press and in the interviews that were conducted during this research (Nédélec, 2013), every single Las Vegan has had to face incredulity and the inability for outsiders to grasp the idea of 'living in Las Vegas'. The value judgments and moral condemnations that dominate analyses dealing with Las Vegas are transferred onto the local population, challenging the line of work and overall respectability of people living there. Two of the respondents expressed this process perfectly:

> When you travel and you say to people you're from Las Vegas, it got this kind of mystique. Everyone assumes you work in a casino or that you're a showgirl.
>
> (Interview #18[9] – 43-year-old female)

People were really curious: 'Where do you live?'. Because people didn't realise that there was anything beyond just the immediate downtown and the Strip. They didn't know that there was anything. 'Where do you go to school?' They had no concept, because they've never gone out and seen it.

(Interview #7 – 66-year-old female)

The contamination of Las Vegas stigma to its inhabitants shaped a 'stained identity' that Las Vegans have to struggle with on a daily basis.

The 'stained identity' of Las Vegans

If the idea that the 2 million people residing in Las Vegas live in hotel-casinos may appear laughable to some, the propagation of Las Vegas stigma takes more serious forms in the discrediting of the local population. The inhabitants that I interviewed (Nédélec, 2013) most often expressed annoyance at negative representations created outside the city, and were systematically frustrated by the need to justify to outsiders that their residing in Las Vegas does not lead to them living any differently than other 'regular people'. The following quotes demonstrate the gap between the perceived vision of the people of Sin City and the mundane everyday life experienced by Las Vegans:

They think we're all gambling all the time, well no we're just families like everybody else.

(Interview #7 – 66-year-old female)

I don't appreciate Las Vegas being, you know, the whole 'What happens in Vegas, stays in Vegas.' You know there's a lot of negativity associated with Las Vegas, and I think drinking and partying, the shows and topless dancers, that has absolutely nothing to do with our lives.

(Interview #12 – 25-year-old male)

When people know that you are from Las Vegas, they expect that you work as a cocktail waitress and you live in a hotel and you work for the casinos and there's that whole image. They don't understand that we have soccer teams, we have PTAs [Parent-Teacher Associations], we have churches, we have schools, and all of this stuff. I think that's the one problem: we get mischaracterised that way.

(Interview #22 – 60-year-old male)

Similar comments have been collected on a wider scale by a study on the social capital of Las Vegas conducted by the Harwood Institute:[10] 'I feel that so many negative images that are shown on television about the Strip have nothing to do with the Las Vegas, the Clark County, and the community I know' (quoted in Harwood and Freeman, 2004: p. 10.)

Gradually, the stigmatisation that is imposed on a city perceived to be lacking any sense of morality has merged with the reputation of local inhabitants. Value judgments and condemnation are transferred from an imagined 'Vegas' to the local population. As one respondent sums up: 'We get blamed for a lot of stuff that we really don't have anything to do with' (Interview #23 – 23-year-old female). Las Vegans have to deal with mischaracterisation, a distorted horizon of expectation from outsiders who seem unable, or unwilling, to disconnect marketing strategies developed by the local tourism industry to promote the Strip establishments and their daily reality, causing them pain and disbelief: 'People think Las Vegas is the Strip. [. . .] That's when the shock, the reality of how others perceive Las Vegas [happens], because when you live here, you think everybody understands' (Interview #11 – 48-year-old female).

Among locals, women are especially targeted by Las Vegas stigma. As a result of the mystique of legal prostitution and the omnipresent commoditisation of female bodies, women in Las Vegas are assigned to degrading social positions, as illustrated by a popular saying: 'Women who work [in Las Vegas] are either making the beds or lying in them' (Kershaw, 2004). This vilifying discourse has an impact on a large part of the population, including the highest elected officials. Interviewed while in charge of the City of Las Vegas city hall,[11] Jan L. Jones was expected to explain to a reporter how she had made it from being a dancer to her position as mayor (Kershaw, 2004). The assumption was that, since she was living in Las Vegas she had to have been a stripper at some point. . . even though she had never worked as one. More than an anecdote of an ill-intentioned or incompetent journalist, this demonstrates that reputation and reality are indistinguishable in Las Vegas, especially when talking about women.

The majority of the population of Las Vegas is regularly belittled and reduced to a group of pathological gamblers and mob gangsters, while women are associated with the apparently tainted profession of exotic dancer – and crucially, this affects those who have nothing to do with the tourism industry and its skills and trades. This symbolic violence is not only a source of personal suffering, but also a cause of contempt. As a consequence, Las Vegans are often not taken seriously which is especially problematic for young Las Vegans aspiring to higher education. Interviewing students from the UNLV (University of Nevada, Las Vegas), being belittled because of diplomas holding the Vegas stamp appeared to be a real concern, as stated in the following quote:

> To carry on that negative connotation: [when looking for grad school] one of the things at the back of my mind was 'will these universities take me seriously?'. Coming from here, I feel like I have to work harder to kind of just shut down this idea, this perception of what the city is.
>
> (Interview #12 – 25-year-old male)

Some interviewees feared that Las Vegas' negative reputation might even prevent businesses and officials from investing locally. Once again, this is a sign that the city and its population are not taken seriously:

We're gonna be the last to come out of all the bad stuff [the economic down-turn] that's been going on in this country because of the stereotypes that people have about this place that are partially true but partially rigid. I think those stereotypes really, really hinder us from making a better impression.

(Interview #16 – 26-year-old female)

For all that, Las Vegans are not dogmatic about their city and do not necessarily try to reduce Vegas' peculiarities, as illustrated by the two following excerpts:

I don't particularly like the idea that people have of Las Vegas in the outer world. It's not a horrible thing, and I don't try to say 'No, no, no, we're a city of churches and we're all nice people', but not all of us are gamblers, not all of us are crooks.

(Interview #24 – 64-year-old male)

If somebody said Las Vegas, as Sin City, [my daughters] sort of roll their eyes, they defend it in that way. They don't defend it as being the absolute most perfect place in the world. But I think they have a fairly realistic idea of how it [Las Vegas] fits into the world cities or the system of cities of the US. [. . .] They would admit its shortcomings and its pitfalls.

(Interview #24 – 64-year-old male)

Having spotlighted Las Vegas stigma and the resulting stigmatisation of its inhab-itants, the question is now whether this trickles down, affecting the sense of pride and weakening social interactions.

Living with stigma in 'Sin City'

Demonyms as signs of a problematic local identity construction

Because of the negative reputation of Las Vegas, it is difficult for inhabitants to reclaim their identity as inhabitants of Las Vegas. In order to highlight their local identity construction, I asked locals about the words they used to introduce themselves to strangers. Studying the use of the demonym 'Las Vegan(s)' gives valuable indications about symbolic territorial appropriation and place attach-ment. Claiming a local demonym is a way to signify to the Other one's belonging to a specific locale, and thus takes place in the symbolic ownership of a territorial identity and the attachment that one has to a certain place.

When asked if the term 'Las Vegans' meant something to them, only a quarter of the interviewees answered positively; and 65 per cent were aware of the term but didn't use it (Nédélec, 2013: p. 303). The vocabulary used by residents when introducing themselves reveals a sort of coping mechanism, allowing them to differentiate themselves from Las Vegas stigma. By not using the 'Las Vegan' demonym, the local inhabitants reject more than a word: they entirely refute the idea of integration in Las Vegas and its related sense of ownership. Therefore, it is a way to distance themselves from the city and of what it stands for in the public imagination. Admittedly, the respondents live in the Las Vegas urban area but they insist on highlighting that they come from elsewhere. This second defence mechanism consists of emphasising the distinction between locals' current place of residence, which is conceived of as temporary, and where they originally come from, where their real 'home' is. The word 'home' implies a condensed emotional attachment accentuating a symbolic rejection of Las Vegas (Blunt and Varley, 2004; Blunt and Dowling, 2006; Brickell, 2012). If Las Vegas is not home, it does not embody one's sentimental belonging, and cannot be a place for which one has much personal affinity. Behind this apparently subtle nuance, one can read the refusal to set up roots in Las Vegas, to make it one's home even at the symbolic level. The clarification 'I live in Las Vegas but my home is elsewhere' transcribes a reluctance to commit locally, and a difficulty to envision their future in Las Vegas. Explaining this type of behaviour is the idea that Las Vegas is only a step on one's personal trajectory, before moving to a more satisfactory place – a place that is easier to live in? – or going back home. This feeling is palpable in the following quote:

> No I wouldn't say I'm a Las Vegan. I'd say I lived here for 13 years but I feel like I'm a Midwesterner or from Chicago. [. . .] No, I don't think I'll ever feel like a Las Vegan [. . .] I think [Las Vegas] is where we live but not where I'm from.
>
> (Interview #20 – 44-year-old female)

During his interviews with local inhabitants, geographer Rex Rowley witnessed the same kind of reaction, trying to symbolically attenuate the impact of territorial stigmatisation:

> [When] asked where I was from, I would always reply, 'Oh, I live in Las Vegas but I'm from Kansas.' I don't know how long I have to live here before I will start answering that I'm from Las Vegas. I think if I were to move to a new city and was asked where I was from, I would still explain that I lived in Las Vegas for X amount of years, but I'm from Kansas. I like living here, but when I think about getting married and raising a family, I don't envision it happening here. Maybe if I did, I would feel more like I was from here.
>
> (Quoted in Rowley, 2013: p. 8)

From individuals to urban community: a weakened sense of community

On a wider scale, refusing to use the Las Vegan demonym echoes a weakened local community, deprived of tight-knit interactions, a strong sense of place and a process of collective identity construction fastened on place. The following quote is based on the opposition between an insufficient community in Las Vegas compared to the interviewee's home state:

[Does the term 'Las Vegans' mean something to you?] Yeah. [*Silence*] I'm not sure what! Because, where I'm from, we're called Hoosiers, and there is a real kind of community: 'I'm an Indiana native, I've been raised here all my life, I'm proud to live here' kind of mentality, kind of a community ethic, and here [in Las Vegas] there just does not seem to be, like, you can call yourself that [a Las Vegan] but I don't know how you would participate in something like that.

(Interview #13 – 30-year-old male)

Place attachment expresses the emotional and affective bond felt by inhabitants towards their place of residence (Scannell and Gifford, 2010; Manzo and Devine-Wright, 2014). Yet, this is precisely what is missing for a majority of Las Vegans: the lack, or weakness, of an emotional commitment towards the city in which they live, and as such the lack of a true sense of belonging and of appropriation. A study conducted by sociologists from the UNLV (Futrell *et al.*, 2010) tried to generalise from individuals to the population of the entire urban area, and statistically measure the fragile sense of belonging and weak collective action in Las Vegas. Through their large-scale survey, they have shown that this deficient place attachment is shared by the majority of the Las Vegan population[12] (Futrell *et al.*, 2010: p. 29): residents of the Las Vegas urban area have the strongest sense of attachment to the United States (66.9 per cent), whereas only 36.9 per cent declared a strong sense of belonging to the city they live in. That number is even lower when switching to a smaller scale: only 33.2 per cent of respondents felt a strong sense of belonging to their neighbourhood. The authors of the survey partially explain those findings through the growth that Las Vegas has experienced since the late 1970s, both demographically and spatially. This entailed both positive and negative consequences on the sense of connection of the inhabitants. Focus groups offered additional material to portray the complicated sense of belonging and attachment in the Las Vegas urban area, as stated in the final report:

> Participants expressed pride in the [Las Vegas] Valley's growth and its status as an international tourist destination. But they also feel that one of the costs of development is transience and impermanence in their neighborhoods, which affects their sense of attachment and belonging. Focus group participants report that they are wary of getting too attached to their neighbors. They say that too many people have come to Las Vegas on only a temporary basis with no interest in establishing roots and giving back to the neighborhood.
>
> (Futrell *et al.*, 2010: p. 32)

The weak sense of place attachment expressed by Las Vegans is not only a source of complaint about the lack of involvement in the community among locals, it also raises important issues about civic involvement on a broader scale. If residents feel only a limited sense of attachment to their neighbours and neighbourhood, then they may be less willing to act together to solve neighbourhood or city problems.

Beyond the negative reputation of Las Vegas: glimpses of place attachment

Despite the above analysis, it would be excessive to say that there is no hope of place attachment for Las Vegans. Even if the 'Las Vegan' demonym is abandoned, I argue that the use of the term 'local' reveals a way to distinguish oneself from tourists and outsiders, and as such it becomes a source of pride (Nédélec, 2013). It is useful to ask how this paradoxical claiming of 'local' versus 'Las Vegan' happens. I analyse this selective use of words as another expression of the coping mechanisms deployed by the local population to deal with Las Vegas stigma. A large number of the people I met are proud to define themselves as being 'from around here', indicating through this formulation that they have a sharp knowledge and understanding of the city and of its customs. By extension, they are thus claiming authority in their command of the city, while at the same time not fully accepting that 'here' means Las Vegas. This explains the preference for the more neutral term of 'local', another self-distancing mechanism from the Las Vegas stigma. Place attachment can therefore be seen as camouflaged as normalised 'localness', while omitting to mention 'Las Vegas'.

One interviewee's acute analysis, expressing his 'conflicted view' about his use of the demonym 'Las Vegan', is worth quoting at length:

> I've lived here longer than I lived anywhere else, I've lived here about a half of my life! [. . .] If I'm not a Las Vegan, I'll be like an Albuquerquian because I've lived there for six years, but I don't feel like an Albuquerquian, and if I'm not a Las Vegan, I feel like a person without a city! I guess I'm a Las Vegan but I don't feel like I say that proudly, I don't feel like I take that on voluntarily, I feel that I say that reluctantly, like I say that 'I guess I'm a Las Vegan because I'm not anything else, that's what I am by default.' And I came to that realisation one day and I almost felt, like, upset or sad. [. . .] I don't have very many things from Las Vegas that I can say are unique and dear to my heart; I have memories but I can't say that I have things, just memories. [. . .] It's with this reluctance that I identify myself as a Las Vegan. And when I go back home. . . Even there, when I say 'back home', I mean go back to Denver! It seems bizarre to me to say I'm a Las Vegan, 'cause I don't feel like I hold Las Vegas in my heart, even though I've lived here almost half of my life.
>
> (Interview #21 – 25-year-old male)

The intention here is not to present this personal point of view as a general case that would be valid for the entire population of Las Vegas. I do, however, deem this quote to be especially emblematic of the conflictual relationship between Las Vegans and the city in which they live. What stands out in the excerpt is the absence of emotional territorial anchoring or of a local sense of place ownership, as featured by the numerous references to the 'heart': Las Vegas hasn't produced elements 'unique and dear to [his] heart'. Moreover, after a Freudian slip, the respondent realised that his 'home' isn't in Vegas, but in Denver where part of his family currently lives. He resides in Las Vegas 'for the moment', but he isn't attached to this place. His choice of words is particularly remarkable. Far from being a source of personal anchoring, being a Las Vegan is for him a 'by default' identity, to the point that he expresses 'reluctance' in referring to himself in that way. He doesn't 'take on voluntarily' the Las Vegan identity, to which he refers 'reluctantly', to the extent that he ends up comparing himself to a stateless person ('a person without a city'). This example argues in favour of a complex identity construction for the inhabitants of Las Vegas, torn between the city's stigma and their personal place attachment process.

The relationship between Las Vegans and their city can thus be summarised as a 'love/hate relationship'. While the majority of respondents expressed an absolute need to detach themselves from Las Vegas stigma, a minority of inhabitants do take pride in its all-pervading reputation, as exemplified in those quotes:

> Now in a way I kind of pride myself anyway in being a part of it, and simply because whenever you go around the country, for conferences or visiting family, when you tell them you're from Las Vegas, people know what you're talking about, you know, it has an image, it has a certain reputation, and not that many cities can say that, you know. [. . .] Immediately, there is a brand here, you have a reputation [. . .] I definitely think the city has an identity and on its own unique way, and I think it's kind of fun.
>
> (Interview #5 – 40-year-old male)

> When you travel overseas, like when I traveled in Paris, London, even Windhoek, Namibia, you tell people you're from Las Vegas, everybody knows where you are from and has an image of you immediately. [. . .] So I guess it's easy to tell people where you're at, since it connotes some images, which I don't think are necessarily bad.
>
> (Interview #9 – 58-year-old male)

Some locals even base their sense of pride on the benefits of being a world-renowned tourist destination, as exemplified by the following comments on the significance of the term 'Las Vegan':

> Las Vegan? I think it's a thing of pride. Yeah, it means something to me: it means that I live in the greatest city on earth, it means that I live in a city that attracts people from all over the world, it means that I'm a Las Vegan, it means that I am proud of my city, that I can say I love Las Vegas and I consider myself a Las Vegan.
>
> (Interview #6 – 63-year-old female)

> It means I'm from a place that people are curious about. They really want to know.
>
> (Interview #8 – 61-year-old female)

Nevertheless, the sense of pride remains in minority among the local population (Nedelec, 2013; Futrell *et al.*, 2010). The overall stigma of Las Vegas has a cost and the majority of residents end up unable to feel pride for a city that is above all portrayed as Sin City.

Conclusion

The three main characters of *The Hangover* movie spend two days searching for their missing friend, lost during an epic night that led to the eponymous hang-over. With their headaches in tow, they hunt him down throughout Las Vegas, and, in the process, they meet and interact with a number of especially colourful locals. The only female that they encounter describes herself in these words: 'I'm a stripper. Well, technically, I'm an escort but stripping is a great way to meet the clients.' This leads the main characters to characterise her as a 'whore' for the rest of the movie – with no departure from the simplistic image of women in Las Vegas. In that similar vein, when the three protagonists have to deal with the local police, they end up becoming test subjects for a demonstration of stun guns, which is performed in front of local kids. Of course, this is a million miles from the professional response that one expects from police officers but this is Las Vegas where anything can happen and even the police acts in unpredictable and quirky ways! The scene also says a lot about the mundane leisure activities of Las Vegas youth. Beyond the exaggeration that can be expected from a comedy, depictions of Las Vegans in *The Hangover* are especially emblematic of the ways in which the stigma that marks out Las Vegas has contaminated its inhabitants.

The Sin City label can be said to act as a double-edge sword for Las Vegas and its residents. It has brought international visibility to the city and led to the staggeringly high numbers of tourists, with people from all over the United States – and the world – wanting to view the dizzying illuminations on the glitzy central boulevards, as well as to visit the renowned hotel-casinos. Yet, the label has also evidently led to Las Vegas being perceived as a place without morals, a city where immorality is the norm, and therefore a city that can only be populated by residents who must somehow be immoral too. These assumptions mean that, on a daily basis, Las Vegans have to endure the constant reactivation of stereo-types and degrading spatial perceptions that originate from tourism imaginaries and marketing strategies. As I have suggested, this has tremendous impacts on identity construction of inhabitants of the city, on their attachment to place and on their sense of community.

Even if other factors need to be taken into consideration,[13] Las Vegas stigma is one key element to understand the deficient nature of social dynamics at the scale of the entire urban area. Though Las Vegas is original in that its stigmatisation comes from the incredible economic success of its tourism industry, the negative impacts of the stigmatic imagery on the local population are nevertheless undeni-able. The complex place attachment of Las Vegans highlights more broadly the hybrid nature of Las Vegas, a city that is torn in half between exceptionalism – as a gaming destination – and urban banality. It demonstrates that the urban area is intrinsically determined by tourism, even if this economic activity is spatially contained in functional enclaves. There are two sides to every coin, and this is also valid for the ways in which tourism has affected the city of Las Vegas, embodying both its strengths and weaknesses. While a formidable asset for the local population, it also brings with it a negative reputation that partially strips Las Vegans of their identity, imposed onto them by outsiders, as a uniformly stigmatising and degrading series of representations. As demonstrated through the extended research conducted in Las Vegas, it seems to me that the majority of the local population does not contest Las Vegas stigma, but rather accepts it, as well as their own stigmatisation, which is internalised as an unquestioned characteristic of local life. Even though some Las Vegans feel that they profit from their city's reputation, most inhabitants seem just too tired of fighting.

Notes

1 IMDb (Internet Movie Database); http://www.imdb.com/title/tt1119646/ (accessed 11 November, 2015).
2 The local tourism agency, Las Vegas Convention and Visitors Authority (LVCVA), started collecting visitation statistics in 1970, hence the absence of solid statistics prior to this date.
3 According to the US Census data, Las Vegas showed an 83.3 per cent growth rate between 1990 and 2000 (Perry and Mackun, 2001); and a 41.8 per cent growth rate between 2000 and 2010 (Mackun and Wilson, 2011) which was the third largest rate of growth for cities in the United States.
4 The toponym Las Vegas is somehow misleading since it does not correspond to any official territorial entity. The so-called 'Las Vegas' urban area is composed of three incorporated municipalities (City of Las Vegas, North Las Vegas and Henderson) and six unincorporated towns under the authority of Clark County (Whitney, Paradise, Winchester, Sunrise Manor, Spring Valley, Enterprise).
5 The Old Spanish Trail dates back to the sixteenth century, but its consolidation did not happen before the 1830s to 1850s.
6 To this day, Utah and Hawaii are the only states where all forms of gaming are forbidden.
7 See Nevada Revised Statutes § 244.345, 'Dancing halls, escort services, entertainment by referral services and gambling games or devices; limitation on licensing of houses of prostitution.'
8 The other two are referred to as 'abominations of the body' and 'blemishes of the individual character' (Goffman, 1963: p. 14).
9 The numbering refers to fieldwork made by the author (Nédélec, 2013: p. 418).
10 The Harwood Institute for Public Innovation is an independent nonprofit association, focusing on urban public policies and community development. Its survey about social capital in Las Vegas relies on an eight-month survey conducted in 2003–2004, composed of 12 focus groups, 75 detailed interviews and three public reunions, thus collecting the words of 275 people.
11 Jan L. Jones served two mandates (1991–1995, 1995–1999) as the City of Las Vegas first female mayor.
12 Respondents were asked if they had a strong, moderate, low, or no sense of belonging to the various locales: the United States, the Southwest, Nevada, the Las Vegas urban area, the city they live in, the neighbourhood they live in (Futrell *et al.*, 2010: p. 29).
13 These factors are a low native population added to a high turnover (see Nédélec, 2013).

References

Blunt, A. and Varley, A. (2004), 'Introduction: Geographies of home', *Cultural Geographies*, 11(1): 3–6.

Blunt, A. and Dowling, R. (2006), *Home*, London/New York: Routledge.

Bourdieu, P. (1991), *Language and Symbolic Power*, Cambridge: Polity Press.

Bourdieu, P. (1993), 'Effets de lieu', in P. Bourdieu (ed.), *La misère du monde*, Paris: Le Seuil.

Brents, B. G., Jackson, C. A. and Hausbeck, K. (2010), *The State of Sex: Tourism, Sex, and Sin in the New American Heartland*, New York: Routledge.

Brickell, K. (2012), '"Mapping" and "doing" critical geographies of home', *Progress in Human Geography*, 36(2): 225–244.

Census Bureau (2015), 'Clark County', Nevada. *Quickfacts*. Available online at: http://quickfacts.census.gov/qfd/states/32/32003.html (accessed 31 July, 2015).

Eumann, I. (2005), *The Outer Edge of the Wave: American Frontiers in Las Vegas*, Frankfurt: Peter Lang.

Findlay, J. (1986), *People of Chance: Gambling in American Society from Jamestown to Las Vegas*, New York: Oxford University Press.

Foster, J. (2009), *Stigma Cities: Dystopian urban identities in the United States West and South in the Twentieth century*, PhD diss., Las Vegas: UNLV.

Foster, J. (2012), 'Stigma Cities: Birmingham, Alabama and Las Vegas, Nevada in the National Media, 1945–2000', *Psi Sigma Siren*, 3(1): Article 3. Available online at: http://digitalscholarship.unlv.edu/psi_sigma_siren/vol3/iss1/3 (accessed 31 July 2015).

Friess, S. (2004), 'A firm hits jackpot on Las Vegas ads. Campaign phrase enters the lexicon', *Boston Globe*, 28 March. Available online at: http://www.boston.com/news/politics/advertising/articles/2004/03/28/a_firm_hits_jackpot_on_las_vegas_ads/ (accessed 31 July 2015).

Futrell, R., Batson, C. D., Brents, B. G., Dassopoulos, A., Nicholas, C., Salvaggio, M. J. and Griffith, C. (2010), *Las Vegas Metropolitan Area Social Survey, 2010 Highlights*. Metropolitan Area Research Team, Department of Sociology, Las Vegas: University of Las Vegas, Nevada. Available online at: https://www.unlv.edu/sites/default/files/24/LVMASS_0.pdf (accessed 4 August 2015).

Goffman, E. (1963), *Stigma: Notes on the Management of Spoiled Identity*, Englewood Cliffs, NJ: Prentice Hall.

Gragg, L. (2013), *Bright Light City: Las Vegas in Popular Culture*, Lawrence, KS: University Press of Kansas.

Harwood, R. and Freeman, J. (2004), *On the American Frontier: Las Vegas Public Capital Report*, Bethesda, MD: The Harwood Institute.

Kershaw, S. (2004), 'They take it off, but they also put on suits, uniforms and blue collars', *The New York Times*, 2 June. Avaialable online at: http://www.nytimes.com/2004/06/02/us/women-s-work-they-take-it-off-but-they-also-put-suits-uniforms-blue-collars.html (accessed 4 August 2015).

Kirkness, P. (2014), 'The *cités* strike back: restive responses to territorial taint in the French *banlieues*', *Environment and Planning A*, 46(6): 1281–1296.

Lippmann, W. ([1922] 2012), *Public opinion*, New York: Harcourt, Brace and Co, and Mineola: Dover Publications.

LVCVA (2015), 'Visitors Statistics'. Available online at: http://www.lvcva.com/stats-and-facts/visitor-statistics/ (accessed 31 July 2015).

Mackun, P. and Wilson, S. (2011), *2010 Census Briefs. Population Distribution and Change: 2000 to 2010*, US Census Bureau, March. Available online at: https://www. census.gov/prod/cen2010/briefs/c2010br-01.pdf (accessed 31 July 2015).

Manzo, C. L. and Devine-Wright, P. (eds) (2014), *Place Attachment: Advances in Theory, Methods and Applications*, New York: Routledge.

Moehring, E. and Green, M. (2005), *Las Vegas, A Centennial History*, Reno: University of Nevada Press.

Moody, E. N. (1994), 'Nevada's legalization of casino gambling in 1931. Purely a business proposition', *Nevada Historical Society Quarterly*, 37(2): 79–100.

Nédélec, P. (2012), 'L'enclave fonctionnelle du Strip à Las Vegas: quand l'insularité façonne la ville', *Espaces et Sociétés*, 150: 49–65.

Nédélec, P. (2013), *Réflexions sur l'urbanité et la citadinité d'une aire urbaine américaine: (dé)construire Las Vegas*, Ph.D. diss., Lyon, France: University of Lyon.

Nédélec, P. (in press), '"What happens in Vegas doesn't stay in Vegas": When tourism imaginaries fashion the scientific discourse', In Gravari-Barbas, M. and Graburn, N. (eds), *Tourism Imaginaries at the Disciplinary Crossroads: Place, Practice, Media*, Farnham, UK: Ashgate.

OED (2015), *Oxford English Dictionary*. Available online at: http://www.oxforddictionaries. com/ (accessed 31 July 2015).

Perry, M. and Mackun, P. (2001), *Census 2000 Brief. Population Change and Distribution, 1990 to 2000*, US Census Bureau, April. Available online at: https://www.census.gov/ prod/2001pubs/c2kbr01-2.pdf (accessed 31 July 2015).

Raento, P. (2003), 'The return of the one-armed bandit: Gambling and the West', in Hausladen, G. (ed.), *Western Places, American Myths: How We Think About the West*, Reno, NV: University of Nevada Press.

Reza, J. (2013), 'Ask A Native', *Vegas Seven*, 25 April–1 May, 163: 18.

Roske, R. J. (1990), 'Gambling in Nevada. The early years, 1861–1931', *Nevada Historical Society Quarterly*, Spring, 33(1): 28-40.

Rowley, R. (2013), *Everyday Las Vegas. Local Life in a Tourist Town*, Las Vegas: University of Nevada Press.

Scannell, L. and Gifford, R. (2010), 'Defining place attachment: A tripartite organizing framework', *Journal of Environmental Psychology*, 30(1): 1–10.

Schwartz, D. (2006), *Roll the Bones: The History of Gambling*, New York: Gotham Books.

Sheehan J. (2008), 'Why Las Vegans can be a tad defensive about how others see their city', *Las Vegas Sun*, 13 April. Available online at: http://m.lasvegassun.com/news/2008/ apr/13/why-las-vegans-can-be-tad-defensive-about-how-othe/ (accessed 31 July 2015).

Slater, T., Pereira, V. B. and Wacquant, L. (2014), *An international bibliography on territorial stigmatisation*, AdvancedUrbanMarginality.net. Available online at: http:// www.advancedurbanmarginality.com/uploads/5/9/6/7/5967617/terstig_bibliography_ may2014.pdf (accessed 31 July 2015).

The Hangover (2009), Comedy, Directed by Todd Phillips, [DVD] US: Warner Bros.

Wacquant, L. (2007), 'Territorial stigmatisation in the age of advanced marginality', *Thesis Eleven*, November, 91: 66–77.

Wacquant, L., Slater, T. and Pereira, V. B. (2014), 'Territorial stigmatisation in action', *Environment and Planning A*, 46(6): 1270–1280.

White, M. and White, L. (1962), *The Intellectual Versus the City, From Thomas Jefferson to Frank Lloyd Wright*, Cambridge, MA: Harvard University Press.

Zukin, S., Baskerville, R., Greenberg, M. and Wissinger, B. (1998), 'From Coney Island to Las Vegas in the urban imaginary: Discursive practices of growth and decline', *Urban Affairs Review*, May, 33(5): 627–654.

3 Imaginary politics of the branded city

Right-wing terrorism as a mediated object of stigmatisation

Lucas Pohl

Introduction

In almost every urbanised part of Germany, one will find examples of city market-ing, place branding and image politics. In a century where 'marketing and branding have become hegemonic in cities across the globe' (Jonas *et al.*, 2015: 203), so-called 'local identities' have become a highly influential aspect of everyday life. Being scalar as well as political units, cities in the common sense of local image politics have become 'separate homogeneous objects' (Boisen *et al.*, 2011: 137) with specific functions on higher scales: 'people understand cities in the same way as they understand brands' (Kavaratzis, 2007: 703). The imagination of cities as brands involves a variety of characteristics, such as 'clarity', 'simplicity', 'memorability' and 'distinctiveness' (Anholt, 2010: 178).

Against this background, a city like Zwickau in Germany can be signified as the 'No. 1 city for the automotive industry in Eastern-Germany' (Zwickau, 2015a).[1] While several scientific approaches have already called for a hetero-geneous understanding of local images (see Jansson, 2003; 2005; Johansson, 2012), the image politics of Zwickau also emphasise a variety of possible identifications, focusing not just on the automobile industry but on Zwickau's museums, theatres, music scene and architectural facets (Zwickau, 2015c). Even though the city calls for the recognition of its multiple colors, Zwickau's 'uniqueness' is defined through its 'diverse economic structure' and its 'good basis for research and development' (Zwickau, 2015b). One claims to be both a 'seller seeking to attract and retain place costumers' (Kerr and Oliver, 2015: 62) and a vehicle for 'appropriate ways of living the brand' (Johansson, 2012: 3613). In this sense, the branded city is more than just a simplified image for outsiders. It is a fetishised way of governing the city as a subject. But even if the city is presented through a coherent and stable portrait, local identities remain none-theless contingent, as they can become hegemonically layered with negative stereotypes and become stigmatised as a result. In the case of Zwickau, this is linked to right-wing terrorism. Following psychoanalytic theories, I seek to analyse the city as a subject of mediated stigmatisation and develop a critical understanding for ways of dealing with right-wing terrorism and extremism through local identities.

The city as the subject of topoanalysis

Grasping cities like Zwickau as some*thing* that can be imagined – a harmonious entity with identifiable actors – is extremely desirable from a political perspective. Without denying the inherent place-fetishism and 'perversity' (Marcuse, 2005) of those self-portraits that constantly miss the resistant and contradictory facets of places, in my analysis of 'sense of place', I will keep the focus on this desire. For a psychoanalytic reading of territorial stigmatisation,[2] my approach acknowledges that place becomes fetishised when power becomes attributed to it (Harvey, 1996: 320), but as images are such a crucial part of local politics, they cannot be grasped as an ideology that scientists can somehow get rid of. As 'the new opium of the people' (Kristeva, 2003: 22), images must be taken seriously. As such, I will question the imaginaries of a desirous city as well as the fragile dimensions and problematic consequences that emerge from governing the city as a unified subject.

One way to capture the genesis of such a portrait of the city is through the lens of Jacques Lacan, whose psychoanalytic work can be read as an attempt to topologise the psyche. Therefore, the mirror plays an important role. According to Lacan (2006: 76ff.), it enables the possibility to assume an image that differentiates between inner and outer parts of the subject. The mirror produces spatial dialectics, such as *here* and *there*, and in this sense leads to the imaginary insight that the ego as a singular and unique unit exists *in* it.[3] By transferring this perspective to the local, it is crucial not to misunderstand the notion of the imaginary as some kind of 'false reality': 'imaginary does not mean that it is illusory' (ibid. 290). Even if it is 'far from inconsequential' and 'not simply something that can be dispensed with or 'overcome' (Evans, 1996: 84), the imaginary – or respectively the ideal-ego for Lacan – means first and foremost 'Alienation' (Lacan, 2008a: 146). It is bound to a constitutive process of 'Othering' (Lacan, 2007: 482), which means that for Lacan the subject is not sovereign, unbound or self-sufficient, but rather a relational, split and fragmented matter that needs an image to stabilise itself. For this reason, just as the mirror has edges, every image has limits. As a result, if one analyses places psychoanalytically – or 'topoanalytically'[4] – there will always be something that lies beyond the limits of imagination. An image enables not just the creation of competitiveness by allowing for the emergence of an ego, but also needs to confirm its status, and therefore it has to be represented in the symbolic.

Media as the big Other

To introduce the Lacanian notion of the symbolic into the city, it is possible to examine how the 'sense' becomes substantial for places. Campelo (2015: 51) highlights the crucial influence of 'meaning and signifiers that create the uniqueness of each place'. For an analysis of local brands, it is important to acknowledge that when a place develops a name – a toponym – there is a direct connection to the individual mind and memory (Medway and Warnaby, 2014: 157). In other words, the city as a singular unit receives a 'sense' at the moment where it is named as such. From that point, one can say that the 'name' is crucially influenced by those who are *naming*. The agency of naming is what Lacan refers to as the symbolic or the 'big Other'. As the 'grammatical rules' (Žižek, 2006: 9) of social life, it can be seen as a powerful instance that exceeds the subject, even if it is spoken, and in this sense realised through it.

As a key instance of the symbolic, of how we see, name and speak about cities, the role of the media is crucial. Being one of the most influential ways to connect local processes to the public, local images are inherently produced and circulated through media: 'most of the information we get about places, and the images produced in our minds by that information, comes to us through one form of mass media or another' (Cromwell, 2011: 107, quoted in Sevin, 2014: 50).[5] Even if a new 'sense of being *inside mediation* is becoming an increasingly significant form of everyday experience' (Jansson, 2005: 1689), Gunder speaks of a missing debate about 'the role of contemporary media in shaping public aspirations as to what is desired for the future of our cities' (Gunder, 2011: 325). In what follows, I will stress this gap by focusing on the media as the big Other. As an instance that (co)constitutes the city, the media opens up the possibility to assume a local image, and in this sense affects how the specular city appears. The media warps and distorts the image, which does not mean that there is an objective reality which the media is falsifying, but rather that it introduces an imaginary point of view into the local reality, whereas it is impossible to know what the big Other – the media – wants. That is what Lacan calls desire: a desire, which is 'always the desire of the Other' (Lacan, 2008b: 38). There is no security about the symbolic, just an imaginary identification and the question '*Che vuoi?*' (Lacan, 2014: 6): what do you, the big Other, want?

Topological limits of the desirous city

In this sense, local politics are neither the product of the media, nor do they have the power to finally control the image that the media reproduces. They are imaginary, and therefore influence how a place is politically and materially shaped as well as they are influenced by it. A toponym is neither only an issue of the residents who are connected to the named place, nor is it just an issue of the ones who are naming a place. It is something that is registered within and beyond the scene. Therefore, the production of a desirous city – a subjectified place, which is constituted through the imaginary and the symbolic – is in the same way embodied and internalised as it is objectified and externalised. Spatially speaking, one could say that the desirous city is organised both internally (perceived and reproduced by the locality) *and* externally (perceived and reproduced by actors, which are not part of the local everyday life)[6] – a dialectic that is topologically represented in the Moebius strip. The Moebius strip, one of the famous topological figures in post-mathematical thinking (Martin and Secor, 2014), is a double-sided spiral strip, which makes it impossible to finally separate between inside and outside. Lacan referred to it several times to capture the structure of the subject (cf. Lacan, 1998: 155f.; 2014: 96f.) and it can become a fruitful concept here if one uses its structure of constituting both difference and identity at one place to analyse and criticise local processes (Secor, 2013).

This way of reading the city topologically does not imply the reduction of its spatiality to an abstract model. The 'imageable city' (Pile, 1996: 220) – a city that is able to imagine – is always at the same time an 'imagineered city'[7] – a city which is *built as* a place for 'strategies and politics of shaping urban identities' (Jonas *et al.*, 2015: 195). Therefore, I do not presume a 'natural' sense of place, but on the contrary, assume that the connection of understanding cities based on its imaginary is related to the mediated society, where 'urban citizens may occasionally experience that the imaginative structure of their home territory is not primarily decided by social dynamics in the local setting, but through public mediation (news, films, advertisements, etc.)' (Jansson, 2005: 1671). From a Lacanian perspective, this topo-logic of a desirous city is, like every process of demarcation, 'un-natural' in its core (Chiesa, 2007: 137). For Lacan, every border that constitutes an interiority and exteriority – and in this context is veiling the Moebius structure of place – is in itself constituted through something that 'makes sense'. 'Sense' here can be understood as the result of the inherent connection between the imaginary and symbolic aspects of the production of place (Lacan, 2005). It provides the possibility of combining issues of place branding and image politics 'from the inside' (as part of the imaginary), with processes of being named and classified 'from the outside' (as part of the symbolic), to follow the logic of a mediated production of place. At the same time, this notion of a 'sense of place' problematises all borders in terms of the distinction between in- and out- side, because what seems to be far beyond the limits of the local imagination can, from this point of view, even take place right in the heart of a city.[8]

Objects of imaginary politics

Assuming that imaginary politics usually seek to draw very clear, simple and memorable portraits of the city, we can now focus on the details of the city's image. While we have pictured the image as a process to differentiate between inner and outer parts of the subject, a desirous city needs image-building factors to emphasise its borders. Therefore, 'local identities' are necessarily bound to a process of transforming particular objects into 'objects of desire'. Economic importance, infrastructures, significant historical facts or well-known personalities, to name just a few highlights of local branding, become necessary objects in order to produce a coherent and identifiable 'sense of place'.

Here it is crucial to mention that for Lacan, such objects cannot be understood as 'natural' or even fully available 'resources'. Rather, the object of desire is a result of the paradoxical need to 'plug the holes' within the city's image, or rather, between the imaginary and the symbolic, to first *create* a 'sense of place'. As I have already mentioned, there is an inherent insecurity or rather unfixable lack, between the imaginary and the big Other. From the imaginary point of view, we cannot know, and we just desire what the big Other is seeing in us. We can only ask: what do you want? Thus the local image is never entirely complete, but even though it lacks, every identification-process follows the fantastic idea to find objects that accomplish identity.[9] When the city marketing of Zwickau claims that an 'appropriate identity has not yet sufficiently developed' (Zwickau, 2008), this is not a temporary, but an inherent demand.

Transferring this logic to place branding in a mediated society, we can assume that because imaginary politics are crucially influenced by 'the voices of the media', city marketing is captured by the idea to find objects that create a 'sense of place', which will be positively recognised as unique and distinctive. But because of the lack, which in this approach can be seen as a 'hauntological' part of any kind of 'local identity', imaginary politics are not able to ultimately fulfill this task. From a psychoanalytical point of view, 'local identities' are never completed or closed (Bullock, 2014: 218f.) and imaginary politics are inherently missing their final goal. Against this background, they have no total control over their objects of desire. The portrait is never fully complete and all the characteristics, like uniqueness, memorability and distinctiveness arc never good enough. There is always a possibility to become more creative, smart, sustainable, resilient etc. Any 'local identity' is constituted according to a general absence, because every kind of particular object is just *one* option to produce *one* kind of place. Objects are substitutable, which becomes obvious when we look at the different strategies of local images in Zwickau as the 'No. 1 City for Automobiles in Eastern-Germany'. But even more so, the temporary and competitive character of 'local identities' becomes apparent when we look at dysfunctions or distortions within those identification processes. Following the notion of imaginary politics, the moment when an object in fact takes place is not something desirable. The moment, when the lack seems to be filled with something – when the object of desire cannot be desired anymore, because it took place already and there is no need to stimulate the search for it – is on the contrary a moment, which has the ability to destabilise the way imaginary politics work. It is this moment, when a presence appears and the 'lack happens to be lacking' (Lacan, 2014: 42) – or, as I will discuss it in the last part of this paper: a placement of an object of stigmatisation.

Right-wing terrorism as an object of territorial stigmatisation in Zwickau[10]

> The fantasy of a utopian harmonious social order can only be sustained if all the persisting disorders can be attributed to an alien intruder, [. . .] through the construction/localisation of a certain particularity which cannot be assimilated but, instead, has to be eliminated.
>
> (Stavrakakis, 1999: 108)

According to imaginary politics, it is a distortion that reveals the problem of territorial stigmatisation. Before I elaborate on my psychoanalytical reading of the city and its imaginary politics with territorial stigmatisation, I will try to give a brief explanation as to why I think it is necessary to go beyond the original understandings of territorial stigma itself. The concept of territorial stigmatisation is associated with Loïc Wacquant who presents one of the most influential frameworks to combine spatial and sociological perspectives on stigmatisation. His major goal – 'to advance our grasp of the ways in which noxious representations of space are produced, diffused, and harnessed in the field of power, by bureaucratic and commercial agencies, as well as in everyday life in ways that alter social identity, strategy, and structure' (Wacquant *et al.*, 2014: 1273) – seems at first glance quite similar to the argument that I make here. However, there are at least three differences. First, Wacquant links the connection between spatial and social practices of stigmatisation primarily to the state. For him, it is the state that 'determines the scope, spread and intensity of marginality' (Wacquant, 2010: 174) and therefore creates the spaces of possible stigmatisation. I would argue that the idea of a 'neoliberal leviathan' (Wacquant, 2012) that is responsible for the victims of stigmatisation ignores the fact that the state itself can become a possible 'victim' of place-based stigmatisation. This is the case in the example that I develop here, of a city that becomes branded as a 'right-wing territory'. Second, Wacquant's notion of 'place'. Concerning the US-American 'black ghetto', Wacquant describes a basic change of their spatial structure. Originally, the ghetto was a 'place' that enabled some form of stability and security (Wacquant, 2007). Today, the ghetto is transformed into a 'space' which is in opposite to former times, far from creating a collective identity defined as 'a socio-spatial contraption geared to naked exclusion that [. . .] offers none of the collective protections and side-benefits of ghettoization' (Wacquant, 2010: 171). I do not want to question his observation on the change of the US-American Ghetto, but my approach seeks to focus on not on the dissolution of place, but on its reproduction of place. By looking at current local image policies, one can perceive the important efforts to create, transform and control local identities. Places do not dissolve in space, but rather receive a central role to new forms of local or imaginary politics. Against this background, Wacquant mentions that the 'dissolution of "place"' is not equivalent with a change of the physical, but rather a change of the social space (Wacquant, 2007: 69f.).[11] As the third point, I would like to stress that territorial stigmatisation is never just an issue of social relations, but always a material phenomenon, which forms and shapes the built environment of a place. The 'social' is not separated from the 'physical', but produces as well as reproduces the matter of place.

One way to move beyond territorial stigmatisation is to understand it as the 'other' side of successful place branding. Territorial stigmatisation, in this sense, goes back to the desire to fix the inherent flaws of the 'local identity'. Instead of the image-building objects of desire, here the object *takes place* and for this reason works not in favor of the desirous city, but against it. One example of a process that employs such an object and has received attention in German and foreign debates, was the revelation of the *Nationalsozialistischer Untergrund* ('National Socialist Underground') or the *Zwickauer Terrorzelle* ('the terror cell of the city of Zwickau') towards the end of 2011.[12] On 4 November, right after two men killed themselves after a robbery, 200 kilometers away in Zwickau, a city of nearly 100,000 inhabitants, a house exploded and Beate Zschäpe who lived there together with those men, turned herself in to the police.[13] One interviewee stated in relation to this that from then onwards it was:

> like an invasion. [. . .] all the curious people [. . .] staring at the people as if all of them would have a swastika on their forehead. And all of this just because three Nazi terrorists have taken root here, with false names, false attitude, false friendliness. Since 4 November, the city is an everyday topic in the news: Zwickau right-wing terror, Zwickau, city of terrorists, Zwickau, a city of murderers [. . .]. You feel stamped in Zwickau, misunderstood.
>
> (SZ 02.02.2012)

After the 'shock that right in the middle of Germany, a right-wing terrorist group could hide whose brutality lays beyond the limits of imagination' – to quote the state chairmen of the German Green party (FP 12.11.2011) – the city of Zwickau was facing the challenge of being branded as *the* heart of right-wing terrorism in Germany. The local media spoke about an 'aura of evil' and suggested that after this event, everyone would now think of 'brutality and death' when gazing upon the ruined meeting place of the former terrorist group (ibid.). From one day to the next, Zwickau's image changed and right-wing terror became the dominating characteristic of the city. As a result, Zwickau gained an imagined idea of what it was, realising what it *did not* want to be.

> Those searching for Robert Schumann on the Internet get 200,000 hits. The '*Zwickauer Terrorzelle*' brings it to 3.3 million.
>
> (FP 25.11.2011a)[14]

Around the same time, with the increasing coverage of local and regional media, the city implemented its first urban policy interventions. On 18 November 2011, the headline of a local newspaper said: 'Zwickau wants to distance itself from right-wing terror.' The city council is promoting a public event to formulate its 'Appeal for Democracy and Tolerance' to overcome the 'speechlessness' (FP 18.11.2011a), as they have called it. During another event which was entitled 'A sign against right-wing extremism and hasty defense reflexes', the media, as well as several public voices spoke about 'remembering the victims' on the one hand, but also the necessity of 'saving the city's reputation' (FP 25.11.2011b). 'All this has nothing to do with Zwickau. [. . .] [The] city is no right-wing stronghold', was the reaction of the chief of the German Trade Union (FP 18.11.2011a). Similar to this, the city council stated that all this was just 'a bad coincidence' and that all this could just as well have taken place in other cities as well (FP 21.11.2011).

While these two techniques of 'undoing' and 'isolating' the issue are symptomatic of the traumatic political reaction in Zwickau,[15] the object nevertheless *signs* the city as what it *is*: the city of right-wing terror. In Lacanian terms, one could suggest that right-wing terrorism becomes the object-cause or the '*objet petit a*' of the city. It obtains a hegemonic status according to the production of 'local identity', which from now on leads to a significant problematisation of the negative effects on the local in general.

'After the shock, now one has to think about the consequences', stated the City Mayor for Finance and Order. 'The term of the '*Zwickauer Terrorzelle*' will damage the city that thrives on tourism and export-oriented industries' (FP 18.11.2011b). 'The city's image as a place of right-wing terror is already delivered abroad', said the chief of the German Trade Union. 'Companies report that they are already being addressed by its foreign partners about what is going on here. For Zwickau, one of the few industrial centers of the country, the damaged image will have devastating consequences' (FP 21.11.2011). As an object of stigmatisation that interrupts the fantasy of the harmonious spatial order – an 'enemy', which is 'internal' and 'out there' all at once (Stavrakakis, 2007b: 199) – the object is able to change both the imaginary portrait of the city 'from the inside' and the symbolic way in which the city is being named and classified 'from the outside'. In other words, it seems to change the 'sense of place'.

To understand how the object of stigmatisation evolves, one can assume that it takes place 'inside' the city, but that its control lies 'somewhere else'. It is what Lacan calls 'extimacy' – an external intimacy. Even if it is constitutive to understand how places can exist as stigmatised places, the object of stigmatisation 'ex-sists'[16] somewhere else (Lacan, 1999: 22) and, in this sense, it is neither just the product of the imaginary nor the symbolic (Lacan, 1991: 177). The object of stigmatisation is connected to what Lacan calls the Real – something which is linked to the imaginary and symbolic as their 'unassimilable' (Lacan, 1998: 55). As real, the object is neither *in* the social reality nor *out* of it, but a marker of its boundaries – a socio-spatial moment that highlights the incompleteness of the local identity, with its symbolic order and imaginary conceptions.

On their way to 'eliminate' the 'alien intruder' (Stavrakakis, 1999: 108), the media and politics stated that the ruined former shelter could have become a place of pilgrimage for some. For so-called security reasons, and with the stated ambition of restoring 'normality', in April 2012, the city council started to demolish the building. To quote the Mayor: 'We prevent the site from becoming a place of pilgrimage for the extreme right [. . .]. Therefore, a demolition of the building is indispensable' (FP 24.04.2012). However, as a materialised proof that the *Zwickauer Terrorzelle* really existed, the ruin as the problem of built presence did not have to be reducible to the possibility of right-wing pilgrimage. Its sustained existence could have functioned not just as an urban landmark for right-wing terrorism, but as a sign that right-wing terrorism is still 'out there' and that politics should keep an open eye to its continued existence. In this context, imaginary politics sought to exclude the traumatic shock of the object of stigmatisation by 'imagineering' and eliminating any built proof of the shock. In its place, one can now find a simple green meadow. In regards to my approach, the practice of placing stigmatising objects is not directed *to* the place, but it is rather an inherent process *of* the production of desirous place. Through a wide range of practices and different kinds of particular objects, imaginary politics try to create an absolute harmonious, and unified 'sense of place'. This can be seen as the general fantasy of current imaginary politics and a crucial aspect of producing desirous cities (Gunder, 2005). Therefore, every kind of particular object has a temporary as well as competitive dimension, and place is no fixed reference point. Hence, it is not entity of which one feels stigmatised from, but rather an imagined, symbolised, as well as materialised process of re-producing place-based identification. The likelihood that territorial stigmatisation might emerge from within a place itself (as opposed to being imposed from outside) is always a terrifying possibility: every territory can become a 'terrortory'.[17]

Conclusion

While outlining the local discourse around the *Zwickauer Terrorzelle*, it is crucial not to point only to the temporary and competitive character of 'local identities', but to highlight the contingent roots of the perception of territorial stigmatisation itself. It is not 'natural' to describe what happened in Zwickau as a traumatic shock, or for those in charge of urban policy to feel that they are responsible for the city's image. This 'normal state' of subjectivation is, like every process of demarcation, un-natural in its core, as it is part of a hegemonic formation, developed to legitimise such patterns of thought, argumentation and action. Even though images become 'the new opium of cities' to displace the inherent 'perversities' of all the consistent narratives developed by city bandings, my attempt has been to follow those local imaginaries to better understand its topo-logic. To regard the city as a singular subject with distinct borders, an inside and outside, does not lessen the alienated process cities go through, but my approach has sought to stress this particular form of fetishism a little bit further in order to question the limits of this kind of governing the city.

To understand the fantasies of local politics, I have tried to capture place branding and image politics topoanalytically. This perspective helps us to analyse political 'negations of the real and the topological structure of the psyche and social that twists the binaries of said/unsaid, near/beyond [. . .] and so on' (Kingsbury, 2015: 341). It thus allows an analysis of the branded city as the subject of research, and therefore leads to an understanding of right-wing terrorism as something excluded by imaginary politics. Therefore, I have followed a Lacanian approach that allows to grasp places, not through an inscribed or fixed 'identity', but by following the political attempt to fix the constitutive lacking structure that is crucial to any kind of spatial 'closure'. As inherent parts of the production of desirous cities and 'local identities', I highlighted that local politics depend on image-building objects of desire to emphasise their borders. At the heart of this topo-logic, the problem of territorial stigmatisation – or, to put it best 'imaginary stigmatisation' – takes place. Depicting the other side of place branding, or more precisely the other side of the Moebius strip, the object of stigmatisation functions as the irrational opposite of the local imaginary. Uncontrollable on the one hand and vital part of any local identification process on the other, the possibility of becoming territorially stigmatised is an inherent part of the fantasy of 'local identities'. Every territory – as a singular controlled demarcation – *needs* objects of stigmatisation to exclude the others who are not part of the imaginary and every territory forms the basis for a 'terrortory', a place of anxiety.

The media play a key role in reproducing the imagination which allows the governing of a desirous and harmonious city. They also have a significant role in locating an object of stigmatisation. To understand the role of the media as the big Other makes clear that descriptions and appellations, such as that the 'aura of evil' (FP 12.11.2011) are more than just the voices of journalists. The connection of news about Zwickau with right-wing terrorism leads to a significant change of the city's 'sense'. The city became a 'synonym for right-wing violence', as a newspaper concludes more than one year after the uncovering of the *Zwickauer Terrorzelle* (SZ 12.05.2013). This means more than just a local rebranding. The change of 'sense' offers the insecure and mutable nature, not just of the toponym (Medway and Wanaby, 2014: 157), but of the place in general. Therefore, the right-wing terror became *the* signifier of the branded city and a 'logical point of identification for inhabitants and visitors alike' (Boisen *et al.*, 2011: 138). But as Lacan (1999: 114) notably stated, the big Other does not exist by itself, but just as a 'hole'. This means that the symbolic itself is lacking, and if we take this seriously, no-one has the control over the city's image. From the local as well as the media perspectives, the problem of right-wing terrorism can be understood as 'an inherently spatial process of psychical apprehension' (Kingsbury, 2007: 253). The object of stigmatisation leads to a modification of the spatial extimacy, a change that is powerfully objectified and externalised (even by official politicians who deny, via the media, that there is really something like right-wing terrorism in the city) as well as embodied and internalised (by city politics, citizens and even by the built environment).

This way of problematising right-wing terrorism as a process that changes local extimacies, allows us to put the focus on the contingency of imaginary politics and the significance of media in general.[18] Since, in this case, right-wing terrorism is limited to a problem of place branding, local politics are mainly acting in accordance with the filled lack of local identification. Terrified by the idea that the *Zwickauer Terrorzelle* could really 'finalise' the portrait of the city once and for all, the imaginary politics produce specific possibilities of activism. Via the tension of competition and factual constraints, politics and citizens seek to fight for a common goal: 'to save the city's reputation' (FP 25.11.2011b). In this sense, we have to question *what* the city is fighting for/against? Every attempt – the appeals, demonstrations, interviews and even the demolition of the ruin – is based on the medial representation of the city image. There is no consistent problematisation of issues such as right-wing extremism in general, the importance of place branding policies or even the local right-wing scene in the city – just an attempt to find solutions for the traumatic object, which seems to fulfill the imagination of Zwickau as being *the* city of right-wing terror in Germany. In this context, although the city claims to 'distance itself from right-wing terror' (FP 18.11.2011a), it is notably 'struggling for its image' (FP 21.11.2011).[19]

But how to get rid of the 'problem' of being locally stigmatised through right-wing terrorism? According to Lacan we could say that, from an imaginary point of view, there is 'no way out'. Following a topological view of the city, local politics – as long as it seeks to draw a unique and distinctive portrait of the city – will never find image-building objects to finally 'fulfill' its own task without running into possible objects of stigmatisation. A first step to 'get rid of the problem' is to find ways to problematise right-wing terrorism and extremism without reducing 'the political' to problem-solving practices. We have to acknowledge that to govern a city inherently means to govern its competitive, lacking and contingent 'sense of place'. From this point of view, we have to question the problem of territorial stigmatisation itself. We should be attentive and ask how and why the stigmatised worry and fight against objects of stigmatisation. Because, as we have seen, a fight against 'right-wing territories' does not necessarily have to be a fight against right-wing terrorism as well.

Acknowledgements

I am grateful to Thomas Bürk, who was a great source of inspiration to write this chapter. Furthermore, I would like to thank Nathan Bullock, Alev Coban, Mona Friedrich, Moritz Herrmann, Paul Kingsbury and Simon Runkel for their responses and advice.

Notes

1 All translations from Germans are by the author.
2 Even if there is intensifying scientific research in the field of 'psychoanalytic geographies (for a recent example, see Kingsbury and Pile 2014), Kingsbury (2015: 330) claims that 'although many geographers are familiar with psychoanalytic ideas, many geographers are unsure about how and why its theories could further their research practices'. Furthermore, Bullock (2014: 213) mentions that 'Lacanian psychoanalysis has been used in film, literature and other areas of social thought, but rarely in the domain of urban studies and human/cultural geography'. In this sense, my approach seeks to contribute to both of these claims.
3 For Lacan the ego is not understood as the subject in a traditional sense. The ego is an imagination of oneself as a whole, which is a result of the splitting between organism and reality in relation to the mirror (Lacan, 2006: 78). On the contrary, the subject for Lacan is a 'subject-with-holes' (Lacan, 1998: 184).
4 The term 'Topo-Analysis' is inspired by Bachelard's *Poetics of Space* where it is defined as part of the field of psychoanalysis (Bachelard, 2014 [1964]: 82).
5 For example, almost every German city is working on a 'corporate identity' – emerging as a unified and coherent local identity. This kind of local branding is systematically produced for the public and often spread visually via brochures, websites or social media platforms.
6 Here, one can draw parallels to already existing understandings of image politics, which in a similar way differentiate between internal and external dimensions of image politics (see Vanolo, 2008: 371).
7 The term 'imagineering' is a semantic combination of the words 'imagination' and 'engineering'. Originally used by the Walt Disney Corporation, 'it suggests a conscious attempt to mould the world through imagination and imaginative work' and is useful for us 'because it brings together the representational and the material aspects of the practice' (Jonas *et al.*, 2015: 195).

8   In this way, my notion of 'sense' has strong similarities to Massey's approach as she claims that a 'progressive concept of place' needs to highlight that places are not static, but processes with necessary boundaries, not as simple divisions, but as something that constitutes an in- and outside; and that places 'do not have single, unique 'identities', but are defined by 'internal conflicts' that continually reproduce the 'specificity of place' (Massey, 1991: 29).

9   This is what Stavrakakis (2007a: 144) called the 'fundamental fantasy' of the utopias of the late-capitalist city: 'the idea that all holes can be filled.'

10 The main part of the empirical data that I use is based on a wider discourse analysis, which I did conducted between the 4 November, 2011 and the 30 July, 2012. Several parts of it are already published in German (Pohl 2013) and now translated in English. The quotes are from the local newspaper *Freie Presse* (FP) and the national wide newspaper *Süddeutsche Zeitung* (SZ). It is crucial to mention that Zwickau is far from being the only case of place-based stigmatization through right-wing extremism. There have in fact been various empirical studies completed especially in peripheral parts of Germany, where phenomena like decline in jobs, lack of social infrastructures and the absence of young and educated, as well as migrated people, have often resulted in xenophobic and even racist urban developments (cf. Bürk, 2012; Döring, 2008).

11 Wacquant's separation between social and physical space goes back to his former teacher Pierre Bourdieu. For him the social space is a 'relation of social positions', characterized by its 'mutual exteriority', relative distance, and rank ordering (Bourdieu, 1998). It can, but does not have to affect the physical landscape.

12 During the first months, the term *Zwickauer Terrorzelle* dominated the media discourse about the right-wing terror group. As the mayor of Zwickau declared in an interview, it was therefore necessary to contact the chancellor 'in order to formulate a different kind of language' (FP 03.06.2015). This is a main reason why today's politics and media speak about the 'National Socialist Underground' instead of the *Zwickauer Terrorzelle* to avoid a negative connotation with the city.

13 After arresting Beate Zschäpe, the police revealed that the three citizens of Zwickau were the planners of a German-wide series of murders of men with an immigrant background. From 2000 to 2006 the terror group killed nine people and since 2013, Zschäpe is accused of murder in legal proceedings, which in the last year became widely discussed throughout the media as well as in politics.

14 Robert Schumann was a famous German composer who was born in Zwickau. For this reason, the image politics of Zwickau also referred to the city as 'the city of Robert Schumann' (see Schumannzwickau, 2015)

15 An interesting correlation here is that Freud describes two techniques of symptom according to obsessional neurosis, namely isolating and undoing (Freud, 2013 [1928]: 40).

16 The term 'ex-sistence' is based on Lacan's reading of Heidegger, who used it in reference to the Greek *ekstasis* and the German *Exstase* (Fink, 1995: 122) to focus on the 'eccentric place' as the location of the subject of the unconscious (Lacan, 2006: 6).

17 For a state theoretical approach to the term 'terrortory', see Hindess (2006).

18 Here I address Gunders' (2011: 325) call for a more precise debate of the role of media according to questions of urban desires in city planning, as well as to develop a discussion about the role of media in regards to urban anxiety produced by imaginary politics that is still lacking in geographical debates.

19 As Pohl (2013) has pointed out, the city of Zwickau framed itself in the same way years before the *Zwickauer Terrorzelle* took place. The place-branding strategies of Zwickau, that I referred to in the introduction of this chapter, have not changed. If one analyses the city's profile on its website today, one will find no word about what happened in Zwickau at the end of 2011.

References

Anholt, S. (2010), 'Place image as a normative construct; and some new ethical considerations for the field', *Place Branding and Public Diplomacy*, 6(2): 177–181.
Bachelard, G. (2014) [1964], *The Poetics of Space*, New York: Penguin Books.
Boisen, M., K. Terlouw and B. van Gorp (2011), 'The selective nature of place branding and the layering of spatial identities', *Journal of Place Management and Development*, 4(2): 135–147.
Bourdieu, P. (1998), *Practical Reasons. On the Theory of Action*, Cambridge: Polity Press.
Bullock, N. (2014), 'Lacan on urban development and national identity in a global city: Integrated Resorts in Singapore: Identity of integrated resorts in Singapore', *Singapore Journal of Tropical Geography*, 35(2): 213–227.
Bürk, T. (2012), *Gefahrenzone, Angstraum, Feindesland? Stadtkulturelle Erkundungen zu Fremdenfeindlichkeit und Rechtsradikalismus in ostdeutschen Kleinstädten*, Münster: Westfälisches Dampfboot.
Campelo, A. (2015), 'Rethinking Sense of Place: Sense of One and Sense of Many', in M. Kavaratzis, G. Warnaby and G. J. Ashworth (eds), *Rethinking Place Branding. Comprehensive Brand Development for Cities and Regions*, Heidelberg: Springer: 51–60.
Chiesa, L. (2007), *Subjectivity and Otherness. A Philosophical Reading of Lacan*, Cambridge, MA: MIT.
Cromwell, T. (2011), 'The East-West Nation Brand Perception Indexes and reports: Perception measurement and nation branding', in F. Go and R. Govers (eds), *International Place Branding Yearbook 2011: Managing Reputational Risk*, Basingstoke: Palgrave Macmillan: 92–101.
Döring, U. (2008), *Angstzonen. Rechtsdominierte Orte aus medialer und lokaler Perspektive*, Wiesbaden: VS Verlag für Sozialwissenschaften.
Evans, D. (1996), *An Introductory Dictionary of Lacanian Psychoanalysis*, London: Routledge.
Fink, B. (1995), *The Lacanian Subject. Between language and Jouissance*, Princeton, NJ: Princeton University Press.
Freud, S. (2013) [1928], *Hemmung, Symptom und Angst*, Hamburg: Nikol.
Gunder, M. (2005), 'The production of desirous space: Mere fantasies of the Utopian city?', *Planning Theory*, 4(2): 173–199.
Gunder, M. (2011), 'A metapsychological exploration of the role of popular media in engineering public belief on planning issues', *Planning Theory*, 10(4): 325–343.
Harvey, D. (1996), *Justice, Nature and the Geography of Difference*, Malden, MA: Blackwell.
Hindess, B. (2006), Terrortory, *Alternatives*, 31: 243–257.
Jansson, A. (2003), 'The negotiated city image: Symbolic reproduction and change through urban consumption', *Urban Studies*, 4(3): 463–479.
Jansson, A. (2005), 'Re-encoding the spectacle: Urban fatefulness and mediated stigmatization in the "City of Tomorrow"', *Urban Studies*, 42(10): 1671–1691.
Johansson, M. (2012), 'Place branding and the imaginary: The politics of re-imagining a Garden City', *Urban Studies*, 49(16): 3611–3626.
Jonas, A., McCann, E. and Thomas, M. (2015), *Urban Geography. A Critical Introduction*, Malden, MA: Blackwell.
Kavaratzis, M. (2007), 'City marketing: The past, the present and some unresolved issues', *Geography Compass*, 1(3): 695–712.

Kerr, G. and Oliver, J. (2015), 'Rethinking place identities', in M. Kavaratzis, G. Warnaby and G. J. Ashworth (eds), *Rethinking Place Branding. Comprehensive Brand Development for Cities and Regions*, Heidelberg: Springer: 61–72.

Kingsbury, P. (2007), 'The extimacy of space', *Social and Cultural Geography*, 8(2): 235–258.

Kingsbury, P. (2015), 'Becoming literate in desire with Alan Partridge', *Cultural Geographies*, 22(2): 329–344.

Kingsbury, P. and Pile, S. (2014) (eds), *Psychoanalytic Geographies*, Farnham: Ashgate.

Kristeva, J. (2003), 'The Future of a Defeat. Julia Kristeva interviewed by Arnaud Spire', *Parallax*, 9(2): 21–26.

Lacan, J. (1991), *The Ego in Freud's Theory and in the Technique of Psychoanalysis. The Seminar of Jacques Lacan. Book II 1954–1955*, New York: W. W. Norton.

Lacan, J. (1998), *The Four Fundamental Concepts of Psychoanalysis. The Seminar of Jacques Lacan. Book XI 1964*, New York: W. W. Norton.

Lacan, J. (1999), *Encore. The Seminar of Jacques Lacan. Book XX 1972–1973*, New York: W. W. Norton.

Lacan, J. (2005), *Le séminaire livre XXIII: Le sinthome*, Paris: Seuil.

Lacan, J. (2006), *Ecrits*, New York: W. W. Norton.

Lacan, J. (2007), *Das Seminar Buch IV. Die Objektbeziehung 1956–1957*, Wien: Turia+Kant.

Lacan, J. (2008a), *The Psychoses. The Seminar of Jacques Lacan. Book III 1955–1956*, New York: W. W. Norton.

Lacan, J. (2008b), *My Teaching*, London: Verso.

Lacan, J. (2014), *Anxiety. The Seminar of Jacques Lacan. Book X 1962–1963*, Cambridge: Polity.

Marcuse, P. (2005), '"The city" as perverse metaphor', *City*, 9(2): 247–254.

Martin, L. and Secor, A. (2014), 'Towards a post-mathematical topology', *Progress in Human Geography*, 38(3): 420–438.

Massey, D. (1991), 'A Global Sense of Place', *Marxism Today*, June: 24–29.

Medway, D. and Warnaby, G. (2014), 'What's in a name? Place branding and toponymic commodification', *Environment and Planning A*, 46: 153–167.

Pile, S. (1996), *The Body and the City. Psychoanalysis, Space and Subjectivity*, London: Routledge.

Pohl, L. (2013), 'Verräumlichung von Stigmatisierungsdiskursen. Zur städtischen Problematisierung von Rechtsextremismus am Beispiel der ‚Zwickauer Terrorzelle'', in H. Kellershohn and J. Paul (eds), *Der Kampf um Räume. Neoliberale und extrem rechte Konzepte von Hegemonie und Expansion*, Münster: Unrast: 58–74.

Secor, A. (2013), 'Topological City', *Urban Geography*, 34(4): 430–444.

Sevin, H. E. (2014), 'Understanding cities through city brands: City branding as a social and semantic network', *Cities*, 38: 47–56.

Stavrakakis, Y. (1999), *Lacan and the Political*, London: Routledge.

Stavrakakis, Y. (2007a), 'Antinomies of space. From the representation of politics to a topology of the political', in BAVO (eds), *Re-Imagining Democracy in the Neoliberal City*, Rotterdam: NAi: 141–161.

Stavrakakis, Y. (2007b), *The Lacanian Left. Psychoanalysis, Theory, Politics*, New York: SUNY.

Vanolo, A. (2008), 'The image of the creative city: Some reflections on urban branding in Turin', *Cities*, 25: 370–382.

Wacquant, L. (2007), 'Territorial stigmatization in the age of advanced marginality', *Thesis Eleven*, 91: 66–77.

Wacquant, L. (2010), 'Designing urban seclusion in the 21st century', *Perspecta: The Yale Architectural Journal*, 43: 165–178.

Wacquant, L. (2012), 'Three steps to a historical anthropology of actually existing neoliberalism', *Social Anthropology*, 20(1): 66–79.

Wacquant, L., Slater, T. and Pereira, V. B. (2014), 'Territorial stigmatization in action', *Environment and Planning A*, 46: 1270–1280.

Žižek, S. (2006), *How to read Lacan*, New York: W. W. Norton.

Sources from media

Freie Presse (12.11.2011), 'Mutmaßlicher Komplize der Terror-Gruppe gefasst.'

Freie Presse (18.11.2011a), 'Zwickau will auf Distanz zu rechtem Terror gehen.'

Freie Presse (18.11.2011b), 'Zwickauer Demokraten planen eine Demo.'

Freie Presse (21.11.2011), 'Zwickau kämpft um sein Image.'

Freie Presse (25.11.2011a), 'Pia Findeiß: „Es hat mich eiskalt erwischt".'

Freie Presse (25.11.2011b), '"Das Gewissen darf mit diesem Abend nicht beruhigt sein".'

Freie Presse (24.04.2012), 'Tulpen statt Terror.'

Freie Presse (03.06.2015), '"Da muss man standhaft bleiben".'

Schumannzwickau (2015), http://www.schumannzwickau.de/en/default.asp (accessed 19 July 2015).

Süddeutsche Zeitung (02.02.2012), 'Im Schutt.'

Süddeutsche Zeitung (12.05.2013), 'Umgang mit Terrorzelle entzweit Zwickauer.'

Zwickau (2008), http://www.zwickau.de/de/wirtschaft/service/publikationen/wfoerderkonzept/Marketing.php (accessed 19 July 2015).

Zwickau (2015a), https://www.zwickau.de/de/wirtschaft.php (accessed 19 July 2015).

Zwickau (2015b), https://www.zwickau.de/de/wirtschaft/wirtschaftsstandort.php (accessed 19 July 2015).

Zwickau (2015c), https://www.zwickau.de/de/tourismus.php (accessed 19 July 2015).

4 Extensive territorial stigma and ways of coping with it

The stigmatisation of the Roma in Italy and France

Gaja Maestri[1]

Introduction

Alexandra is a young girl who lives in a Roma camp in Rome. But she does not tell her schoolmates that she is Roma[2] for fear that they might change their attitudes towards her. This case is not isolated and sadly represents the effects of the current general feelings towards the Roma in France and Italy, where there has been an increase in negative attitudes towards this group. This chapter focuses on two main questions: first, it interrogates this increasingly anti-Roma feeling and asks what led to this exacerbation; second, it discusses the strategies developed to cope with this form of stigmatisation. By unpicking the different aspects contributing to the process of stigmatisation of the Roma, I aim to show that what Loïc Wacquant has called 'territorial stigma' can also apply to people not living in negatively represented places. In this rather specific case, that I call *extensive* territorial stigma, geographic mobility is no longer sufficient to escape it. In fact, the strategies adopted by Roma living and not living in stigmatised camps, rather than constituting a form of escape, are based on new and reinforced feelings of place attachment, solidarity and resistance. Furthermore, the chapter emphasises the capacity of resistance of the most marginalised Roma migrants who, despite their highly precarious living conditions, are consolidating the emergence of a new political subjectivity.

In the first part, I conduct an analysis of both the historical and political dynamics that have produced this hostility towards the Roma people. I examine the historically rooted anti-Roma attitudes and how these have been increasingly coupled with representations of Roma informal settlements as an expression of Roma otherness. In both Italy (in 2008) and France (in 2010), repressive measures targeting Roma migrants living in informal settlements played a crucial role in the production of anti-Roma feelings, extending the effects of the territorial stigmatisation of informal settlements to the whole Roma community. In the second part, I analyse the different – and ambivalent – strategies developed by Roma people to cope with this extensive territorial stigma: submission to stigma yet attachment to the settlement community, solidarity between Roma living in informal settlements and the rest of the Roma population and, finally, resistance uniting Roma migrants and squatters. This section draws on the fieldwork conducted in Rome from September to December 2013 (and a few other cities throughout 2013 and 2015, including Bologna and Brescia) and in Paris from January to June 2014 and part of 2015. During the fieldwork I collected more than 40 in-depth interviews but I also entered into a large number of informal conversations with both Roma people and members of pro-Roma advocacy groups.

Increasing negative attitudes towards the Roma and the process of territorial stigmatisation

During the last decade both Italy and France have witnessed an increase in anti-Roma attitudes. According to a recent survey conducted by the Pew Research Center,[3] Italy and France are the European countries where negative views about the Roma are at their highest (Pew Research Center, 2015). The proportion of Italians who perceive the Roma negatively has increased from 84 per cent in 2009 to 86 per cent in 2015 (Pew Research Center, 2009, 2015). Similarly, in France, a report by the *Commission Nationale Consultative des Droits de l'Homme* (National Consultative Commission on Human Rights, CNCDH) published in 2015,[4] concluded that unfavourable attitudes towards the Roma increased by 16 per cent between 2011 and 2014 (CNCDH, 2015). These results are supported by a study of the European Union Agency for Fundamental Rights (FRA),[5] which shows that 66 per cent of the Roma in Italy and 52 per cent in France feel discriminated on the basis of their ethnicity (FRA, 2012).

Interestingly, the studies conducted by the Pew Research Center and by the FRA have also shown that anti-Roma feelings have actually decreased in Spain, where Roma also feel less discriminated against.[6] This demonstrates that negative attitudes and discrimination are neither linked to the size of the Roma population, nor to the recent arrival of Roma migrants from Eastern Europe. Figures of the Council of Europe indeed show that in both France and Italy the Roma constitute less than 1 per cent of the population. While Italy's Roma population is about 150,000 (i.e. 0.25 per cent of the population, which makes it one of the countries with the lowest percentage of Roma in Western Europe) and in France is around 400,000 (i.e. 0.62 per cent), in Spain, there are nearly one million Roma. As for the presence of Roma migrants from Eastern Europe, it is estimated that they approximately amount to 20,000 in France, 30,000 in Italy and 50,000 in Spain (Doytcheva, 2015).

Survey data need be carefully contextualised as, otherwise, they risk underplaying the political dynamics at the origins of the opinions that they reveal (Vitale and Claps, 2011). The different levels of anti-Roma feelings in these countries show how both the presence of Roma and the recent arrival of Roma migrants living in informal settlements have been framed differently through processes of stigmatisation and articulated within specific historical and political contexts, which produced in some cases a widespread territorial stigma. Territorial stigma relates to one's place of residence and the term was coined by the French sociologist, Loïc Wacquant (2007) by drawing both on Goffman's account of stigma (Goffman, 1963) and on the Bourdieusian theory on symbolic power and space

(Bourdieu, 1985). As Wacquant argues, this type of stigmatisation has become increasingly prominent in contemporary urban spaces and, in theory, can be dissimulated and attenuated relatively more easily than other forms of stigma through geographic mobility (Wacquant, 2007). Wacquant's work has informed several studies, but as Slater observes, the literature on territorial stigma is often limited to the analysis of its '*consequences* rather than *causation*' (Slater, in press, p. 6). In fact, today there is the need of more research on stigmatising agents, institutions and policies. Following Slater's call, the next sections will examine the historical and political dynamics involved in the articulation of the territorial stigma of the Roma in Italy and France respectively.

The historical roots of the territorial stigmatisation of the Roma

The Roma people have historically suffered from discrimination and constituted the target of various attempts to control their mobility and spatial practices. Okely (1983) argues that the Roma were local groups in Western Europe who adopted an itinerant lifestyle at the end of Feudalism, primarily for economic reasons. Because at that time workers were not entitled to free mobility but were attached to the land on which they were working, the Roma started being stigmatised and targeted by the first attempts to control labour migration (Lucassen, 1998). The discrimination of the Roma has been historically framed in different ways: initially they were portrayed as vagabonds, then they were treated as an untrustworthy racial group, and more recently, they have been depicted as nomads (Okely, 1983). However, all the negative representations of this group justified repressive policies and forms of control aimed at their mobile lifestyle and places of residence, reinforcing the historical stigma towards the Roma, and even leading to see them as inferior race. For example, in 1912 France obliged the Roma to carry a *carnet anthropometrique d'identité* (anthropometric identity card) reporting their civil status and biometric information with the aim of enforcing control over them (Noiriel, 1988). In Italy, despite the existence in the past centuries of laws banishing the Roma (Clough Marinaro, 2009), there was no official documentation for them. The proposed introduction in Italy of a document similar to the French *carnet* failed but this did not result in a less discriminating situation, and in both countries the Roma were persecuted and detained in camps during the Second World War.

From the 1970s onward, the discourse on nomadism replaced the more explicit racist one dominant in the first half of the twentieth century.[7] In the 1960s, in France, the term *gens du voyage* (i.e. Travellers) replaced the negatively connoted *nomades*, and the *carnet anthropometrique* was changed to a compulsory *livret de circulation* (circulation card), focusing on the control of mobility rather than on a racial categorisation. The Roma became the object of policies aimed at both their forced sedentarisation and confinement. For instance, the 1969 law that established the obligation of residency, also fixed a cap on the presence of Roma in each French municipality. Italy started developing a discourse on nomadism in the 1960s under the pressure of associations supporting the Roma living in informal settlements, that advocated the protection of the right to roam. As a result of this pressure, in the 1980s several Italian regions adopted laws for the protection of the Roma minority, which also established the creation of official Roma camps for nomadic groups. Although these camps were created as a protection of nomadism, they led to the further exclusion of the Roma, many of whom today live in these segregated spaces, separated from the rest of the population (European Roma Rights Center, 2000). In these two countries, the discourse on nomadism and the policies adopted to manage the mobility of nomadic populations resulted in a forced sedentarisation and fed into a process of stigmatisation of the places of residence of these groups, and also of newly arrived Roma migrants.

The current exacerbation of anti-Roma feelings in Italy and France needs to be read within the context of this historically rooted anti-Roma feeling and the current stigmatisation of the newly arrived poor Roma from Eastern Europe. In contrast with other Western European countries experiencing similar historical racism and recent migration – such as Spain (Manrique, 2015) – France and Italy were the only two Western European countries in which the migration of poor Roma from Eastern Europe was framed as a national emergency discourse (Magazzini and Piemontese, 2015). Following a series of highly mediatised clashes and events involving Roma communities, the Italian government promulgated in 2008 a *Decreto Emergenza Nomadi* (Nomads Emergency Decree) (Consiglio dei Ministri, 2008) aimed at the removal of most of the Eastern Roma informal settlements. Two years later, the French government enacted a series of controversial repatriations of Eastern European Roma. In the next two subsections I will illustrate how the territorial stigma of the Roma settlement was generalised and extended to the whole Roma population.

The 2008 Nomads Emergency in Italy and the creation of the threatening nomads

The city of Rome is an interesting entry point to unpack the discourse on the Roma and informal settlements as the political debate in the Italian capital often strongly influences other cities (Clough Marinaro, 2009). Roma informal settlements started being seen as a political problem during the early 1990s with the arrival of several Roma asylum seekers escaping the conflict in ex-Yugoslavia. At their arrival in Italy, many of the Roma asylum seekers did not receive any protection – partly because they were seen as nomads by policymakers (Sigona, 2003) – and they therefore started squatting in abandoned buildings and on plots of land (Sigona, 2015). In order to tackle this so-called 'Roma problem', the municipality of Rome adopted in 1994 a Nomads Plan, consisting of a set of policy measures that would supposedly solve the 'critical' situation by clearing the informal settlements and relocating the evicted Roma in official camps run by the city of Rome. However, the problem was far from being solved, and since then until today, every administration of Rome has claimed to be facing an emergency with regard to the Roma informal settlements, portrayed either as constituting a threat to public health because of their unhealthy and unhygienic conditions (therefore justifying a humanitarian intervention), or as threatening the national and public security (followed by repressive measures).

The measures against the Roma aggravated in 2008 when the national government adopted the so-called Nomads Emergency Decree, arguing that the presence of several 'irregular nomad camps' constituted a threat to public health, order and security. The prefects of the cities of Naples, Rome and Milan were thus given the role of extraordinary commissioners for solving this alleged 'emergency situation' through increased monitoring powers, the right to clear Roma settlements and the ability to collect the personal data of Roma and to repatriate undocumented migrants.

The Nomads Emergency Decree condemned the 'settlements of nomadic communities', and reported that the 'emergency situation' stemmed from the presence of 'several non-EU citizens, irregular immigrants and nomads that have stably settled in urban areas' (Consiglio dei Ministri, 2008). Although the text begins by mentioning non-EU citizens and undocumented migrants, the rest of the declaration only refers to solutions destined to solve the 'problems' linked to 'nomadic communities' and to the critical situation affecting provinces where there was a higher concentration of informal settlements. Similarly, the regulation of official camps, adopted in Rome as part of the national emergency (Regione Lazio, 2009), mentions nomadic communities as the main target of these measures. When the emergency started in Rome, the local political debate was mainly focused on Roma originating from Romania, whose presence and mobility fell within the interpretative framework of nomadism, lumping together the stigmatisation of recent Roma migrants living in informal settlements and the historical negative stereotypes of the Roma people.

The state of emergency was initially set to last one year, but it was eventually renewed three times, being officially annulled in May 2013. The consequences on the stigmatisation of the Roma people can still be felt today. The *Associazione 21 Luglio* recorded a continuous stigmatisation in the national media and newspapers with almost one episode of hate speech a day on average in the Italian media in the first six months of 2015 (Associazione 21 Luglio, 2013, 2014). In addition to this, over the course of the few last months, there has been a growing amount of hate speech proffered by senior politicians and directed towards this group – which also found its way into the media. One example was highlighted by *Associazione 21 Luglio*, when Matteo Salvini, a member of the far-right Northern League, reignited the debate against the Roma informal settlements by launching the slogan 'Bulldozers against Roma camps'. This received high media coverage and has bolstered negative stereotypes about the Roma.

The 2010 Roma deportations in France and the construction of Roma deviance

In France, popular discussions about Roma informal settlements began to emerge during the early 2000s (Cahn and Guild, 2010). The combination of the arrival of a high number of Roma migrants and the accession of Romania and Bulgaria to the European Union in 2007 led to the deployment of a series of measures that explicitly targeted the Roma migrants from Eastern Europe. These measures mainly focused on the demolition of Roma settlements and repatriations. In 2010, the situation of stigmatisation of Roma migrants was exacerbated by a series of highly mediatised violent episodes which had allegedly implicated members of French Travellers communities and led to rioting against the police (Barbulescu, 2012; Parker, 2012).

The situation eventually brought about the Declaration on Security, presented in a speech delivered by President Nicolas Sarkozy on 30 July, 2010, in Grenoble. Sarkozy opened the speech by defending the actions of the police and blaming the 'gangsters involved in the rioting' (Parker, 2012, p. 477). He stressed the importance of the fight against criminality and outlined some measures to tackle crime (Carrera and Atger, 2010). Despite not being directly involved in the violent episodes, the second part of the speech mainly concerned the situation of the Roma from Eastern Europe and informal settlements, considered as 'lawless zones that', he continued, 'we cannot tolerate in France.' Sarkozy promised that 'in three months, half of these untamed settlements will be erased from French territory', and he referred to 'various kinds of delinquents in France, including immigrants that have failed to integrate and, in particular, the Roma' (Parker, 2012, p. 478).

By linking the category of Roma (especially those from Eastern Europe) to topics such as illegal immigration, Sarkozy was 'transforming every Roma into an illegal immigrant' (Barbulescu, 2012, p. 287). While a communiqué from the French Presidency, published on 28 July, 2010, restated that only a small minority of the French Travellers constituted a threat to the public order, it explicitly targeted all the Roma migrants coming from Eastern Europe. The French President stated that he could not accept the unlawful situation of 'the Roma populations arriving from Eastern Europe into French territory', and who he associated to activities of illegal trafficking, 'shameful' living conditions, child exploitation and delinquency. Sarkozy's speech framed the problems affecting informal settlements (i.e. low sanitary standards, squatting, tensions with the neighbours) within the lens of imagined cultural specificities of the Roma minority.

From the outset, France's policing operations received strong criticisms, mainly from the European Commission (Carrera and Atger, 2010; Atger, 2013). Concerns about ethnic discrimination originated from a series of ministerial memorandums explicitly referring to the dismantling of Roma settlements and repatriations (European Roma Rights Center, 2000). One of these memorandums, dated from 5 August, 2010, has been particularly criticised for an 'explicit targeting of the Roma as ethnic group by the French authorities' (Parker, 2012, p. 479). The existence of these administrative circulars had been constantly denied by the French government, until they were published in the press following an information leak (Carrera and Atger, 2010; Barbulescu, 2012), after which the Interior Ministry declared that they would remove all the references to the Roma (Parker, 2012).

The election of a centre-left government in 2012 has hardly improved the situation. In fact, the number of deportations and evictions of Roma migrants has remained stable (Martin, 2013) and there have been no substantial changes in the political debate. For instance, in 2013 Manuel Valls, the Interior Minister, stated that the Roma are 'people that have a way of living extremely different from ours' (*Le Monde*, 2013). The associations *Mouvement contre le racisme et pour l'amitié entre les peuples* and *La Voix des Rroms* pressed charges against him for inciting hatred against the Roma but both cases were closed without further process.

Coping with extensive territorial stigma: new forms of place attachment, solidarities and coalitions

The strong anti-Roma feelings illustrated at the beginning of the chapter are due to the enhancement of a long-lasting stigmatisation of Roma people and nomadism through the current discrimination of Roma living in informal settlements. Through the two events in Italy and France, the mediatised stigmatisation of specific places, i.e. Roma migrants' informal settlements, combined with the historical discrimination of the Roma minority. As a consequence, the informal settlement has been interpreted as a Roma cultural feature, and its negative connotation has been extended to the representation of the whole Roma population, bearing effects also on those who are not living in these stigmatised places.

After having investigated the causes of these anti-Roma attitudes in France and Italy, this section aims to understand its consequences. If, as Wacquant (2007) argues, territorial stigma is relatively easier to deal with as compared to other forms of stigma, since it can be escaped by geographic mobility, how do Roma people, including those who do not live in stigmatised neighbourhoods, deal with it? From the interviews I conducted with Roma individuals and association members it emerged that, in these cases, geographic mobility and the strategy of 'exit' (Wacquant, Slater and Pereira, 2014), are not enough. Rather to the contrary, it is through 'voice', solidarities and new forms of place attachment that the Roma cope with this type of stigma, troubling the distinction between submission and resistance.

Coping with stigma I: submission strategies and retreat into the community life

Strategies of submission, such as 'mutual distancing' and 'lateral denigration', imply that someone suffering from territorial stigma starts to look at the other members of their communities as those who are responsible for the bad reputation of the neighbourhood. For example, mutual distancing was adopted by some Sinti towards the Roma at a demonstration held in Bologna on 16 May, 2015 (Figure 4.1).[8] Although both Roma and Sinti associations were involved in the organisation of this rally against the growing negative attitudes towards the Roma population, the main organisers were Sinti groups. The demonstration conveyed an ambivalent message since, despite constituting a moment of solidarity, the parade was divided into two main groups: the Sinti in the front, singing the Italian anthem and claiming their Italian identity, and other associations with banners on the solidarity between Roma, Sinti and other migrants, at the back. For the Sinti, stressing their Italian identity was a way to distinguish themselves from Roma migrants and, therefore, to distance themselves from anti-Roma feelings.

In France too, there have been cases of ambivalent relationships between French *gens du voyage* and Roma migrants. For instance, in 2013 the president of the association *France Liberté Voyage* (an association of *gens du voyage*) stated that:

> We highly respect the Roma and we have to help those who are bound to live in horrible slums. But our problems are different. We reject all these misunderstandings that only add the prejudices against the Roma to all the discriminations that we are already subject to.
>
> (*Dépêches Tsiganes*, 2013)

The main concern of this group of *gens du voyage* is that the discriminations against the Roma migrants living in informal settlements could further worsen the already difficult situations that they experience. For this reason *France Liberté Voyage* did not participate in the French Roma Pride parade that took place on 6 October, 2013, distancing itself from pro-Roma mobilisations.

Figure 4.1 Protesters attending the demonstration in Bologna wave a banner stating, 'With the Roma and Sinti, against all forms of racism, to defend a common humanity.'

Source: Gaja Maestri, 2015.

From interviews, it also emerged that Roma recourse to the tactic of mutual distancing. For instance, Imer, a Kosovan Roma man living in an official settlement in Italy, drew a clear distinction between his Roma family group and his neighbours, who are Sinti:

> They are Sinti, Italian gypsies! [. . .] The gypsies are people who fiddle around [. . .]. There are gypsies also in Kosovo, and here there are Italian gypsies. I know how they are because also in Kosovo there are people that steal, that are stopped by the police. When we visit my wife's sister in Pisa they're everywhere, stealing!

In this case, Imer (re)affirmed the negative stereotypes of the 'gypsy' as thief and untrustworthy. At the same time, he stressed that his family – and the community to which he feels they belong – were very different from the negatively portrayed Sinti families, thus articulating a form of attachment to his community:

> But I like it here [. . .] I mean, we would also enjoy living in a flat, of course, but everybody are here, my sister is here [. . .]. Everybody are from my village, my sister, my cousin, my grandpa, and then there is my dad, his brother [. . .]. We all grew up together like a big family. We never fight, and we help each other.

Other examples collected during the fieldwork also reveal strategies of 'dissimulation' and 'retreat into the private sphere' (see Wacquant, Slater and Pereira, 2014). When doing interviews in an official settlement in Rome, I met Alexandra, a 15-year-old Roma girl, and asked her if her schoolmates knew that she lived in a camp. She replied:

> No, no! Well, some of them yes, some understand, but most don't. . . You know, I don't tell everyone that I live in a camp! I'm a bit ashamed. I hate living here; I really look forward to leaving this place.

Jelena, another girl that I encountered during an associational meeting, shared similar feelings. The association was shooting a video for a television broadcast but Jelena did not want to appear in it because she was afraid that some of her classmates might recognise her and discover that she is Roma. She had never told her classmates about her ethnicity because she was afraid they would change their opinion about her, associating her with the negative media representations of informal settlements. This strategy of dissimulation is quite common among people from these communities both in Italy and in France. During a meeting with pro-Roma associations in Paris, Gary, a French Traveller man, told me about when he discovered that his children did not tell their classmates they are Travellers. He asked me:

> How would you feel as a father if your kids tell you that they feel ashamed of saying they're Travellers?

These types of reaction constitute an effort to escape the effects of this stigma without, however, challenging the stigma itself. But at the same time they strengthen the solidarity to the group and also the attachment to the stigmatised informal and official Roma settlements, as emerged in the interview with Imer, the Roma man from Kosovo. Also Alexandra, despite stating that she felt uncomfortable discussing her ethnicity and her place of residence with her school friends, added that nevertheless she feels attached to the community of the camp:

> I hate it here. But not because of the people, in a way I also enjoy living here because I like the community.

Although the exacerbated stigmatisation of the Roma, strongly linked to the negative representation of informal settlements, can generate submission to stigma at first sight, from the interviews it emerged that it also generates forms of place attachment. The interviewees quoted above underscored the solidarity within the informal settlement community as a strong important and positive aspect of the place in which they live. But there are also other cases of resistance to territorial stigmatisation based on the solidarity between those living in the informal settlements and Roma (and non-Roma) not living in these stigmatised places.

Coping with stigma II: resisting through solidarities

In both Italy and France there are Roma associations that campaign against misguided and stereotypical representations of Roma people as either nomads or adherents to a deviant culture, fostering the solidarity between Roma people. For instance, the *Fondazione Romanì* in Italy aims to promote positive images of the Roma through social awareness advertising and through supporting research and public debates around the issue of Roma discrimination. The *Fondazione Romanì* and other associations are taking part in a popular initiative called *Se Mi Riconosci Mi Rispetti* (i.e. If You Know Me You Respect Me) advocating the official recognition of the Roma minority. However, this initiative is controversial and some pro-Roma advocacy groups believe that it actually risks perpetuating the same ethnic approach that justifies the Roma segregation in camps. Other criticisms were put forward by some of the interviewees during the Roma and Sinti demonstration in Bologna, who feared that the petition could imply a dismissal of the differences between Roma and Sinti.

In France, there is one Roma association that draws a distinction between, on the one hand, the Roma as ethnic group and, on the other, the poor and marginalised Roma migrants that live in informal settlements, advocating the solidarity between the two. During an interview held in Paris in June 2015, Edi, a member of this group, explained the association's main message:

> You see, there are ethnic Roma and 'sociological Roma'. I am an ethnic Roma, but, sociologically, I've never been Roma. [. . .] Today, if you live in an informal settlement, you are Roma [. . .] even though maybe you've never been ethnically Roma!

This Roma association promotes solidarity between these two categories of Roma but remains distant from groups of French Travellers (*gens du voyage*), since they believe these groups to suffer from alternative forms of discrimination.

This emerging solidarity between Roma living in informal settlements and those who are not living in negatively represented places also create new forms of place attachment to the strongly stigmatised informal and official Roma settlements. For instance, some of the members who later founded the association *La Voix des Rroms*, as well as the association Parada, were involved in the fight for the housing exclusion of a group of Roma living in Saint-Denis, in the Paris region. In 2004 this Roma community settled in an informal settlement called Hanul (later acknowledged by the municipality that agreed to provide drinking water and waste collection services), and was eventually evicted in 2010. Several pro-Roma associations supported the Hanul community and fostered activities in this settlement, supported the families and children and also built, thanks to the help of volunteering architects, a tent in the camp for various types of activity, from leisure to political meetings.

Another non-Roma association, called PEROU, similarly supported a Roma community living in an informal settlement in Grigny, in the department of the Essonne, at the south of Paris. Their activities mainly aimed at making both informal settlements and the official ones more enjoyable places. For instance, they built

Figure 4.2 Residents and volunteers of the informal settlements in Grigny work together at the construction of the shack.

Source: Gaja Maestri, 2014.

a shack for an artist-in-residence programme in the Grigny settlement (Figure 4.2), and also a temporary structure in an official settlement in Ris-Orangis (not far from Grigny) for having meetings and parties with the residents (Figure 4.3).

These activities aimed to 'humanise' (a term used by a member of PEROU in an informal conversation) these places. The people living there took active part in the construction of the shack and of the tent, and these moments were used to socialise and to meet volunteers and activists living outside the settlements.

PEROU drew inspiration from an Italian association of architects, called Stalker, that promoted similar activities in Roma settlements in Rome. For instance, their last project in 2009, called Savorengo Ker (i.e. Everyone's House in Romani language), entailed the construction of a wooden house in an informal settlement with the help of the residents. The participation of the community helped to develop a feeling of belonging to the place where they lived, which is often lacking in informal settlements as the Roma are continuously forcibly evicted from one place to another.

The actions of these associations constitute a voice against Roma stereotypes and stigmatisation through the enabling of solidarities that develop between non-Roma and Roma, whether the latter are residing in settlements or not. However, although there has been a rise in the number of associations engaging in this type of resistance, the world of pro-Roma associations is still highly fragmented.

Figure 4.3 Residents of the official settlement in Ris Orangis gather for a party.
Source: Gaja Maestri, 2014.

As pointed out by Wacquant (2007), this is a problem affecting many of the organisations that represent and fight for the rights of the dispossessed and the stigmatised as they lack 'a common idiom around and by which to unify themselves' (Wacquant, 2007, p. 72). In this case, stigmatising the residents of informal settlements, newly arrived Roma from the East of Europe as well as Roma communities whose presence in the country is long established (such as the Italian Sinti or the French *gens du voyage*) weakens the claims of associations promoting solidarity among these groups. Despite being subject to the same type of stigmatising discourses, they constitute different groups and the attempts of coalitions against this widespread stigmatisation remain deeply affected by their different histories. But there are new forms of solidarity that bypass these differences, uniting the Roma with other non-Roma groups.

Coping with stigma III: Roma and squatters in-between 'exit' and 'voice'

In the last five years, a new type of strategy has emerged, combining both forms of exit and submission through the establishment of new solidarities and resistance practices. A series of political squats, set up by urban movements fighting against housing exclusion, encouraged the participation of Roma communities. In Rome, the first squat where this strategy was enacted is called Metropoliz, that was joined in 2009 by a group of 150 Roma after the demolition of their settlement. There were already cases of Roma squatting in abandoned buildings before, but they were not supported by a political claim or social movement. Metropoliz was set up in an abandoned factory in the eastern periphery of Rome in 2009 by the BPM group (*Blocchi Precari Metropolitani,* i.e. Metropolitan Precarious Blocks) with the help of Popica, an association working with Roma in informal settlements, that bridged the Roma community's and the squatters' claims. In Metropoliz, squatters, Roma, and non-Roma, present themselves as part of a new 'urban lumpenproletariat' composed of people that are left at the margins of the formal economic and political system. The Metropoliz group negotiated with the City of Rome that in case of evictions the occupants will be entitled to social housing.

In France, two Roma families are living in Dilengo, a squat in Ivry-Sur-Seine, just outside Paris. The squat was set up after the occupation of an abandoned building in August 2013, and was joined by two Roma families coming from an informal settlement in the same area. Notwithstanding an eviction order in June 2015, the squat residents managed to receive an extension for staying three more years. Similarly to the cases in Metropoliz, in the Dilengo the Roma did not make claims based on their ethnicity but based on the socio-economic conditions they share with the other occupants.

These cases combine a strategy of exit with a form of submission to stigma. The Roma that moved out of the informal settlement (in these cases, to political squats), choose to identify themselves primarily as squatters instead of making claims on the base of their ethnicity. These new shifting identities also generate new types of place attachment to the place of the squat, where new solidarities are nourished. Although in Metropoliz there were some difficulties at the beginning between the Roma and the rest of the squatters because of the widespread negative stigma towards the Roma groups and also because the Roma preferred to live in an isolated hangar in front of the occupied factory, over time these tensions disappeared and the co-habitation has continued peacefully. After the hangar in which the Roma initially lived was demolished, the Roma moved into the main building of the factory, strengthening their attachment to the squat (Figure 4.4). The squat Metropoliz is also a space of exhibitions where artists can work and organise art and music events. These initiatives reinforce Metropoliz as a space of encounter between different people, including both Roma and non-Roma groups, activists, artists and citizens, that increase their feeling of attachment to this place.

In addition to this, there is also an emerging solidarity between these squatting movements and pro-Roma associations. For instance, members of Dilengo set up a stand at the *Fête de l'insurrection gitane* (Festival of Gypsy insurrection) held in May 2015 in Saint-Denis, a municipality in the North of Paris, underscoring the importance of this emerging solidarity between Roma and squatters.

Although there are also problematic aspects in these squats, mainly regarding the cohabitation between different groups and the potential further stigmatisation of the Roma as squatters (which is an increasingly stigmatised category), there have been important positive achievements. The shift from being seen as Roma to being seen as squatters enabled the Roma to escape the policies for Roma living in informal settlements (e.g. relocation to official Roma camps) and to be included in the negotiations and solutions offered to squatters (see Maestri, 2016). While these types of solution (such as the extension of occupation and inclusion

Figure 4.4 The apartment of a Roma family in the main building of Metropoliz.
Source: Gaja Maestri, 2013.

in social housing in case of eviction) are not usually offered to the Roma living in informal settlements, the Roma who participated in these squatting movements benefited from the negotiation power of other squatters. These achievements reveal that also the most oppressed by this exacerbated stigmatisation, i.e. poor Roma migrants living in informal settlements, have a strong power of resistance and can turn a moment of crisis into an opportunity for building new places of solidarities (Maestri, 2014). Wacquant (2008) argues that the weakening of the labour integrative capacity is one of the properties of advanced urban marginality, in which social fragmentation and the emergence of the 'precariat' leads to class decomposition instead of class consolidation. However, the cases of Metropoliz and of Dilengo show that the 'lumpenproletariat' does not necessarily imply a depoliticisation but can constitute the basis for new solidarities and resistance.

Conclusion

This chapter has investigated the causes that led Alexandra, the young Roma girl, to conceal her ethnicity to her classmates. However, it has also shown how this girl, together with other Roma, is resisting this historically rooted and currently exacerbating stigma. The current stigmatisation of the Roma groups in Italy and France comes from a generalisation of the negative representations of poor Roma informal settlements, which have been framed as a threatening aspect of the Roma people. While in Italy informal settlements have been presented as expressions of the nomadism of the Roma, in France they have been interpreted as symptoms of an alleged Roma deviant culture. Both these two framings have culminated in the security discourse targeting the Roma informal settlements, producing a peculiar kind of intense and overwhelming form of territorial stigmatisation extended to the entire Roma population. As a consequence, this type of stigma not only affects the life opportunities of the people living in stigmatised places but, as I have argued in the empirical section of this chapter, it also impacts on the perceptions of those people *who are not* living in stigmatised areas but to which they become associated.

To escape the effects of this extensive territorial stigma, geographic mobility is not enough. As shown by the strategies that I have described here, as I encountered them during trips to France and Italy, this specific type of stigma is not one that is easily escapable. Yet, instead of surrendering to it, the Roma people and associations engage with new creative forms of resistance and place attachment: strategies of submission to stigma, such as 'mutual distancing' and 'lateral denigration', are counterbalanced by attachments to the informal settlement communities in which new solidarities are constantly emerging, such as between Roma and squatters, thanks to the recent involvement of new urban movements.

Notes

1 Special thanks go to Nicolò Palazzetti, Paul Kirkness and Andreas Tijé-Dra for their help in improving the quality of this work. I would also like to thank Tommaso Vitale for supporting part of this research. This work was funded by the Economic and Social Research Council [ES/I007296/1].

2 Following the terminology adopted by the Council of Europe, I employ the term Roma to refer to 'Roma, Sinti, Kale and related groups in Europe, including Travellers and the Eastern groups (Dom and Lom)' (Council of Europe, 2012, p. 4). Although I am aware of the differences between these groups, for the sake of convenience I decided to follow the COE terminology, using Roma as noun and adjective to refer to all of them. The chapter mentions two main Roma groups from France: French Travellers, an official category called *gens du voyage,* of French nationality, and Eastern European Roma, mostly arrived to France from the 2000s. As for the Italian case, there are the *Sinti* (indigenous Roma groups mainly present in Northern Italy), while those generally called Roma arrived from ex-Yugoslavia in the 1990s and, more recently, from Eastern Europe.

3 The Pew Research Center, an American think tank, conducts a series of surveys monitoring global attitudes and trends since 1991 (on issues spanning from religion, to politics, economy and immigration). Surveys are based on a sample of 1,000 interviewees in both France and Italy. In France interviews were telephone-based and in Italy they were face-to-face.

4 The *Commission Nationale Consultative des Droits de l'Homme* is a French governmental organisation charged with monitoring the respect of human rights. Since 2002 they have published an annual report on racism, xenophobia and antisemitism. The 2015 report is based on a survey conducted in 2014 with a sample of 1,020 people.

5 The European Union Agency for Fundamental Rights, with the support of the United Nations Development Programme, conducted a large-scale survey to study the situation of the Roma in 11 European countries in 2012 (an earlier version of the survey was conducted in 2008 but did not include France and Italy). In France, the sample included 3,617 people and covered sedentary and nomadic French Roma and Roma migrants both living in informal settlements and houses. In Italy, the FRA survey considered 2,670 Roma both living in houses and in informal settlements (FRA, 2012, 2013).

6 In Spain, while 41 per cent of people expressed negative attitudes towards the Roma in 2014, in 2015 this figure decreased to 35 per cent (Pew Research Center, 2014, 2015). The FRA survey shows that in Spain 32 per cent of Roma feel they suffer discrimination because of their ethnicity.

7 However, today the majority of the Roma in France and Italy is sedentary (Liégeois, 2007; UNAR, 2012)

8 The data collected on the Roma and Sinti demonstration are part of the RONEPP project (Sciences-Po, Paris).

References

Associazione 21 Luglio (2013), *Antiziganismo 2.0. Rapporto Osservatorio 21 Luglio (2012–2013).*

Associazione 21 Luglio (2014), *Antiziganismo 2.0. Rapporto Osservatorio 21 Luglio (2013–2014).*

Atger, A. F. (2013), 'European citizenship revealed: sites, actors and Roma access to justice in the EU', in Isin, E. and Saward, M. (eds), *Enacting European Citizenship*, Cambridge: Cambridge University Press, pp. 178–194.

Barbulescu, H. (2012), 'Constructing the Roma people as a societal threat. The Roma expulsions from France', *European Journal of Science and Theology*, 8(1): 279–289.

Bourdieu, P. (1985), 'The social space and the genesis of groups', *Theory and Society*, 14(6): 723–744.

Cahn, C. and Guild, E. (2010), *Recent Migration of Roma in Europe*, OSCE and Council of Europe.

Carrera, S. and Atger, A. F. (2010), *L'Affaire des Roms. A challenge to the EU's Area of Freedom, Security and Justice*, CEPS Liberty and Security in Europe.

Clough Marinaro, I. (2009), 'Between surveillance and exile: Biopolitics and the Roma in Italy', *Bulletin of Italian Politics*, 1(2): 265–287.

Commission Nationale Consultative des Droits de l'Homme (2015), *La lutte contre le racisme, l'antisémitisme et la xénophobie (2014)*, Paris: La Documentation française.

Council of Europe (2012), *Council of Europe Descriptive Glossary of Terms Relating to Roma Issues*. Available online at: http://a.cs.coe.int/team20/cahrom/documents/Glossary%20Roma%20EN%20version%2018%20May%202012.pdf (accessed 6 June 2013).

Consiglio dei Ministri (2008), *Decreto del Presidente del Consiglio dei Ministri, 21 maggio 2008. Dichiarazione dello stato di emergenza in relazione agli insediamenti di comunita' nomadi nel territorio delle regioni Campania, Lazio e Lombardia*.

Dépêches Tsiganes (2013), *Français itinérants : solidarité avec les roms réels : oui/amalgame avec les préjugés non !* Available online at: http://www.depechestsiganes.fr/francais-itinerants-solidarite-avec-les-roms-reels-oui-amalgame-avec-les-prejuges-non/ (accessed 21 July 2015).

Doytcheva, M., (2015), Roms et Tsiganes en Europe méditerranéenne: l'actualité d'une question. *Confluences Méditerranée*, (93), pp. 9–25.

European Roma Rights Center (2000), *Campland. Racial segregation of Roma in Italy*. Country Report Series.

European Union Agency for Fundamental Rights (FRA) (2012), *The Situation of Roma in 11 EU Member States. Survey Results at a Glance*, Luxembourg: Publications Office of the European Union.

European Union Agency for Fundamental Rights (2013), *Roma Pilot Survey. Technical Report: Methodology, Sampling and Fieldwork*. Luxembourg: Publications Office of the European Union.

Goffman, E. (1963), *Stigma: Notes on the Management of Spoiled Identity*. Englewood Cliffs, NJ: Prentice-Hall.

Le Monde (2013), Le MRAP porte plainte contre Manuel Valls pour ses propos sur les Roms. [online] 12 November. Available online at: http://www.lemonde.fr/politique/article/2013/11/12/le-mrap-porte-plainte-contre-manuel-valls-pour-ses-propos-sur-les-roms_3512261_823448.html (accessed 21 July 2015).

Liégeois, J.-P., (2007), *Roms en Europe*. Strasbourg: Council of Europe Publishing.

Lucassen, L. (1998), 'Eternal vagrants? State formation, migration and travelling groups in western Europe, 1350-1914', in Lucassen, L., Willems, W. and Cottaar, A. (eds), *Gypsies and Other Itinerant Groups*, London: Macmillan, pp. 55–73.

Maestri, G. (2016), 'From nomads to squatters. Towards a deterritorialisation of Roma exceptionalism through assemblage thinking', in Lancione, M. (ed), *Re-thinking Life at the Margins: Assemblage, Subjects and Spaces*. Avebury, UK: Ashgate, pp. 122–135.

Maestri, G. (2014), 'The economic crisis as opportunity: How austerity generates new strategies and solidarities for negotiating Roma access to housing in Rome', *City: analysis of urban trends, culture, theory, policy, action*, 18(6): 808–823.

Magazzini, T. and Piemontese, S. (2015), 'Modèles de gestion de la diversité en Europe et migrations roms: le cas espagnol', *Roms et Tsiganes en Europe méditerranéenne*, (93): 51–62.

Manrique, N. (2015), 'Les Gitans d'Espagne: une catégorie sui generis?', *Roms et Tsiganes en Europe méditerranéenne*, (93): 63–72.

Martin, M. (2013), 'Expulsion of Roma: The French Government's Broken Promise', *Statewatch journal*, 22(4): 1–10.

Noiriel, G. (1988), *Le Creuset Français; Histoire De L'immigration (19ème-20ème Siècles)*. Paris: Seuil.

Okely, J. (1983), *The Traveller-Gypsies*. Cambridge: Cambridge University Press.

Parker, O. (2012), 'Roma and the politics of EU citizenship in France: Everyday security and resistance', *Journal of Common Market Studies*, 50(3): 475–491.

Pew Research Center (2009), *End of Communism Cheered but Now with More Reservations*. The Pew Global Project Attitudes.

Pew Research Center (2014), *A Fragile Rebound for EU Image on Eve of European Parliament Elections*. The Pew Global Project Attitudes.

Pew Research Center (2015), *Faith in European Project Reviving*. The Pew Global Project Attitudes.

Regione Lazio (2009), *Commissario Delegato per l'emergenza nomadi nel territorio della Regione Lazio. Regolamento per la gestione dei villaggi attrezzati per le comunità nomadi nella regione Lazio.*

Sigona, N. (2003), 'How can a 'nomad' be a refugee? Kosovo Roma and labelling policy in Italy', Sociology, 37(1): 69–80.

Sigona, N. (2015), 'Campzenship: reimagining the camp as a social and political space', *Citizenship Studies*, 19(1): 1–15.

Slater, T. (in press), 'Territorial stigmatization: symbolic defamation and the contemporary metropolis', in Hannigan, J. and Richards, G. (eds), *The Handbook of New Urban Studies*, London: Sage.

UNAR, 2012. *National Strategy for the Inclusion of Roma, Sinti and Caminanti Communities*. European Commission Communication no. 173/2011.

Vitale, T. and Claps, E. (2011), 'Not always the same old story: Spatial segregation and feelings of dislike against Roma and Sinti in large cities and medium-size towns in Italy', in Rovid M. and Stewart, M. (eds), *Multi-Disciplinary Approaches to Romany Studies*, Budapest: Central European University Press, pp. 228–253.

Wacquant, L. (2007), 'Territorial stigmatization in the age of advanced marginality', *Thesis Eleven*, 91(1), pp. 66–77.

Wacquant, L. (2008), *Urban Outcasts: A Comparative Sociology of Advanced Marginality*. Cambridge: Polity Press.

Wacquant, L., Slater, T. and Pereira, V. B. (2014), 'Territorial stigmatization in action', *Environment and Planning A*, 46(6): 1270–1280.

5 Redlining or renewal? The space-based construction of decay and its contestation through local agency in Brixton, Johannesburg

Christoph Haferburg and Marie Huchzermeyer

Introduction

Hidden practices of space-based exclusion are prevalent in the residential property market in several countries where financial institutions 'redline' selected neighbourhoods, labelling them as non-credit worthy due to perceived risk. As a consequence of this space-based strategy of risk reduction, prospective home-owners in such areas find it difficult or impossible to secure mortgage finance. This effectively prevents further investment and creates a self-fulfilling trajectory towards crime and grime (Aalbers, 2006). The rationale for banks to employ this instrument, and the potentially detrimental effects for neighbourhood development have been examined in urban studies literature over a number of decades (Tomer, 1992; Holmes and Horvitz, 1994; Hillier, 2003; Aalbers, 2005, 2006).

Actual and aspirant residents of redlined neighbourhoods are subject to a practice of territorial stigmatisation that is based on a powerful normative framing of orderly urban society. Although a common response by property owners is disinvestment or extractive forms of investment, redlining of certain areas may trigger efforts to counter this framing; for instance, by demonstrating desirability and bankability of usually very diverse areas. This chapter takes a look at spatialised lending practices in the real estate sector in Johannesburg, South Africa's financial capital, in order to highlight the consequences of redlining. It focuses on Brixton, a neighbourhood on Johannesburg's desegregating east-west axis. It is precisely in these well- located, desegregating or integrating areas, in which middle-class consumers seek conventional mortgage products, that redlining is rife. Yet they epitomise the lived reality of a more inclusive post-apartheid city more successfully than the few engineered 'inclusive' (yet micro-segregated) post-apartheid housing projects.

The spatial stigmatisation of areas such as Brixton by the financial and property market industry is directly related to the social diversification, which clashes with the perception of normality established by the extensive mortgage market for uniform and exclusionary suburbs. However, the chapter also demonstrates the possibility of local contestations of this particular form of territorial stigmatisation. The case that we present here demonstrates the huge efforts required by spatially discriminated individuals to challenge the spatial ideology that defines the decision-making of financial institutions. There are certainly limits to individual strategies but the example of Brixton serves to illustrate how some residents successfully challenge their exclusion from the property market.

Despite some opposition to redlining during the transition to post-apartheid up to the early 2000s, it has remained legitimate in the toolbox of financial institutions and is currently not in the spotlight of urban research in South Africa. The structural effects of this practice, however, are anything but negligible. The way in which this practice is constituted is of equal importance – it is through a combination of spatial and social perceptions with financial management techniques applied in a 'liberal' legal environment that a problem of exclusionary urban subdivisions is created. We will try to unravel this with the help of conceptual references on space and society after reviewing the use and meaning of redlining.

Redlining in international context and conceptual framing

The term 'redlining' stems from the United States, where it is outlawed, although its illicit practice remains a concern there. Hillier (2003, p. 395) provides one definition of the term: 'Redlining refers to lending (or insurance) discrimination that bases credit decisions on the location of a property to the exclusion of the characteristics of the borrower or property.' Lending decisions under what is referred to as redlining relate to 'where to make loans and what type of loans to make' (ibid., 396). As this affects 'all residents from one area', it is referred to as 'a form of place-based exclusion' (Aalbers, 2005, p. 100). There has been a tendency to legitimise a certain form of decision-making that is included in Hillier's definition of redlining, namely decisions on the type of loan, by referring to this as 'yellowlining'. Aalbers (2005) shows how, in the Netherlands, 'yellowlining' has been used to refer to areas in which loans are granted for only a percentage of the value of the property, whereas 'greenlining' delineates areas where loans were granted well above the value of the property. The red-, yellow- and greenlining terminology has begun to be applied by housing market players in South Africa (Rust, personal communication, 2015).

As we have already suggested, redlining refers to 'perceived risks to real estate investment' related to location of the property in a particular area (Hillier, 2003, p. 395). As such, it has been met with opposition. The term itself was coined late in the 1960s in a Chicago community directly affected by such discrimination, after evidence was found of insurance companies and lenders literally drawing red lines around areas that they decided not to service, but evidence of such practices date back to the 1940s (ibid.). However, since the 1970s the practice of redlining has been outlawed in the United States. This is underlined in particular by the Community Reinvestment Act of 1977 (ibid.). The practice of sub-prime lending in the United States, which alongside securitisation infamously triggered a global economic meltdown in 2008, has been referred to as 'reverse redlining'; this has perpetuated the practice of exclusion, with the geographic concentration of sub-prime mortgages in areas that were formerly redlined (Hernandez, 2009, p. 291).

Tomer (1992, p. 78) relates redlining practices in the US with what he terms 'micro social forces', namely the

> predominate housing market ideology widely shared by housing market participants, such as: real estate brokers, appraisers, home builders, life and fire insurance companies, governmental housing-related agencies, mortgage lenders and others . . . This ideology reflects prevailing prejudices about what types of neighbourhoods, housing, and socio-economic groups are desirable and where decline is likely to take place.

This ideology of players in the housing market may sit at odds with the ideal of preventing decay and impoverishment. The implication of the housing market ideology, when driving decisions not to invest or fund investment in entire areas, certainly defies such ideals and instead 'contributes to the deterioration of housing and other conditions in the area' (Tomer, 1992, p. 79). Aalbers (2006, p. 1064) invokes Henri Lefebvre's notion of abstract space when engaging the way 'governments and real estate actors think about space for political and economic gain'. Lefebvre's concept of abstract space and its location in his spatial triad of lived space (or 'spatial practice' which relates to social space) versus perceived space (or 'representational space' linked to symbolisms and the arts) and conceived space (or 'representations of space' linked to scientific knowledge and ideology) is complex. However, Lefebvre (1991, pp. 33, 40, 57) writes very fittingly of abstract space as being underpinned by a very particular ideology: 'Abstract space, the space of the bourgeoisie and of capitalism, bound up as it is with exchange (of goods and commodities, as of written and spoken words, etc.) depends on consensus more than any space before it'. He also writes of 'differentiation induced within existing abstract space', that is, difference that serves the same converging interests and is supported by the same consensus (ibid., p. 64). Aalbers (2006) provides the example of a Rotterdam neighbourhood in the Netherlands, which banks defined as loss-making and in which they refused to issue mortgages. As a result of the inability of buyers to obtain mortgages, properties lost value and were sold to landlords (people not requiring a conventional mortgage), often exploiting or milking properties for short-term gain. The 'self-fulfilling prophecy' or 'chain reaction' was of 'an even stronger decline' in the area (ibid., pp. 1074, 1081).

While within abstract space an area might be 'written off' or redlined as a result of a spatially stigmatising consensus among market players, Aalbers (2006) also invokes Lefebvre's 'social space' when discussing that a redlined area may still be desirable as a place in which to live. He argues that '[t]he dynamics of social space are such that they have to react to the dynamics of abstract space. This reaction can possibly counteract abstract space dynamics' (Aalbers, 2006, p. 1065). The case study in this chapter provides an example of such a reaction from within social space. The chapter examines both the market based, spatially stigmatising consensus that determines investment patterns in Johannesburg in its transition beyond apartheid, and the reaction and agency from within social space.

The point reflects the fact that financial mechanisms (as well as government policies) are enacted in close correspondence with social subjects – not only do they affect everyday urban life, but they are also met by a variety of local reactions. Interpreting social space with Bourdieu (1984) as an arena in which this interaction plays out helps to grasp the range of social inequalities resulting from this, as well as to identify the different capacities 'at hand'. As such, in order to counter spatial stigmatisation, social actors are understood as being equipped with different portfolios of economic (income, wealth), cultural (education, professional skills) and social capital (networks of contacts). By confronting the structural and cognitive principles at work in the South African real estate sector in our case study, we highlight that it might be relevant to combine the two conceptual perspectives of Lefebvre and Bourdieu to understand how spatial differentiation plays out as a reflection of structure and agency.

Redlining in South Africa

Since the end of apartheid, South African cities have maintained high levels of 'racial' and socio-economic segregation. The transition out of apartheid opened South Africa to market processes that are widely acknowledged as having contributed to the vexed problem of socio-spatial continuities (CoGTA, 2014). Several factors have contributed to this. One is the role of the private-sector in the property market and close ties with the international economy of real estate and finance: investment rationales are no longer embedded in the local urban context (cf. Parnreiter *et al.*, 2013). Another is the increasingly unequal distribution of wealth and income within South African society (Leibbrandt *et al.*, 2012; Everatt, 2014). One should also note the existence of a (local) government system that lacks the appropriate policy, legislation and capacity to address urban inequalities (Sihlongonyane, 2014). All of these factors intersect with an unevenly growing urban economy, a growing number of residents, and a severe housing backlog (Todes, 2014). In this context, market-driven stigmatisation of neighbourhoods plays into the hands of real estate developers and slum landlords, with spatially polarising effects.

It is disconcerting that in South Africa, a country with one of the most extreme spatially discriminating histories stemming from colonial and apartheid-based segregation policies, and a country struggling to overcome this legacy as economic and spatial inequality continue to rise, redlining has not been outlawed. Awareness of this practice has waned, as we will show; over the last decade, the term 'redlining' is used loosely, with little questioning of its precise meaning or the legitimacy of the procedures that it describes. And yet, redlining has a particularly negative impact on renewed efforts to desegregate and restructure South Africa's cities.

Redlining became an accepted practice among banks in South Africa in the late apartheid years. During this time, the big South African cities experienced a substantial increase of population growth based on 'black African' rural-urban migration. This trend was caused by severely limited livelihood opportunities in the 'reserved territories' in the countryside, the so-called 'homelands'. When the wave of migrants to the cities could no longer be banned, the national government took the decision to 'legalise' and control the process through the policy on Orderly Urbanisation of 1986. This included gradually lifting restrictions on home own-ership for 'black Africans' in urban areas outside of 'homeland' areas, allowing banks to lend to an emerging 'black African' middle class (Huchzermeyer, 2004). However, up to the repeal, in 1991, of the Group Areas Act and the Population Registration Act – two key apartheid statutes enacted in 1950 – the new mort-gage financed housing developments were, with few exceptions, on vacant land in racially segregated 'black African' group areas (see Haferburg, 2007). By definition, these were areas of low infrastructural standards and socio-politically marked (though unevenly so) by resistance to the state.

Several factors, but primarily poor workmanship of the bonded houses, resulted in a portion of consumers in these areas boycotting their bank payments (Bond, 2000, p. 132). Efforts to make South African towns and cities ungovern-able remained a key strategy in the anti-apartheid struggle throughout the 1980s, and banks perceived non-payment of bonds as part of this strategy, and used this argument to justify the introduction of redlining (ibid.). While authors such as Bond (2000), close to the civic struggle at the time, are at pains to point out that government commissions and banks were mistaken in their explanation of bond boycotts as being part of orchestrated 'unrest', authors close to the banking indus-try such as Tomlinson (1998, p. 8), an authority in the housing finance discourse in South Africa, uncritically refer to the bond boycotts as 'politically inspired', thus legitimising the practice of redlining politically unstable areas.

In the policy-making process during the transition out of apartheid, a Record of Understanding in 1994 between the newly elected interim government and the private sector included a commitment by banks to lend 'down' to the previously excluded and a commitment by the state to stabilise or normalise township areas (Huchzermeyer, 2001). The Bill of Rights in the post-apartheid Constitution of 1996 (Republic of South Africa, 1996, s. 9(4)) disallows all forms of unfair dis-crimination and obliges the state to draw up legislation to this end. A Promotion of Equality and Prevention of Unfair Discrimination Act 4 of 2000 and a Home Loan and Mortgage Disclosure Act 63 of 2000 were approved four years later. The latter had its commencement date only in 2007 after the delayed enactment of regulations (Republic of South Africa, 2007). In the United States, mortgage disclosure requirements 'formed a strong incentive against redlining practices' (Aalbers, 2005, p. 100). In South Africa, however, the Office of Disclosure, housed in the national Department of Human Settlements, is jokingly referred to as 'Office of non-disclosure', in effect a toothless institution (housing consultant, personal communication).[1]

In the transition beyond apartheid, banks continued to experience a level of non-payment. In the late 1990s, 12.4 per cent of mortgage loans made on 'township' properties were in arrears for over three months (Tomlinson, 1999). The township market, however, represented only a fraction of the banks' mort-gage lending portfolio, at a mere 6.4 per cent (Tomlinson, 1998). Redlining practices remained in the spotlight into the early 2000s, the Human Rights Commission and the Banking Council being tasked with ensuring an end to this practice (Randall, 2000).

By 2002, the national Department of Housing had drafted a Community Reinvestment Bill, going beyond the requirement of disclosure to set 'specific minimum targets by financial institutions in lending to low and medium income level households for housing purposes' (Department of Housing, 2002, p. 5). Banks viewed the bill as 'too onerous' (February, 2002) – if enacted, the bill would require of financial institutions, among other actions, 'to refrain from refusing home loan finance to borrowers purely on the grounds of the current or future expected socio-economic characteristics of the residents in the neighbourhood in which the home is located' (Department of Housing, 2002, s. 4(1)(a)). Interestingly, the bill did not set out to outlaw redlining as such, merely requiring banks to 'refrain from the practice of redlining other than where dictated by safe and sound business practices' (ibid, s. 4(1)(b)). As Tomlinson (2005) points out, the bill differs from the community reinvestment regime in the United States in that is does not compel lending in geographical areas. Its wording supports the idea that geographic areas may, in certain circumstances, be used to determine the risk associated with a loan, irrespective of the applicant's proven socio-economic standing.

This legitimisation of redlining in certain areas in the 2002 bill was insensitive to the processes of desegregation, socio-economic mixing and tenure diversification underway in certain older neighbourhoods. These processes of change occurred particularly in areas whose location and built fabric lent themselves to affordable living, be it through lower purchase prices than in the suburban market, or more dramatically through informal subdivision and renting out of rooms. In the highly segregated South African housing market, in which legislation and planning regulation during apartheid had ensured both racial and socio-economic homogeneity in the mortgage sector over four decades, such diversification ought to have been welcomed in the post-apartheid years.

Despite its lenient approach towards redlining, the Community Reinvestment Bill was hotly contested by the banking sector and never enacted. Instead, the financial institutions committed themselves to a voluntary Financial Services Charter in 2004. Tomlinson (2005, p. 32) summarises the charter as 'transforming the entire financial sector, including increasing investment and the extension of lending, into the low-income housing market'. Under the overarching goal of Black Economic Empowerment (BEE), the charter which was operational from 2004 to 2014 sought to transform ownership of, and participation in the financial sector. It sought to achieve inclusivity by reaching out to a greater range of customers, and to enhance its ability to provide access to a 'greater segment of the population' (ABSIP *et al.*, 2003). It made no area-based commitments. In so doing, it effectively ignored the specific challenges of the fragmented urban context in which the charter was to be applied.

The banking industry could prove relative success in collectively reaching the affordable home loan targets set by the Financial Services Charter. This was achieved through new loan products rather than sub-prime lending, which the

National Credit Act 34 of 2005 seeks to prevent. This success has drawn attention away from the question of redlining or place-based discrimination. All the while, redlining continues to be justified by financial institutions with reference to 'political unrest', particularly given the context of recurring street protests across South African low income urban areas (Pernegger, 2014). In an interview with a banking sector representative (June 2015), frequent mention was made to 'community action' preventing vacant possession, thus occupants or tenants of the former owner refusing to vacate the property when a sale is transacted, and holding banks 'hostage'. According to this interviewee, banks resort to redlining if such actions have occurred even just once in an area. Reference was also made to the grading of areas according to property value increases, in turn related to the nature of demand in an area. Approval of loans is based on a 'strong system' which is centralised and 'computer driven' and in which the credit profile of the applicant, the nature of the property and the neighbourhood are taken into account (interview 1, 2015). Banks extract data on a particular suburb from the deeds registry and from estate agents, but may also 'inspect' the property and area with the assistance of various Google tools (ibid.). In addition, property companies such as Lightstone Property provide toolkits which produce data sets that unquestioningly include 'suburb ranking' (Lightstone Property, 2015a).

Place-based decision-making around the issuing of mortgages occurs particularly in older, formerly 'white' suburbs that have diversified. It is a common experience of applicants that banks cite the suburb or neighbourhood (perhaps among other factors) when providing an explanation for declining a home loan, or when approving a loan for only a percentage of the value of the property. However, it is specific types of diversified neighbourhoods that are prone to be classified in this way. The socio-economic change in these areas is often enabled by a shift from owner-occupation to tenancy arrangements, and this contributes to the integration of different apartheid era population groups.[2] A municipal official in an entity that tracks urban trends (interview 2, 2015) cited knowledge of redlining in the inner city and in Yeoville, a densely built-up area to the east of the central business district (CBD) – both areas that experienced a substantial post-apartheid socio-economic transition alongside a deterioration of buildings and infrastructure.

The banking sector representative (interview 1, 2015) in turn advised that banking decisions may be challenged by invoking, for instance, the Consumer Protection Act 68 of 2008. However, the advice was that it was preferable that customers attempted to negotiate with their banks though this needed to be with the head office, given the centralised nature of the approval system (ibid.). This suggestion assumes a particular type of confidence and professional knowledge among consumers. However, empirical evidence on the reproduction of social hierarchies has demonstrated that only comparatively few subjects within a society have the cultural capital to encounter established institutions at eye level (Bourdieu, 1984). Cultural capital and the associated professional confidence might then enable some actors to navigate a complex system and to compile the necessary evidence and argument. After contextualising Johannesburg with the areas in which redlining and consumer-bank negotiation around redlining occurs, we provide an example from the neighbourhood of Brixton that illustrates the extent of cultural (and professional) capital in a Bourdieusian sense required to successfully challenge spatial decision-making by banks, or to challenge the collective ideology of the market players which marks abstract space and lends it concrete power in a Lefebvrian sense.

Contextualising Johannesburg's diversifying neighbourhoods, including Brixton

Most private residential extensions to Johannesburg post-1994 have been developer-initiated and gated, with an exponential increase in the land consumption and exclusivity of individual estates (Murray, 2014). Notwithstanding, Johannesburg's residential landscape remains dominated by large swathes of socio-economically uniform apartheid-era suburbs, too easily labelled as 'normal' in comparison to the city's apartheid era townships (Mabin, 2014, p. 403). This normality for financial institutions was and continues to be underpinned by exclusionary zoning schemes, with standards that determine the socio-economic class of the inhabitants (Berrisford, 2011). Post-apartheid municipal strategies have treated these areas as an important tax base, and have sought to prevent degradation, ensuring continuity with uniform standards (City of Johannesburg, 2003, p. 225).

This protectionist approach did not apply to the city's east-west axis, a series of formerly whites-only suburbs on opposite sides of Johannesburg's extensive CBD. For one of these suburbs, Brixton (to which we turn in more detail below), the City of Johannesburg admits to 'relatively lenient regulations guiding land-use' (Harrison, 2007, p. 5). As in other parts of the east-west axis, 'there is drug-dealing, residential decay, deteriorating facilities, illegal activities and uses' (ibid.). Yet the City, in contrast to the profit-oriented thinking of market players, recognises areas like Brixton for the co-existence of 'people of diverse backgrounds' – students, young professionals, artists and musicians, 'long standing residents' and 'ordinary working people . . . of all races, cultures, and sexual orientation' (ibid.). The City celebrates Brixton as 'one of the few places in Jo'burg that has sustained diversity during transition' (ibid.). Key to this success has been the fact that the area as a whole, unlike nearby Melville to the north, has not been gentrified (ibid.). This condition of (transitory) diversity is not unrelated to the socio-spatial ideology of financial institutions. In a sense, this could be considered as the flip side of redlining, but it would be short-sighted to conclude with this assessment, as will become clear when considering Johannesburg's post-apartheid urban context.

The east-west axis of Johannesburg runs immediately to the north of the historical Main Reef Road which traverses the city along the northern side of the gold reef and mining belt (see Figure 5.1). This, in turn, separates Soweto and Johannesburg's southern suburbs (formerly white working-class and later middle-class extensions to the south of the CBD from the remainder of the almost exclusively formerly

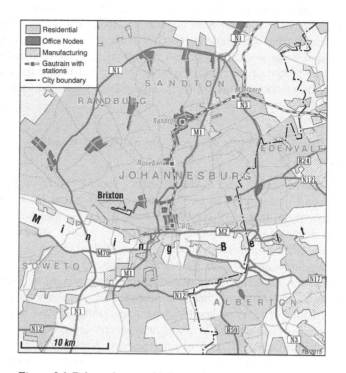

Figure 5.1 Brixton is part of Johannesburg's diversifying east-west axis, north of the
        mining belt.

Source: Own design based on Haferburg *et al.,* 2014; cartography by Thomas Böge.

'whites-only' urban fabric to the north (Harrison and Zack, 2014). Crankshaw's
(2008) analysis of the 2001 census data for Johannesburg shows neighbourhoods
on the east-west axis, along with the CBD forming a strip of areas in which only
up to 23 per cent of residents are 'middle-class' (Crankshaw, 2008, p. 1701). Our
own analysis of census data of the distribution of income groups in 2011 appears in
Figure 5.2. This shows that Brixton is neither part of the concentration of suburban
wealth, nor of the poverty concentration of the south and the northern periphery.
Brixton's attraction both to the middle classes and very low income earners is
evident in its location in relation to the segregated city. Neighbourhoods on the
east-west axis, along with some old areas to the south of the mining belt eastward of
Soweto also stand out as being most diversified in terms of former apartheid population group categories (ibid., p. 1704; Harrison and Zack, 2014).

Johannesburg's east-west axis also follows a geographic divide (the conti-
nental watershed): the northern suburbs slope down to the north with higher
rainfall and less wind than southern Johannesburg, which slopes down to the
south (Storie, 2014). While Johannesburg initially developed informally and
haphazardly, sprawling from the current CBD and around the nearby gold
prospecting sites, its segregationist and later apartheid planners erased any

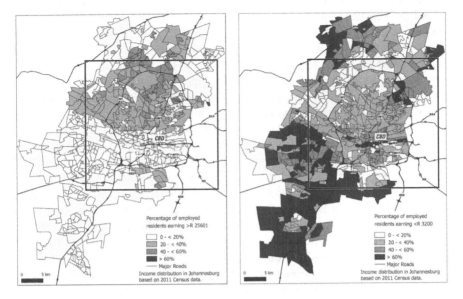

Figure 5.2 High- and low-income distribution across Johannesburg, 2011.

Source: Own construction from census data accessed through Quantec; cartography by Miriam Maina.

socio-economic and cultural mixing that emerged. They planned the city's expansion such that the climatically pleasant north was reserved for those who the mining economy and segregationist system privileged, while the harsh south was largely staked out for those who both the economy and the system of privilege needed to exploit (Beavon, 2004). The 'Old South' formed a pocket of white working class neighbourhoods to the south of the mining belt, built here due to their proximity to the city centre and mines, despite the climatic disadvantage (Harrison and Zack, 2014).

East-west expansions of Johannesburg were surveyed and proclaimed in the first decade of the 1900s (Beavon, 2004, p. 89), mostly as white working class residential areas with small plots on an open grid layout. A mere three kilometers to the west of the CBD, Brixton was proclaimed as a residential area in 1903. Along with other neighbourhoods on this east-west strip, Brixton developed over the following decades mostly with one and two-bedroomed semi-detached and small freestanding houses It also includes a limited number of mansions, constructed by the randlords[3], whose wealth primarily led the city's expansion with lush suburbs to the north (Mabin, 2014). The relatively compact east-west expansions were well endowed with public amenities; the western tramline reached Brixton in 1917 serving the area until 1948 when a decision was made to discontinue the city's tram system (Beavon, 2004, pp. 90,161).

Johannesburg saw a sequential implementation of the segregationist Native (Urban Areas) Act of 1923, which could be used to remove the right of residence of Africans but not of other groups – this was only made possible later, under the

Group Areas Act of 1950 (Beavon, 2004). Initially the Native (Urban Areas) Act paved the way for 'slum clearance' in the immediate surrounding of the business district. In the late 1920s its provisions were applied to northern suburbs (and an enclave of suburbs to the south east), and only in the early 1930s to much of the east-west axis, including Brixton (Beavon, 2004, p. 99). The Afrikaaner working class, 'a struggling minority grouping in English dominated Johannesburg' and ranked below the English working class, managed to establish itself on the western side of the CBD, including Brixton (Harrison and Zack, 2014, p. 4).

During 46 years of apartheid rule, Brixton became an icon of Afrikaanerdom, home to key Afrikaaner political and cultural figures and also to several key institutions of the apartheid government (see Figure 5.4). The late 1950s through to 1962 saw the construction of the 247m high, iconic broadcasting tower on the Brixton ridge. The Albert Hertzog Tower (now Sentec Brixton Tower) was named after then Minister of Post and Telegraph, a conservative Afrikaaner nationalist (Harrison, 2007). Brixton was also home to the Brixton Murder and Robbery Squad, one of the apartheid state's infamous death squads (O'Brien, 2011), based at what is now simply the Brixton Police Station.

A 'white' group area throughout apartheid, census figures show that already in the mid 1990s Brixton was in the fairly rare category of neighbourhoods with almost equal share of 'white' and 'non-white' residents ('non-white' referring to former population categories 'African', Indian' and 'Coloured') (see Table 5.1). By 2001, the change in the composition of apartheid-era population groups had continued towards the increase in members of the 'black African' and drop in members of the 'white' and 'Indian/Asian' groups, respectively. In the year of the latest census, in 2011, the relative shares of the groups in Brixton have consolidated to fall more or less in line with population mix in Johannesburg's metropolitan region as a whole.

Over the period 1996 to 2011, Brixton's population more than doubled, from 1,746 to 4,068, as per the census figures presented in Table 5.1. This went hand-in-hand with a change in the distribution of tenure forms, namely a significant increase in tenants (see Figure 5.3). There is some evidence that the population increase was absorbed in part through slumlordism, if defined as the overcrowding of tenants facilitated by contravention of land use regulations. An unpublished sample survey of 142 properties evenly spread across Brixton, which we conducted in collaboration with the Brixton Community Forum in 2010, showed 12 per cent had occupation rates of over 10 people, these being mostly unrelated to one another (53 per cent of the sample households were renting, and only 39 per cent owning) (Haferburg and Huchzermeyer, 2010). Land use regulations permit up to eight unrelated people in a single residential property, any higher number requiring rezoning and the requisite upgrading of buildings (City of Johannesburg, 2009). Given the immediate proximity of Johannesburg's two public universities, a dominant approach among landlords has been the conversion of single residential houses into often cramped and overpriced student accommodation, a practice that occasioned the City of Johannesburg, together with the Brixton and neighbouring Community Forums as well as landlord organisations, to introduce a Commune Policy (ibid.).

Table 5.1 Residents according to apartheid population groups in Brixton as compared to Johannesburg as a whole.

BRIXTON (SUBPLACE)	1996		2001		2011	
African Black	427	24%	1.112	56%	3.074	76%
Coloured	107	6%	153	8%	323	8%
Indian Asian	255	15%	115	6%	249	6%
White	920	53%	591	30%	373	9%
Unspecified/Other	37	2%	not designated		49	1%
Total	1.746	100%	1.971	100%	4.068	100%
JOHANNESBURG (METRO)	1996		2001		2011	
African Black	1.848.050	70%	2.368.043	73%	3.395.636	77%
Coloured	171.121	6%	206.172	6%	249.959	6%
Indian Asian	96.757	4%	134.071	4%	212.746	5%
White	492.343	19%	515.403	16%	541.198	12%
Unspecified/Other	24.760	1%	not designated		35.289	1%
Total	2.633.031	100%	3.223.689	100%	4.434.828	100%

Source: census data accessed via Quantec.

Census data shows that Brixton's income distribution was already diverse in 1996. While only 12 per cent of households earned an income that banks would consider for mortgage purposes (earning upwards of R6 000/month at the time), 38 per cent earned little enough to qualify for fully subsidised housing from the state (earning up to R1 500/month at the time). By 2011, taking substantial inflation into account, 13 per cent of households earned a mortgageable income (earning upward of R12 000), and just over 42 per cent would qualify for fully subsidised state housing (earning up to R3 500/month).[4]

Brixton's proximity to two universities, the public broadcaster, and the revitalising inner city business district Braamfontein, have helped maintain its attractiveness across the different socio-economic strata. In the sample survey of 2010, an overwhelming majority of the 142 respondents indicated that they chose to stay in Brixton because of its location in relation to facilities and the city centre (Haferburg and Huchzermeyer, 2010). This includes a small discerning middle class not wishing to reside in Johannesburg's uniform 'leafy suburbs' or in its gated estates (ibid.). A property owner profile for Brixton, drawn on the Lightstone Property data facility, shows that since June 2014 buyers are predominantly in the mid-thirties to late-forties age bracket (Lightstone Property, 2015b). These may be home owners, or investors who target the student market either by turning family houses into so-called 'student communes' or by purchasing houses already converted for this purpose. Estate agents capitalise on this market, a typical caption being:

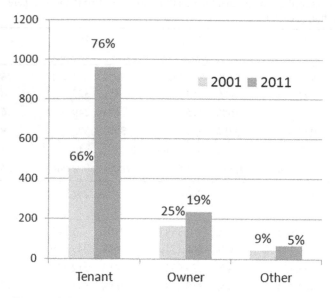

Figure 5.3 Tenure status in Brixton, 2001–2011.
Source: Census data accessed via Quantec.

'Calling all investors: Student Accommodation 3 streets behind. . .' (Pam Golding, 2015). In the 2010 sample survey, the most cited aspect about Brixton's future (mostly noted with concern) was the expansion of student communes (Haferburg and Huchzermeyer, 2010). The aforementioned *Commune Policy* stipulates that the goal is to maintain a mix of home owners and communes across all sub-areas within Brixton and the surrounding neighbourhoods (City of Johannesburg, 2009). As the student market saturation is reached, and noting the concerns voiced across Brixton in the 2010 survey, the Brixton Community Forum regularly lobbies estate agents to drop the call for student commune investors from their advertising.

To contextualise Brixton's mortgage market, it is worth noting that in 2007, before the global economic downturn sparked by the sub-prime lending and securitisation crisis in the United States, mortgage arrears in South Africa, meaning defaults of over three months, were at a low of 3.2 per cent of total value; mortgage arrears rose steadily and peaked in 2010 at 9.4 per cent of value, and have since declined, by 2014 standing at 3.9 per cent (Melzer, 2015). In Brixton, the home loan market is dominated by South Africa's four largest mortgage banks. ABSA takes the lead, followed by Standard Bank of South Africa. First National Bank and Nedbank have comparatively small loan books in Brixton, followed by South African Homeloans, Investec and others. According to Lightstone Property, the number of mortgage bonds in Brixton was at an all-time high in 2007, dropping steeply in 2008 and hovering at the same level through to 2015. (Lightstone Property, 2015b). It is in this period that the mortgage case in Brixton, which follows, unfolds. It is based on our acquaintance with the mortgage seekers, and a formal interview conducted with them in 2012 (anonymous interview 3, 2012). The case is an example of how a territorialised and stigmatising approach to the financing of the built environment through redlining is met by a determined response on the part of mortgage seekers.

Local agency: challenging redlining in Brixton

In 2009, a self-employed couple with one small child, who were already residing in Brixton as homeowners, sought to relocate to a larger property within the neighbourhood. They identified an historic property with a visual presence, and large enough to be adapted to a family home along with offices for their small architectural firm (see Figure 5.5). As the property was unsuitable for use as a rental establishment or commune the couple were able to induce the absentee owner to sell. The next challenge was to raise finance for the purchase and redevelopment of the property.

When seeking mortgage finance from several banks, the couple found all their applications declined. When asked to explain, an Investec employee explicitly referred to the property's location in Brixton; the bank had also checked Google Earth's street view facility which had confirmed their fears – people were loitering in the streets. As could have happened in any other area, some other banks also cited the couple's self-employed status as reason for declining the bond.

The couple did not give up. Instead, they embarked on a counter-intervention. They were convinced that the banks' perception of Brixton was mistaken. In order to build their case, they approached the Brixton Community Forum. The plan was to use the Forum's contact list of over 100 residents in order to conduct an email survey on 'professionals' living in the area.[5] The survey asked 'professionals' to name their occupations and indicate the location of their home in Brixton. The response was supportive, with many willing to help prove the bankability of the couple's project. The results were represented on a map, showing 'professionals' dotted across most of Brixton. However, there were no responding professionals in the two southernmost streets (the commercialised part of the suburb) and in the most western part of Brixton (predominantly residential were the community forum did not have significant reach). The map included a few properties to the north of Brixton's boundary (see Figure 5.4), homeowners in the neighbouring upmarket suburb of Auckland Park who identify with Brixton due to the activities of its community forum. Of particular interest were the comparatively high values of a few such properties bordering directly onto Brixton. A compelling picture of the neighbourhood emerged, not one of crime and grime, but of an area in which significant personal wealth was invested, and with a connection to the bankable neighbouring suburb Auckland Park.

Armed with an empirical argument, spatialised on a map (see Figure 5.4), the couple had the confidence to reapproach banks. ABSA, the dominant mortgage lender in Brixton, agreed to reconsider its decision and finally approved a bond for the property and its conversion. The couple's plan could go ahead: the property was purchased, renovated and extended, and a year later the family as well as the creative firm could relocate (Figure 5.5).

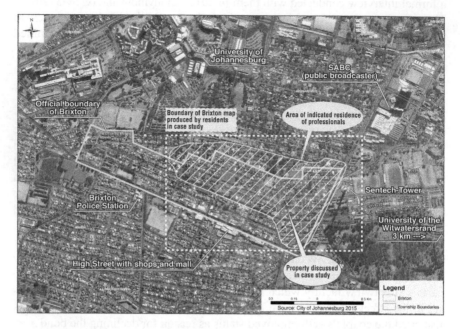

Figure 5.4 Map of Brixton's key facilities and areas in which professionals' homes were identified.

Source: own construction based on the architect couple's survey in 2010; background: 2012 aerial photography from City of Johannesburg, with permission.

Figure 5.5 A century-old building, earmarked 'off-limits' by credit-lending institutions due to redlining (left), and the same building after credit was finally granted – renovated and extended to the back (right).

Source: photographs by the authors, 2010 and 2012.

The socio-spatial composition of Brixton, as the case demonstrates, contradicts the property industry's 'norm' which is determined by uniform, orderly middle-class suburbs. The banks translate a perception of Brixton into a fixed representation for decision-making (in Lefebvre's abstract space): Brixton must be redlined. The stigmatisation of Brixton as an unbankable suburb is further deepened through a practice that deems it necessary to remotely check (on Google

Earth's street view) what kind of people might be visible. Streets in a racially representative, mixed income neighbourhood such as Brixton, with high tenancy rates and student communes, inevitably contain diverse groups of people. With the property industry's socio-spatial norm being one of orderliness, those visible in the streets of Brixton aligned with stereotypes of what the banks perceive as problematic (whether in assumed intention, assumed poverty, or racial stereotyping). When the aspirant lenders questioned the representational basis of the bank's decision, the bank employee's immediate response was to tap into the negative stereotype of loitering.

However, the case also demonstrates banks' openness to accept challenges to their redlining decisions, as was suggested by our interviewee from the banking sector (interview 1, 2015). Indeed, as social space, areas that contravene the property industry's norm and are therefore stigmatised, may be a desirable place for private investment, middle class home and professional work. In our case of Brixton, there is no evidence as yet that this desirability is driving gentrification, as the concentrated locational choice of the creative class is prone to do. The couple's success in changing a bank's perception in relation to one property and securing a loan has not unlocked mortgage finance for others – the strategy would have to be repeated on a case-by-case basis. Our case shows an attempt at countering, in relation to one property, the powerful socio-spatial ideology that drives the built environment through financing decisions. The counter rationale is that it is 'ok' for banks to finance a 'middle class' or 'professional' property aspiration in an area where home ownership and investment by 'professionals' is present. Interestingly, this alternative representation did not cover the entire Brixton, in particular leaving out the most western portion, in which the community forum has little or no reach (see Figure 4). It represents a loose network of homeowners that also crosses Brixton's boundary to the more affluent north, all of who clearly identify with Brixton as a 'community'. Given the diverse composition of Brixton, it therefore represents only one of many overlaying social networks and forms of identification with the area (some of which are transient), but one which the mortgage seekers deemed relevant to their cause. The effective communication of the alternative rationale through the spatialisation of the 'professionals' in the area – pushing for a new/alternative representation – led one of the banks to bend its lending policy.

This counter rationale could be successfully presented given collective support. The bond seekers drew on social capital in the form of the community forum and the closely associated network of home owners willing to identify themselves as working in 'professional' occupations. The response of these individuals (who are well represented on the community forum) also has to be understood in the light of their mutual interest in presenting Brixton in a positive light – the financing of any future investment decision, and indeed the property values of home owners, clearly depend to a considerable extent on the area's reputation.

Over and above the collective support or social capital, the couple was also able to draw on their own creative professional and cultural capital. They successfully planned and professionally carried through the mapping and the establishment of an alternative representation of Brixton. The couple was able to combine these two forms of capital to gain access to financial capital (see Bourdieu, 1984). In this sense, the case is also about two possible trajectories for the area. One is the maintenance of the uniquely diverse status quo (representative of the city as a whole) which relies on mortgage finance for middle-class investment dotted between often run-down and non-compliant rental establishments. The other, in the absence of mortgage finance for homeowners, is an ongoing shift towards dominance of low-income rental accommodation. This would be underpinned by *rentier*-style extractive forms of investment – the realisation of the trajectory inherent in the stigma banks impose on the area.

However, the extent to which the couple had to go in challenging this spatial discrimination, assumed trajectory for Brixton and financing decision, and the extent to which they relied on collective support, as well as on their own cultural and professional skills, implies that this is not a strategy that the regular mortgage seeker can or will employ. Challenging the powerful structural determinants of Johannesburg's built environment, manifesting in the very uneven distribution of financial capital through investment circulation, market dynamics and political as well as planning decisions by government, will take more than individual or neighbourhood agency. And yet, the case of Brixton, if brought into wider public debate, may help undo perceptions of social difference and related assumptions on wealth or poverty. With a careful analysis of the conditions that allow for gentrification in a socio-spatial context such as Johannesburg, steps would need to be taken not to unleash the takeover of entire areas by the middle class.

Conclusion

Within a global trend of growing urban inequality related to the way incomes are distributed, and with the private financial and built environment industries driving the urban property market, South African cities have remained alarmingly segregated. A vicious circle of value is attached to the pre-apartheid, apartheid and post-apartheid creation of uniform and segregated suburbs which epitomise a long-established norm for the property industry. One implication is the stigmatisation areas that are undergoing processes of integration and transformation, thereby appearing not to fit a so-called norm. Financial sector criteria for the bankability of suburbs and their relevance as a municipal tax base, tie bank and government fiscal interests together in a way that frustrates the process of diversification, desegregation and cosmopolitanisation of South African cities and neighbourhoods. As demonstrated in this chapter, spatial stigmatisation in the form of redlining by financial institutions is not only a form of space-based economic exclusion, but it is de facto a significant contributor to the reproduction of spatial segregation in the post-apartheid condition. More specifically, the practice is detrimental to the many potential local initiatives of individual and collective investment, upgrading and bottom-up urban renewal, be it middle class or not: as soon as credit would be required, the territorialising effects on the urban fabric set in.

In South Africa today, redlining remains a legitimate practice, not challenged in public discourse nor on the radar of legislators or policy makers. From this point of view, it is notable that individual mortgage seekers are motivated to challenge this system, and may employ alternative spatial representations to do so.

In further research on the countering of spatial stigmatisation by the property industry and financial institutions, it would be relevant to examine the roles and relationships of different actors. The 'modes of engagement' (see Haferburg and Huchzermeyer, 2014) of banks and the supporting industry of estate agents, property advisers and aspirant consumers that were examined in this chapter, need to be complemented with an analysis of the rationales and approaches of state and political actors. Thus we conclude this chapter with a hypothesis that stigmatisation is not only the purview of private-sector actors, but might well be found in mutually reinforcing rationales across different spheres of actors that shape the city at both micro- and macro-scales.

Acknowledgements

We would like to acknowledge assistance from Kate Tissington and Miriam Maina at the University of the Witwatersrand, Johannesburg, and Thomas Böge, Benno Hankers, Luca Sommer and Simon Pommerin at the University of Hamburg.

Notes

1 The Office of Disclosure is inaccessible to the public and has issued no reports that are in the public domain. There are no media reports that cover its findings.
2 These categories continue to be monitored in South Africa for purposes of redress.
3 Randlords are described as 'that legendary or infamous class of mine owners and financiers' (Mabin, 2014, p. 396).
4 Other qualification criteria for subsidised housing, some of which would disqualify many of Brixton's tenants, are that the applicant must be a South African citizen or permanent resident, have dependents, be over the age of 18, not benefited before, and fit to contract (Department of Housing, 2009). Citizenship status had also diversified. Census data shows that whereas in 1996 only 0.9 per cent of people in Brixton were citizens of other African countries, in 2011 this figure had risen to 9 per cent.
5 In South Africa, the term 'professional' is strongly associated with a loose use of the term 'middle class', i.e. decently paid white-collar work and bankability.

References

Aalbers, M. (2005), 'Place-based exclusion: redlining in the Netherlands', *Area*, 37(1): 100–109.
Aalbers, M. (2006), '"When banks withdraw, slum landlords take over": the structuration of neighbourhood decline through redlining, drug dealing, speculation and immigrant exploitation', *Urban Studies*, 43(7): 1061–1086.
ABSIP et al. (2003), *Financial Services Charter*. Association of Black Securities and Investment Professionals (as mandated by the Black Business Council); Association of Collective Investments; Banking Council of South Africa Bond Exchange of South Africa; Foreign Bankers Association of SA Investment Managers Association of SA;

Institute of Retirement Funds JSE Securities Exchange South Africa; Life Offices' Association of South Africa South African Reinsurance Offices' Association; South African Insurers Association, available online: http://www.banking.org.za/docs/default-source/default-document-library/financial-sector-charter.pdf?sfvrsn=10 (accessed 22 December 2015).

Beavon, K. S. O. (2004), *Johannesburg: The Making and Shaping of the City*, Pretoria: Unisa Press.

Berrisford, S. (2011), 'Unravelling apartheid spatial planning legislation in South Africa: a case study', *Urban Forum*, 22: 247–263.

Bond, P. (2000), *Elite Transition: From Apartheid to Neoliberalism in South Africa*, London: Pluto Press.

Bourdieu, P. (1984), *Distinction: A Social Critique of the Judgment of Taste*, Cambridge, MA: Harvard University Press.

City of Johannesburg (2003), *Integrated Development Plan 2003/2004*, City of Johannesburg.

City of Johannesburg (2009), *Commune Policy*, Development Planning and Urban Management, City of Johannesburg.

CoGTA (2013), *Towards an Integrated Urban Development Framework: A Discussion Document*. Pretoria: Department of Cooperative Governance and Traditional Affairs.

Crankshaw, O. (2008), 'Race, space and the post-Fordist spatial order of Johannesburg', *Urban Studies*, 45(8): 1692–1711.

Department of Housing (2002), *Community Reinvestment (Housing) Bill*, No. 23423, 17 May. Pretoria: Government Gazette.

Department of Housing (2009), *Housing Code*, Pretoria: Department of Housing.

Everatt, D. (2014), 'Poverty and inequality in the Gauteng city-region', in Harrison, P., Gotz, G., Todes, A. and Wray, C. (eds), *Changing Space, Changing City: Johannesburg After Apartheid*, Johannesburg: Wits University Press, pp. 63–82.

February, J. (2002), 'Redlining to be outlawed', *Mail and Guardian*, 1 January. Available online at: http://mg.co.za/article/2002-01-01-redlining-to-be-outlawed (accessed 22 December 2015).

Haferburg, C. (2007), *Umbruch oder Persistenz? Sozialräumliche Differenzierungen in Kapstadt*, Hamburg: Institut für Geographie.

Haferburg, C. and Huchzermeyer, M. (2010), Unpublished sample household survey of Brixton, conducted in collaboration between the Brixton Community Forum and the School of Architecture and Planning, University of the Witwatersrand, Johannesburg.

Haferburg, C. and Huchzermeyer, M. (2014), 'An introduction to the governing of post-apartheid cities', in Haferburg, C. and Huchzermeyer, M. (eds), *Urban Governance in Post-apartheid Cities: Modes of Engagement in South Africa's Metropoles*, Stuttgart: Borntraeger, pp. 3–14.

Harrison, P. (2007), 'Let's save Brixton', *MATASATASA*, Official Staff Newsletter for Region B's Urban Management, City of Johannesburg.

Harrison, P. and Zack, T. (2014), 'Between the ordinary and the extraordinary: Socio-spatial transformations in the "Old South" of Johannesburg', *South African Geographical Journal*, 96(2): 180–197.

Hernandez, J. (2009), 'Redlining revisited: Mortgage lending patterns in Sacramento 1930–2004', *International Journal of Urban and Regional Research*, 33(2): 291–313.

Hillier, A. (2003), 'Redlining and the Home Owners' Loan Corporation', *Journal of Urban History*, 29(4): 394–420.

Holmes, A. and Horvitz, P. (1994), 'Mortgage redlining: Race, risk, and demand', *The Journal of Finance*, 49(1): 81–99.

Huchzermeyer, M. (2001), 'Housing for the Poor? Negotiated Housing Policy in South Africa', *Habitat International*, 25: 303–331.

Huchzermeyer, M. (2004), *Unlawful Occupation: Informal Settlements and Urban Policy in South Africa and Brazil*, Trenton, NJ: Africa World Press/The Red Sea Press.

Lefebvre, H. (1991), *The Production of Space*, Oxford: Blackwell Publishing.

Leibbrandt, M., Finn, A., and Woolard, I. (2012), 'Describing and decomposing post-apartheid income inequality in South Africa', *Development Southern Africa*, 29(1): 19–34.

Lightstone Property (2015a), Sample Suburb report. Johannesburg: Lightstone Property. Available online at: http://www.lightstoneproperty.co.za/SampleReports/SuburbReport.pdf (accessed 10 October 2015).

Lightstone Property (2015b), Suburb Report: Brixton, June 2015. Johannesburg: Lightstone Property.

Mabin, A. (2014), 'In the forest of transformation: Johannesburg's northern suburbs', in Harrison, P., Gotz, G., Todes, A. and Wray, C. (eds), *Changing Space, Changing City: Johannesburg After Apartheid*. Johannesburgh: Wits University Press, pp. 395–417.

Melzer, I. (2015), 'Segmenting mortgage loan performance in South Africa', Centre for Affordable Housing Finance: Johannesburg. Available online at: http://www.housingfinanceafrica.org/blog/segmenting-mortgage-loan-performance-in-south-africa/

Murray, M. (2014), 'City of layers: The making and shaping of affluent Johannesburg after apartheid', in Haferburg, C. and Huchzermeyer, M. (eds) *Urban Governance in Post-Apartheid Cities: Modes of Engagement in South Africa's Metropoles*, Stuttgart: Borntraeger, pp. 179–196.

O'Brien, K. (2011), *The South African Intelligence Services: From Apartheid to Democracy, 1948–2005*, Oxford: Routledge.

Pam Golding (2015), Online advert for property in Brixton, Pam Golding Properties, Johannesburg. Available online at: http://www.pamgolding.co.za/property-search/residential-properties-for-sale-brixton/848 (accessed 15 July 2015).

Parnreiter, C., Haferburg, C. and Oßenbrügge, J. (2013), 'Shifting corporate geographies in Global Cities of the South: The Cases of Mexico City and Johannesburg', *Die Erde*, 144(1): 41–62.

Pernegger, L. (2014), 'The agonistic state: Metropolitan government responses to city strife post-1994', in Haferburg, C. and Huchzermeyer, M. (eds), *Urban Governance in Post-apartheid Cities: Modes of Engagement in South Africa's Metropoles*, Stuttgart: Borntraeger, pp. 61–78.

Randall, E. (2000), 'Banks tackled for 'red-lining' by colour', IOL, 2 September. Available online at: http://www.iol.co.za/news/south-africa/banks-tackled-for-red-lining-by-colour-1.46349#.Vg7cHSsXfFA (accessed 22 December 2015).

Republic of South Africa (1996), *The Constitution of the Republic of South Africa*, Act 108 of 1996, Pretoria: The Government Printer.

Republic of South Africa (2002), Department of Housing, *Community Reinvestment (Housing) Bill*, No. 23423, 17 May, Pretoria: Government Gazette.

Republic of South Africa (2007), *Government Notice 1372* in Government Gazette 21900, dated 15 December, 2000. Commencement date: 2 July, 2007 [Proc. 15, Gazette No. 30050, dated 13 July, 2007].

Sihlongonyane, M.F. (2014), 'A critical overview of the instruments for urban transformations in South Africa', in Haferburg, C. and Huchzermeyer, M. (eds), *Urban Governance in Post-apartheid Cities: Modes of Engagement in South Africa's Metropoles*, Stuttgart: Borntraeger, pp. 40–59.

Storie, M. (2014), 'Changes in the natural landscape', in Harrison, P., Gotz, G., Todes, A. and Wray, C. (eds), *Changing Space, Changing City: Johannesburg after Apartheid,* Johannesburg: Wits University Press, pp. 137–153.

Todes, A. (2014), 'The external and internal context of post-apartheid urban governance', in Haferburg, C. and Huchzermeyer, M. (eds), *Urban Governance in Post-apartheid Cities: Modes of Engagement in South Africa's Metropoles*, Stuttgart: Borntraeger, pp. 15–35.

Tomer, J. (1992), 'The social causes of economic decline: Organizational failure and redlining', *Review of Social Economy*, 50(1): 61–81.

Tomlinson, M. (1998), 'The role of the banking industry in promoting low income housing development', *Housing Finance International*, 12: 8–13.

Tomlinson, M. (1999), 'Ensuring mortgage payment in South Africa', *Housing Finance International*, 13: 40–43.

Tomlinson, M., (2005), 'South Africa's Financial Sector Charter: where from, where to?', *Housing Finance International*, 20(2): 32–36.

Interviews and personal communication

Interview 1: banking sector representative, interview conducted by Kate Tissington, 17 June, 2015.

Interview 2: municipal official, interview conducted by Kate Tissington, 7 July, 2015.

Interview 3: Graupner, A. interview: resident and architect, 26'10 South Architects, Brixton, interview conducted by Christoph Haferburg, 18 October, 2012.

Personal communication: housing consultant, personal communication by email, 5 October, 2015.

Rust, K., personal communication by email, Centre for Affordable Housing Finance in Africa and Secretariat to the African Union of Housing Finance, Johannesburg, 11 November, 2015.

6 The 'not so good', the 'bad' and the 'ugly'

Scripting the 'badlands' of Housing Market Renewal[1]

Lee Crookes

Geography was not something already possessed by the earth but *an active writing of the earth by an expanding, centralizing imperial state.* It was not a noun but a verb, a *geo-graphing*, an earth-writing by ambitious endocolonising and exocolonizing states who sought to seize space and organise it to fit their own cultural visions and material interests.

(O' Tuathail, 1996, p. 6, original emphasis)

No one, wise Kuublai, knows better than you that the city must never be confused with the words that describe it.

(Calvino, 1975, p. 61)

Introduction

Places and spaces are constructed materially, socially, emotionally *and* discursively. Indeed, as Harvey has noted, 'Discursive struggles over representation are as fiercely fought and just as fundamental to the activities of place construction as bricks and mortar' (1996, p. 322). But, just as discourse can play an important role in 'place-making', that is, in the planning, promotion and positive representation of places, it can also be equally instrumental in their wilful decline and destruction (see for example Graham, 2004; Mele, 2000a, 2000b; Beauregard, 1993). As Graham reminds us: 'Cities are destroyed, unmade and annihilated discursively and through symbolic violence, as well as through bombs, planes and terrorist acts' (2004, p. 44). In particular, with respect to urban development, 'states discursively constitute, code and order the meaning of place through policies and practices that are often advantageous to capital' (Weber, 2002, p. 524) and research from several different contexts provides evidence of city governments playing an active role in the production of stigmatising narratives of place to facilitate urban redevelopment (Kallin and Slater, 2014; Allen, 2008; Pfeiffer, 2006; Goetz, 2003; Beauregard, 1993). In some instances, such processes of 'territorial stigmatisation' (Wacquant, 2007) have quite literally helped to prepare the ground for new-build gentrification by serving to legitimise extensive housing clearance programmes such as Housing Market Renewal (HMR) in northern England and HOPE VI in the United States.

As this chapter will demonstrate, territorial stigmatisation is integral to such processes of 'gentrification-by-bulldozer' (Crookes, 2011).

Using carefully-crafted, oft-repeated and consistently applied narratives, powerful policy elites are able to *problematise* selected 'target' neighbourhoods and their residents in particular ways that make certain, favoured solutions (such as demolition) seem natural, logical and inevitable (Goetz, 2013). Through a relentless, linguistic offensive of stigmatisation, demonisation and othering, neighbourhoods are variously discredited, tainted, exoticised and subjected to damaging, homogenising, socio-spatial misrepresentations. More pertinently, through this new 'urban Orientalism' (Wacquant, 1997) the state is then able to delineate the boundaries of an othered space – a housing estate, *banlieue* or housing project – that warrants distinctive 'special treatment':

> Once a place is publicly labelled as a 'lawless zone' or 'outlaw estate', out-side the common norm, it is easy for the authorities to justify special meas-ures, deviating from both law and custom, which can have the effect – if not the intention – of destabilizing and further marginalizing their occupants . . . and rendering them invisible or driving them out of a coveted space.
>
> (Wacquant, 2007, p. 69)

In this way, we can readily see how territorial stigmatisation can be invoked to symbolically and physically turn much-loved, meaningful places to dirt, emptying places of people, homes and meaning to create the *tabula rasa* of a development-ready 'brownfield' site. As Porter has argued in relation to the redevelopment of Eastside, Birmingham:

> to attract developer interest, the Council must create a 'blank slate' to offer to the market – a place with no memory, no people, no community, no spirit. Just a flat, uniform, empty patch of dirt, with all traces of earlier life expunged (except where they might usefully turn a profit).
>
> (2009, p. 144)

Territorial stigma is integral to such practices of erasure and this chapter shows how the symbolic defamation of place operated as a precursor to the housing clearance that formed a major component of the HMR programme, serving to legitimise the demolition of mixed-tenure terraced housing and social housing to meet developers' demands for inner-urban, brownfield sites suitable for residential development in the former manufacturing heartlands of northern England and the West Midlands.

The chapter begins by examining the relationship between place marketing and territorial stigmatisation, arguing that boosterism and spatial defamation often work alongside each other – if sometimes rather uneasily – as part of a city's broader redevelopment efforts. Developing this further, I argue that territorial stigmatisation is an intrinsic feature of urban policy and that it has, in some guise or another, long been implicated in a 'deficit model' approach to urban policy whereby policy officials and academics have been fixated with finding and constructing urban 'problems' (Baeten, 2002).

The next two sections then provide an overview of the HMR programme and an analysis of the spatial discourses that were constructed and invoked by those who were involved in the development and implementation of the programme. Drawing on work in critical geopolitics that examines the politics of 'geo-graphing' spaces (O'Tuathail, 1996), I then describe how the areas targeted by the programme were represented or 'scripted' in three principal ways.

Finally, reflecting on the extent to which academics were involved in the formulation and implementation of the HMR programme, the chapter concludes by offering some suggestions for how academics might take greater responsibility for challenging territorial stigma in their own research and teaching.

Urban policy and research = territorial stigmatisation?

Writing in the late 1980s, David Harvey was one of the first commentators to detect a significant shift in the role and nature of city government. Where urban governments had once been primarily engaged in the provision of services and amenities to their local populations, Harvey (1989) observed that local authorities were beginning to adopt a much more entrepreneurial outlook, competing with other cities to attract private investment, tourism and public funding in support of new public-private ventures. In this shift from managerialism to a new regime of outward-looking urban entrepreneurialism, place itself becomes a source of competitive advantage. What we then see is the emergence of a preoccupation with a city's 'symbolic economy', that is, 'the intertwining of cultural symbols and entrepreneurial capital' (Zukin, 1995) where the symbolic and material processes involved in the production of place are inextricably linked. With the re-imaging of the city emerging as a key factor in inter-urban competitiveness, 'The production of urban images, of advertising and communication about the city has become a field of public policy in its own right, with dedicated budgets, organisations and experts' (Colomb, 2012, p. 11).

But the project of city building is not just about place promotion and 'accentuating the positive'. As Friedmann (2007) observes, the incessant forces of 'creative destruction' are dependent upon the twin processes of place-making *and* place-breaking but the critical focus on civic boosterism may have distracted attention from the routine defamation that attends the latter. Parker and Long (2004, p. 55) offer a useful synthesis:

> The dominant storylines of urban renewal are animated by the politics of vision and the politics of erasure. The politics of erasure is a discursive ground-clearing, the framing of a spectacle of shame. . . The politics of vision defines an aspirational city imaginary for the twenty-first century – a horizon of expectation towards which are projected hopes for renewal.

In short, urban development requires some combination of both place marketing *and* its opposite, that is, territorial stigmatisation. As Smith (1979, 1996) demonstrates, uneven development and the unfolding spatial dynamics of 'rent gaps' are related to processes of devalorisation and re-investment and it is encouraging to see further work exploring the links between territorial stigmatisation and gentrification (Kallin and Slater, 2014; Wacquant, Slater and Pereira, 2014; Slater and Anderson, 2012; Gray and Mooney, 2011; Wacquant, 2007).

Urban imagineers, those responsible for producing discourses about the city must, to use everyday language, be just as adept at 'talking places down' through spatial defamation as they are at 'talking places up' through their attempts at place promotion. But city officials and academics should not be overly challenged or troubled by this pressure to devalue and discredit since much of the day-to-day work of urban policy and scholarship is essentially concerned with finding and analysing 'problems' in the city, with problematising urban places and phenomena and/or producing negative representations of selected, 'target' neighbourhoods. Much urban policy, arguably, can therefore be conceived as territorial stigmatisation writ large.

At this point it might be instructive to draw on some of the classic scholarship on policy analysis, since much of this work interprets policy-making as a process that is concerned with the collective definition or social construction of a problem (Blumer, 1971; Stone, 1989; Wildavsky, 1979). In particular, 'the first maxim of problem definition is the recognition that problems do not exist 'out there', are not objective entities in their own right, but are analytic constructs, or conceptual entities' (Dery, 2000, p. 37). Policy-making, Hajer argues, is very much a question of problem-making: 'Hence, policies are not only devised to solve problems, problems also have to be devised to be able to create policies' (1993, p. 15). Extending this discursive framework to the context of housing policy, Jacobs and colleagues (Jacobs, Kemeny and Manzi, 2003, p. 430) suggest that: 'the construction of each problem draws heavily upon negative stereotyping and rhetorical strategies that undermine the status of certain marginalised social groups whilst privileging others.'

What makes urban policy distinctive is that its problem definition begins with an area (Cochrane, 2000) but this basic fact is frequently overlooked. As Dikeç argues, 'social constructionist approaches, while helpfully focusing on the construction of urban problems and policy discourses, neglect the role space plays in such constructions' (2005, p. 6). For Dikeç, urban regeneration 'is a place-making practice that spatially defines areas to be treated, associates problems with them, generates a certain discourse and proposes solutions accordingly' (2005, p. 41). Whether it be through language, statistics, imagery or cartographic representation, we can begin to see how territorial stigmatisation plays a fundamental role in constituting the spaces that form the 'targets' of urban policy. In this way, urban policy can then be reconceptualised as the practice by which urban elites inscribe spatial stigma, imprinting blemishes upon places that serve to literally mark them out as different or somehow 'deviant' from some arbitrarily defined 'norm'.

Some might argue that we have been acutely aware of the role of territorial stigmatisation in urban policy for some time: the social construction of problematic urban spaces for political and/or development purposes is not new. For example,

Beauregard (1993) shows how cities in the United States have long been the subject of a discourse of decline. Fullilove and colleagues (Fullilove, Hernandez-Cordero and Fullilove, 2010, p. 202) recount an interview with Dennis Gale, Professor of Urban Studies at Rutgers who described how 1950s urban renewal would begin with the city designating certain areas as blighted: 'You label it as bad, you clear it all out. You have a featureless plain and call it urban renewal. There is no longer any bad, there is nothing. And then you build from scratch.'

We see similar processes at work again and again, for example, in New York's Lower East Side (Mele, 2000b), in Minneapolis (Goetz, 2003), in the redevelopment of Cabrini Green in Chicago (Bennett and Reed, 1999; Pfeiffer, 2006), in Vancouver's Downtown Eastside (Blomley, 2004) and in relation to the public housing targeted by the HOPE VI programme, where a major study by the National Housing Law Project suggested that it played 'upon inaccurate stereotypes about public housing to justify a drastic model of large-scale family displacement and housing redevelopment' (NHLP, 2002, p. iii). Goetz (2013, p. 342) similarly identifies a wildly exaggerated, near apocalyptic 'discourse of disaster' whereby those promoting HOPE VI represented US public housing communities 'as deviant, dysfunctional, or obsolete'. Examples such as these and the discussion of HMR presented below help us to expose the inherently discursive nature of urban policy. The more we recognise this, the easier it becomes to see that territorial stigmatisation is something that is routinely practiced by the powerful urban elites who shape urban discourse and policy. This simple recognition is vitally important, for, as Pickles (2004, p. 18) contends, 'Recognising the socially constituted nature of identity claims (our concepts, categories and practices) is a first step to a deconstructive retrieval of other possible worlds, spaces and mappings.'

Jacobs *et al.* (2003, p. 431) conclude their paper by arguing that housing researchers 'need to question far more the construction of 'problems' commonly advanced by government policy makers if the discipline is to retain critical and independent modes of enquiry'. The issue here, however, is that many academic researchers are themselves wedded to a 'deficit model' of urban phenomena, having been schooled in what Baeten (2002, p. 107) describes as the 'peculiar epistemological framework of 'problems', a singularly dystopian approach to the city that he sees as a pervasive feature of wider urban scholarship. Lees (2004, p. 4) has also noted how 'the relentless focus on urban problems has tended to reinforce longstanding narratives of perverse and pathological urbanism. . . The prevailing mood in urban studies has become one of doom and gloom.' But all is not lost because this 'deficit model' provides university research institutes and private consultancies with a lucrative stream of income. As Dileonardo and Reed (1989) remind us, 'poverty research is a huge academic business' (see also Nicolaus, 1968). With academics under increasing pressure to secure research income (Allen and Imrie, 2010), the quest to find or *create* urban 'problems' grows ever stronger. In the contemporary neoliberal university many academics are therefore increasingly complicit, either inadvertently or intentionally, in practices of territorial stigmatisation.

In the case of HMR, the territorial stigmatisation was unequivocally intentional. Moreover, what made HMR particularly interesting was that it re-scaled the practice of territorial stigmatisation. Previously, much of the attention of policy-makers, academics and urban policy had been focused on specific 'problem' neighbourhoods within cities; HMR expanded the scale of spatial defamation to much bigger, housing market areas that covered entire sub-regions. Those who were involved in the formulation of HMR therefore had to find ways of stigmatising large swathes of northern England and the West Midlands. This was achieved through the discursive construction of extensive areas of so-called 'low demand' housing, 'housing market failure' and the associated designation and delineation of these areas as housing market renewal areas.

Borrowing terminology from the field of critical geopolitics, we might say that those involved in constituting HMR's areas of intervention were effectively engaged in a form of 'earth-writing', that is, in the construction of stigmatising 'geographs' or 'scripts'. Geographs, according to O'Tuathail are 'simplified descriptions of the earth' that transform specific places and regions into the objects of politically influential narratives' (1996, p. 476). O'Tuathail also uses the term 'foreign policy script' to describe 'a set of representations, a collection of descriptions, scenarios and attributes which are deemed relevant and appropriate to defining a place in foreign policy' (1996, p. 156). Such scripts, O'Tuathail (1996) argues are employed as a form of 'shorthand' that serves to render complex socio-spatial realities readily comprehensible in ways that appeal to common-sense understandings whilst simultaneously marginalising alternative understandings of these spaces. Urban policy is, in several respects, markedly different from foreign policy but both frequently seek to produce simplified representations of spaces that serve their respective ends, othering identified spaces (and people) in readiness for particular forms of intervention at different scales. In the discussion below, I therefore extend O'Tuathail's ideas to the urban scale and, specifically, to HMR, substituting 'urban policy scripts' for his 'foreign policy scripts'.

Having spent some time theorising the central role of territorial stigmatisation in urban policy and urban regeneration schemes, the remainder of the chapter examines the HMR programme, a clear example of how a 'discourse coalition' (Hajer, 1993) of various public and private interests colluded with academics to create a new problem of 'low demand' housing. In support of this low demand framing, the 'architects' of HMR constructed three, principal 'scripts' that sought, to varying degrees, to reimagine places as empty, problematic or in the throes of a rapidly unfolding disaster, thereby legitimising demolition on a massive and unprecedented scale.

The context: Housing Market Renewal

The HMR programme was introduced by the Labour government in 2002, ostensibly in response to evidence of so-called 'low demand' housing and 'housing market failure' in former industrial areas across the north of England and the West

Midlands (Cole and Nevin, 2004). An imprecise concept, variously defined, 'low demand describes housing which few people want to move into, or remain living in . . . It applies to areas where overall demand is low relative to supply, suggesting an emerging surplus of housing. The areas affected can be small neighbourhoods, estates, cities. . .' (Power and Mumford, 1999, p. vii). With particular concerns about the apparent extent of empty and low demand housing in the north-west of England in the late 1990s, the North West Housing Forum,[2] a newly formed housing lobby group, commissioned research to provide new evidence on changing housing markets and the prevalence of low demand in Manchester, Liverpool and other settlements in the M62 motorway corridor that connects the two cities (Nevin, 2001). The 'M62 study', as it became known, was a crucial moment in raising the profile of the low demand issue, both regionally and nationally.

Introducing the neologism of 'Housing Market Renewal' to the urban policy lexicon, Nevin (2001) suggested that the problem they had identified could only be tackled with a new, long-term, sub-regional approach to regeneration that was focused on the renewal of housing markets. At much the same time, the 2002 House of Commons Select Committee inquiry into Empty Homes was recommending the need for 'radical intervention' in the form of a long-term, conurbation-wide, housing market renewal approach (House of Commons, 2002). In April 2002, having considered the Select Committee's recommendations and the emerging research evidence, the Government announced that it was creating Housing Market Renewal Pathfinders in nine sub-regional areas in northern England and the West Midlands. These HMR Pathfinder partnerships were established to coordinate 'radical and sustained action to replace obsolete housing with modern sustainable accommodation, through demolition and new building or refurbishment. This will mean a better mix of homes, and sometimes fewer homes' (ODPM, 2003). Home to around 1,860,000 people, the nine 'Pathfinder' areas incorporated a total of just over 846,000 houses and flats, of which around 61,000 were vacant. Initial plans to demolish around 167,000 of these dwellings (Northern Way, 2005) were later revised down to 57,000 (Audit Commission, 2006). Spending on HMR between 2002 and 2011 amounted to £2.2 billion but, with limited evidence of its effectiveness (see National Audit Office, 2007; House of Commons, 2008), the programme was eventually wound up in April 2011 as part of the Coalition's budget cuts.

In a critical review of evidence-based policy in housing, Jacobs and Manzi (2013, p. 7) suggest that 'HMR provides perhaps the clearest example of the use (and some would argue abuse) of evidence'. Indeed, as Leather, Cole and Ferrari (2007, p. viii) emphasise: 'It is doubtful whether any previous regeneration programme has been developed from a sounder evidence base.' 'Policy makers and practitioners argue that objective, academic evidence demonstrates the need for housing market renewal' but, as Webb (2010, p. 3) contends, 'this evidence is actually the construct of a highly political environment and demonstrates the utility this evidence has for specific interests involved in urban renewal.' Webb (2010, 2012), Cameron (2006) and Allen (2008) have all argued that HMR was not the evidence-based policy that it was so frequently claimed to be; Jacobs and

Manzi (2013, p. 5) contend that the deployment of evidence in support of HMR was 'akin to "epistemic legitimisation", evoked to demonstrate superior knowledge to undermine opponents by proving their "impartiality" and "rationality" to a wider audience'.

In seeking to add to this line of critique, I would argue that the dominance of the official discourses of low demand and housing market failure also served to divert attention from one important aspect of the context in which HMR emerged. Back in February 1998, as part of its efforts to promote an urban renaissance, the Blair government introduced a target requiring 60 per cent of all new housing development in England to be built on brownfield land. Broadly defined, brownfield land can include abandoned, disused or vacant land that was formerly given over to industrial uses but, significantly, it can also include land that is currently used for residential purposes. Now, the cost of remediating and decontaminating former industrial land can be extremely high and housebuilders will often demand public subsidies to develop such areas. The cheaper, more straightforward alternative, of course, is to build on residential land, to in-fill or, preferably, to actively produce larger sites through the demolition of existing housing. But city authorities and developers cannot simply set about demolishing houses without proper justification. Following Hajer (1993), a problem needs to be formulated to justify a policy of mass demolition on a scale that will satisfy developers' demand for residential brownfield sites. It might be pure coincidence but it is precisely at this juncture that a coalition of support for HMR emerges, advocating clearance on a massive scale.[3] Just as Neil Smith (2002, p. 444) had predicted, Blair's brownfield target would likely be 'aimed at older urban areas that have undergone sustained disinvestment'.

Whilst many studies of territorial stigmatisation have focused on areas of public housing, it is worth adding here that around 50 per cent of the housing in the HMR areas was privately owned. The respective authorities therefore had to acquire a significant number of houses at market value using their compulsory purchase powers. The precise nature of this relationship merits further research but one would surmise that an ongoing, orchestrated process of territorial stigmatisation would help to keep house prices low in the target neighbourhoods. This, in turn, serves to both widen the rent gap in anticipation of redevelopment *and*, crucially, it reduces the cost of the acquisitions to the state. This all lends further weight to the view that HMR was little more than a 'land grab' that was artfully disguised by an appeal to the technical rationality of a low demand discourse, or, as Macleod and Johnstone (2012) have characterised it, a prime example of 'accumulation by dispossession'.

Scripting the 'badlands' of HMR: empty spaces, problem areas and disaster zones

Whilst Allen (2008) and Webb (2010) have made an important contribution to the idea that discourse played a major role in the development of HMR, this chapter's main concern is the *spatiality* of the HMR discourse. Combining Hajer's insight that policy-making is, in essence, a matter of problem-making with Dikeç's (2005, p. 7) contention that 'any analysis of urban policy has to critically analyse the ways in which policies constitute their spaces of intervention', this section examines the 'scripts' that were employed by HMR, drawing on the author's doctoral research (Crookes, 2011). The analysis presented here is based on an extensive review of several key HMR policy documents and participant observation in two public inquiries that were held in Liverpool in 2006 and 2008 to consider the compulsory purchase of around 900 homes in the Merseyside *New Heartlands* HMR Pathfinder area.

Most studies of territorial stigmatisation have, arguably, focused on language as the dominant mode of (mis)representation and this may naturally reflect the reality of how territorial stigmatisation is enacted in a range of different contexts. Certainly, with respect to the politics of New Labour in the UK, language and 'spin' played a significant role in their approach to government (see Fairclough, 2000; Hall, 2003; Levitas, 1998) and HMR was no exception to this, as we shall see. But it is important to remember that the discursive constitution of HMR's spaces of intervention also employed other modes of representation. Using a combination of language, statistics, photographs and maps, the three 'scripts' described here constructed simplified geographies that sought to reimagine the places being targeted as empty spaces, problem areas, disaster zones or, very often, as some combination of all three.

'Empty', 'dying' or 'going to waste'

Across the HMR programme as a whole, the areas targeted for intervention were generally scripted as anachronistic, 'obsolete' or in terminal decline in ways that made change and modernisation – 'transformation' even – appear necessary and inevitable (Allen, 2008). One does not have to look far to find references to obsolescence and/or evidence of Allen's 'end of history' narrative:

> A fundamental issue is for policy to recognise the need for the North to readjust its housing stock to one that responds to the economic challenges of the 21st century, making it competitive. In simplistic terms, whilst the North still has a significant stock of housing intended for accommodating the workforce of an early 20th century industrial economy, it is positioned poorly in terms of a residential offer attractive for the workers that will drive a post-industrial 21st century economy.
>
> (Ove Arup/ Innovacion, 2006, p. 7)

But before the residents of the HMR areas could be physically displaced to make way for this influx of new, post-industrial workers, they first had to be displaced discursively. In order to construct an imagined *terra nullius*, this particular script, informed by what Allen (2009) describes as a 'politics of emptiness', sought to represent the HMR areas in ways that emptied them of meaning, people and value so as to naturalise and legitimise demolition. The apparent spiral of decline, depopulation and abandonment is perhaps best summarised in the introduction to the 2005 document, *Sustainable Communities: Homes for All*, which set out the Government's five-year plan to create sustainable, mixed communities (ODPM, 2005, p. 1):

> In the closing years of the last century and the first few years of this one, the prices of houses in parts of some of our northern cities and towns fell to prices well below their apparent worth. Those who could do so moved out. Homes were abandoned and some of the most disadvantaged in our society became trapped in neighbourhoods characterised by dereliction, crime, anti-social behaviour and poor services.

This sense of dereliction and abandonment can be found throughout the academic research on HMR and it permeates the extensive body of 'grey' literature that accompanied the programme. Reviewing this literature, one finds neighbourhoods being repeatedly characterised as de-populated, de-humanised spaces or as 'Order Lands',[4] that is, as sequentially numbered plots and parcels of land that were seemingly devoid of life. This vocabulary of emptiness spoke variously of 'voids', 'abandonment' and 'brownfield sites' whilst, in Liverpool, the HMR areas of intervention were described in the de-humanising and frontier-like language of 'Zones of Opportunity' (New Heartlands, 2006). Far from being empty, however, it is important to remember that the nine Pathfinder areas were actually home to a total of around 1.9 million people.

The HMR areas are frequently viewed as being in 'terminal decline' and, in the tradition of 'urban triage', housing renewal guidance from the early 2000s (ODPM, 2003) refers to the 'termination of areas'. With HMR alternately invoked as a sort of place-based euthanasia or an 'intervention' that will 'bring life back to these areas' (ODPM, 2003, p. 23), the 'architects' of HMR would lead us to us believe that major cities like Liverpool and Manchester were abandoned 'ghost towns', that were, if not already 'dead', then at least very close to expiration:

> Both major centres contain considerable areas of derelict land and premises, and in some areas the roads, pavements and general environment reflect economic and social activity of a long gone era. Additionally, the physical infrastructure of the core of the two conurbations was constructed for much larger populations raising issues relating to the sustainability of the neighbourhoods.
> (Nevin *et al.*, 2001, p. 7)

The 2002 Select Committee on Empty Homes is similarly pessimistic, concluding that without HMR, 'our northern cities will consist of a city centre surrounded by a devastated no man's land encompassed in turn by suburbia' (House of Commons, 2002, para. 161).

The visual representation of HMR areas reiterates this sense of emptiness and abandonment. Although the HMR areas continued to be home to large numbers of people who did their best to go about their lives whilst living under the threat of demolition, photographs of the HMR areas invariably presented images of empty, boarded-up housing, cleared brownfield sites and/or dereliction. By way of example, the Urban Task Force report, commonly known as the 'Rogers Report' (DETR, 1999), which had a major influence on HMR, includes an aerial photograph of East Manchester (at page six) that shows a mix of urban land-uses, with the foreground dominated by a large social housing estate, comprising a mix of tower blocks and low-rise housing. The photograph is simply captioned as 'East Manchester: Land going to waste'. Matter-of-factly, the caption exudes a frustration that the land under its current use falls short of highest and best use: land, and by implication, potential profits are going to waste. To my mind, this simple statement is hugely revealing, communicating the essence of how urban elites apprehend the urban realm in terms of privileging exchange value over use value, abstract space over lived space and, ultimately, profit over people.

Problem places

Using examples drawn from some of the research that informed the development of the HMR programme, this next section discusses how statistical analysis and mapping often play a critical, if frequently taken-for-granted role in territorial stigmatisation. This body of research was concerned with identifying and spatially delineating those areas that were either experiencing 'low demand' or at risk of 'low demand'. It was about *marking* certain places, differentiating them as 'other' and separating them out from an arbitrarily constructed 'norm'. For instance, in the M62 study this was achieved principally through the calculation of a standardised composite risk score for every Census enumeration district in the Merseyside/Lancashire and Greater Manchester M62 corridor area. These scores were then mapped to highlight those enumeration districts that scored more than 1.51 (the median score) on the combined risk index. As Nevin *et al.* (2001, p. 47) explain: 'Whilst this is an arbitrary threshold, we wanted to focus on those parts of the M62 Corridor that have the highest coincidence of a range of factors associated with low or changing demand.'

With the enumeration of districts that were computed to have the highest risk scores shaded in red, the resultant map showed that the housing at risk of changing (low) demand was spatially concentrated in Manchester/Salford, inner Liverpool and some smaller towns on the edge of the Liverpool city boundary. This modelled approach was later extended across England and provided the basis for the designation and delineation of the Pathfinder areas in 2002:

> The boundaries for the HMR areas did not follow established administrative contours but were largely shaped by the scale and incidence of market failure, as identified through the raft of studies . . . in which the results of data analysis defined the territory.
>
> (Cole and Nevin, 2004. p. 19)

As objective as the research described here might first appear, the presentation of the results of this weighty 'raft of studies' was inevitably a conscious, political act, with decisions about what should be included (or omitted) very much tied to the audience and the message that the researchers wanted to emphasise. Highlighting particular enumeration districts in red served to *mark* them out, perhaps intuitively warning the reader of 'danger': in effect, it served to imprint the blemish of place onto the map. The work of Harley (1988, 1989) and Pickles (1995) shows how maps frequently do powerful rhetorical work in very subtle ways: they are not simply innocent representations of reality. As Pickles notes, maps often play a key role 'in enabling repeated rounds of dispossession, enclosure and colonization' (2006, p. 348).

Developing this further, the drawing of the Pathfinder boundaries actually served several rhetorical functions. Typically, the maps showing the Pathfinder areas of intervention would shade the electoral wards included in the Pathfinder boundary as a solid block of grey. This sounds fairly straightforward and unproblematic. Yet if we think about such maps in the more reflexive and considered way that Harley (1989) suggests, other subtle meanings come to light. Firstly, the delineation of an outer boundary and the shading establishes difference. It puts a boundary around the 'problem' and fixes it in space: 'Discipline sometimes requires enclosure, the specification of a place heterogeneous to all others and closed in upon itself' (Foucault, 1979, p. 141). The wards that are shaded are, in effect, 'othered' as being somehow different from their neighbours, as somehow abnormal. At the same time, this differentiation also establishes a superior 'us' and an inferior 'them', picking out an area of 'failure' or 'weakness' that requires some form of intervention or correction in order to become 'normal' like the areas that lie outside of the boundary. Putting a boundary around an area-based regeneration initiative is, in many ways, akin to a form of 'redlining'.

Inevitably, such maps also convey the impression of an *extensive* and *contiguous* 'problem' zone that demands a 'joined-up', inter-authority response that extends beyond the conventional neighbourhood to an entire sub-region. Such a form of representation adds value to the underlying statistical indices and tables by conveying visually the existence of a continuous zone – a problem that has an apparent geographic logic. It suggests a coherent, bounded problem that is amenable to bounded, area-based intervention, a problem of the delineated area rather than one produced by the structural inequalities and injustices of the wider society. At the same time one also gets a sense of 'contagion' that must be contained and arrested: as noted above, the HMR areas in Liverpool were officially designated as 'Zones of Opportunity' or, more curtly, 'ZOOs', whilst the adjoining neighbourhoods were designated as 'holding areas' (New Heartlands, 2006).

Finally, we might additionally contend that cartographic representation simplifies, hides or renders invisible what is already there, adding to the impression that the enclosed area is simply a 'regeneration space' – a surface amenable to intervention – rather than an array of unique and interesting people and places. No longer a collection of communities it is, in effect, redefined as a Housing Market Renewal Area, delineating an expanse of territory where housing market failure is uniformly evident. Understood in this way, we can see how mapping is one of the methods by which territorial stigma is subtly inscribed onto the pages of an authoritative document, the map allowing its author 'to control what exists by selecting what is depicted and thus officially recognised' (Herb, 2008, p. 30).

Disaster zones

If policy-making is problem-making, then the principle 'the bigger the problem the better' can be an effective strategy for attracting the attention of policy-makers. Creating a large-scale (in this case, sub-regional) 'disaster' scenario was certain to mobilise a large cross-section of interests and create the sort of powerful discourse coalition which ensured that HMR was accelerated to the top of the political agenda (see Webb, 2010). Drawing on Klein's (2007) work, which demonstrated how the state manufactures or uses 'disaster situations' to open up new markets for capital, Allen and Crookes (2009) argue that the research informing HMR created the impression that housing markets in certain parts of northern cities had either failed or were on the verge of failure, thereby creating the potential for a 'housing market disaster': HMR is essentially a 'disaster recovery' programme. This is disaster capitalism par excellence, with state-sponsored gentrification in northern cities being legitimised by the academic manufacture of a 'disaster myth'.

The figures reported in a key 2005 document, the *Northern Way Growth Strategy*, were suitably apocalyptic:

> The Centre for Urban and Regional Studies (CURS) estimates that around 1,500,000 homes are at risk and perhaps up to 400,000 should be replaced. . . Based on current rates, over the next ten years some 167,000 homes will be cleared. This is well below the rate required.
>
> (Northern Way, 2005, p. 53)

Reiterating the trope of emptiness and inexorable depopulation, Brendan Nevin presented some big, jolting numbers in his evidence to the 2008 public inquiry in Liverpool (Nevin, 2007a). He was subsequently questioned on this by one of the principal objectors, Bill Finlay:

> Finlay: 'Can you explain why, in a report about housing, you talk about population?'
>
> Nevin: 'Why? Because it's so striking. To bring the magnitude of it to people's attention: an American-style population loss. . . I think the housing market in Liverpool has increased exclusion. . . People have left the city and this has left behind a social and economic – I don't know how to say this – disaster area.'
>
> (Crookes, personal field-notes, 15 January 2008)

In this brief exchange, we have 'American-style population loss' leaving behind a 'disaster area'. Continuing to draw parallels with the United States, the HMR

94   Lee Crookes

literature makes repeated reference to Detroit, a city that is often held up as the emblem of dystopian urbanism and decline:

> The spectre of abandoned properties (as raised by Lowe *et al.*, 1998) recounted, albeit briefly, the potential of wholesale US-style urban abandonment of the sort that has been exemplified in response to structural changes in places like Detroit.
>
> (Leather *et al.*, 2007, p. 47)

Similarly, in an angry response to Jenkins (2007), Nevin retorts, 'If readers wish to assess how the do-nothing option would work in practice, I recommend a trip to Chicago or Detroit, cities that have seen vast population loss and dereliction. Many of their neighbourhoods have been in decline for 50 years, and show no signs of recovery' (2007b). Presented with such alarming scenarios, it was perhaps inevitable that politicians would readily be persuaded that a step-change in the Government's approach to regeneration – namely HMR – was essential. Exemplifying what Slater (2006) has described as the 'false choice' between gentrification and continued disinvestment, those living in certain parts of northern England were presented with a choice between HMR or a disaster zone: 'What's the alternative? Do we allow the urban core of our big, vibrant, Northern cities to decay? If we do nothing, we leave people in really awful situations' (Boaden, 2003).

But evidence from Finlay (2007), Cameron (2006) and the author's own research (Crookes, 2011) suggests that the magnitude and extent of empty housing and housing market weakness in the HMR areas were greatly exaggerated. Finlay had been a manager of Liverpool City Council's Research, Policy and Strategy Unit in the late 1990s, at the time the Council commissioned an extensive programme of research into the sustainability of the city's inner-urban neighbourhoods. In his written evidence to the Edge Lane public inquiry in Liverpool in 2008, Finlay describes how he and his colleagues became aware that the draft reports being produced by the researchers were overstating the level of vacancies in the social rented sector; soon after, Finlay was suspended from his post and, in his evidence, he claimed that this was because he had questioned the figures presented in the research (Finlay, 2007). His argument at the public inquiry, in essence, was that 'Liverpool City Council and its key housing partners developed and subsequently implemented policy from 1999 onwards. . . in the full knowledge that it rested upon data analysis which was unreliable' (2007, p. 5). Elsewhere, in the Northeast of England, some of the interviews with local government officers reported by Cameron (2006) also suggest that the same researchers may have overstated the severity of the low demand problem in the region.

Similarly, examining the data for a neighbourhood in another HMR area that was targeted for extensive demolition,[5] Crookes (2011, p. 240) found that vacancy rates in the area were actually less than half of the national average, that is, 1.4 per cent compared to 3.2 per cent, with the average across all HMR areas standing at 7.5 per cent. Despite these reported figures, the introduction to the Masterplan for the area argued that the estate, predominantly comprised of social housing, 'exhibits some of the most acute signs of housing market decline and low demand in the housing market renewal area' (NMBC *et al.*, 2005, p. 2). Given that the actual vacancy figures were so low, why was this estate targeted for demolition? Some clues can be found elsewhere in the Masterplan: 'Hilltop has a dramatic setting. Stunning views can be seen from many locations overlooking the countryside. The Masterplan has the opportunity to frame and use many of these to the best advantage' (NMBC *et al.*, 2005, p. 25). In short, Hilltop would provide developers with a very attractive site that the Council could clear relatively cheaply, given that most of the residents were council tenants. Thus, the Hilltop scheme was never about low demand but rather about securing a prime piece of land with 'stunning views'.

Conclusion

Given the displacement, 'symbolic violence' and 'accumulation by dispossession' that attended HMR (see Macleod and Johnstone, 2012; Allen and Crookes, 2009; Allen, 2008) it perhaps comes as little surprise that it has been described as one of the most controversial urban regeneration programmes of recent times (Allen, 2008). It is also one that met with considerable opposition from local residents who were living in the areas that were targeted for demolition. Whilst some residents had what Vale (1997) describes as an ambivalent, 'empathological' relationship with the places where they lived, many residents were keen to stay put and challenge the systematic defamation of their homes and neighbourhoods with more authentic counter-representations that were based on their lived experience. However diminished or deteriorated the conditions of their homes and environment might have seemed to the casual observer, the 'problem' areas targeted by HMR arguably provided the same sort of protections and reassurances, social networks and practical and cultural resources for their inhabitants that the middle-classes seek in their communities (see, for example, Mee, 2007). Whilst there is not the space here to consider how residents contested HMR, further details can be found in Allen (2008), Crookes (2011) and Leeming (2010). All of these works provide an important counterpoint to the official, stigmatised representations of place that were being circulated by those intent on demolition.

As outlined above, territorial stigmatisation played a key role in the formulation, development and implementation of HMR. Other scholars have noted the links between territorial stigmatisation and gentrification and HMR can be understood as a particular form of state-led, new-build gentrification that is predicated on housing demolition. But where conventional forms of gentrification tend mostly to re-imagine places on an upwards trajectory, HMR and other forms of 'gentrification-by-bulldozer' must first represent places in ways that rule out the possibility of any solutions apart from demolition: an initial phase of defamation must necessarily precede the revalorising of the site in a way that naturalises its clearance and redevelopment. Territorial stigmatisation can therefore be seen as having an essential role in processes of new-build gentrification that require the clearance of a site through prior demolition. This is important to note since many accounts of newbuild gentrification frequently assume a pre-existing, unpopulated brownfield site that has already been cleared (see, for example, Davidson and Lees, 2005).

In making the arguments in this chapter, I acknowledge that some of the neighbourhoods characterised as areas of 'low demand' housing or 'housing market failure' were faced with a very difficult set of challenges. But the response to these challenges could have been very different: Hajer's work (*op. cit.*) highlights how the manner in which a problem is constructed serves to delimit the possible policy responses. In constructing a problem of such unprecedented scale and threat, the discourse of low demand and the scripting of the HMR 'badlands' helped to win support for a 'new' sub-regional approach to urban regeneration that advocated mass demolition on a scale not seen in the UK since the 1960s.

The crucial point here is that HMR and the areas it targeted were social constructs, the products of a discourse that rested, to a large degree, on the practice of territorial stigmatisation, as does much urban policy. As Dikeç puts it, urban policy is 'a process of actual boundary shaping' (2007, p. 280) and policy-makers, informed by research evidence, define what come to be understood as 'good', 'bad' or 'ugly' spaces through a variety of statistical, linguistic, visual and cartographic forms of representation. At the very least, as already noted, researchers should do more to question the construction of 'problems' advanced by government policy makers (Jacobs *et al.*, 2003). But, most importantly, the more we recognise these 'problems' as harmful social constructs, the more readily we can challenge them. As Massey puts it: 'changing discourses is part of changing "the world"' (2001, p. 10). If policy-makers and academics can construct discourses that inexorably lead to demolition, then surely we can begin to develop discourses that allow for other possibilities?

Finding these alternative, more constructive narratives, however, requires academics to exercise greater responsibility in terms of how they themselves approach, research and discuss urban neighbourhoods. Baeten's (2002) assertion that urban scholarship is framed by a 'peculiar epistemological framework of problems' suggests that territorial stigmatisation has seeped into our understanding of the urban condition and is already, at least to some extent, taken for granted, naturalised even. This is often done entirely unintentionally as we seek to critique the unjust social and economic dynamics that shape urban inequalities. As Patillo (2009, p. 859) contends, it is a tricky balance to get right:

> All of us who study disadvantaged people and places want to use the strongest language possible in order to powerfully show the human consequences of policy assaults and neglect. But how do we name the phenomenon without ourselves repeating the stigmatizing labels that further fuel the cycle of disregard?

If we are to move beyond Baeten's (2002) 'problem epistemology' – the 'deficit model' of urban scholarship – and adopt a more sympathetic, 'glass half-full' perspective, towards urban places, then one way forward might be to embrace a research framework that mirrors Kretzmann and McKnight's (1993) 'asset-based' approach to community development (ABCD). Their ABCD approach seeks to build on the existing strengths and positive attributes that can be found within any community. Where, for the purveyors of territorial stigma, the glass is only ever half-empty, an assets-based approach inverts this negative perception. What might an asset-based approach to urban scholarship look like? My hope is that it would encourage greater (and closer) engagement with marginalised communities not just in research, through greater use of participatory and collaborative methods, but also in teaching, where greater engagement and personal contact with stigmatised communities would seek to dis-abuse students of their media-driven perceptions of particular places and their residents. The challenge for students and researchers is to engage with places up-close, to augment the cold, detached remote sensing of space with an involved, 'intimate sensing' of place (Porteous, 1986). In turn, this requires a shift towards more engaged forms of scholarship (Boyer, 1996) and a rethinking of research ethics governance in the manner described by Beebeejaun *et al.* (2015).

Finally, if, as Harvey argues, 'The freedom to make and remake our cities and ourselves is. . . one of the most precious yet most neglected of our human rights' (2008, p. 23), then this freedom to remake the city should extend to the right to be a part of the collective re-imagining of the city. This is already starting to happen to some extent. Just as new information and communication technologies have transformed how cities engage in place marketing they are similarly equipping ordinary citizens with the tools to produce their own analyses, maps and representations of the city, thereby helping to democratise and subvert the representation of the not-so-good, the bad and the ugly.

Notes

1 With thanks to Mustafa Dikec, whose book, *The Badlands of the Republic* (Dikec, 2007), provided the inspiration for this title.
2 This coalition-lobby group was formed in 1999 and included the Housing Corporation, the 18 local authorities who administer the M62 Corridor, the National Housing Federation, the National House Builders Federation and a number of Registered Social Landlords (RSLs).
3 It is perhaps also worth recalling that the National House Builders Federation, the principal voice of the home-building industry in England and Wales, was a prominent member of the North West Housing Forum, the group that commissioned the influential M62 study.
4 Areas in England that are subject to compulsory purchase orders are customarily referred to as 'Order Lands'.
5 The neighbourhood and local authority were respectively anonymised as 'Hilltop' and 'Northerly' Metropolitan Borough Council (NMBC) and I have accordingly omitted the citation details of the relevant Masterplan from the bibliography.

References

Allen, C. (2008), *Housing Market Renewal and Social Class*, London: Routledge.
Allen, C. (2009), *Remaking the North of England? Urban Regeneration, Economic Growth and the New Politics of Emptiness*, paper presented to the conference 'Unmaking England? Is London Delinking from its peripheries?' University of Salford, 15–16 January.

Allen, C. and Crookes, L. (2009), 'Capitalist disaster', *Morning Star*, 23 July.

Allen, C. and Imrie, R. (2010), (eds) *The Knowledge Business*, London: Ashgate.

Baeten, G. (2002), 'Western utopianism/dystopianism and the political mediocrity of critical urban research', *Geografiska Annaler*, B 3–4, pp. 143–52.

Beauregard, R. (1993), *Voices of Decline: The Post-War Fate of US cities*, Oxford: Blackwell.

Beebeejaun, Y., Durose, C., Rees, J., Richardson, J. and Richardson, L. (2015), 'Public harm or public value? Towards coproduction in research with communities', *Environment and Planning C*, 33(3): 552–565.

Bennett, L. and Reed, A., (1999), 'The new face of urban renewal: The Near North Redevelopment Initiative and the Cabrini-Green Neighbourhood', in Reed, A., (ed.), *Without Justice For All: The New Liberalism and our Retreat from Racial Equality*, Boulder, CA: Westview Press.

Blomley, N. (2004), *Unsettling the City: Urban Land and the Politics of Property*, New York: Routledge.

Blumer, H. (1971), 'Social problems as collective behaviour', *Social Problems*, 18: 298–306.

Boaden, J. (2003), 'The way to resurrect a city', *Building*, 17 April.

Boyer, E. (1996), 'The scholarship of engagement', *Journal of Public Outreach*, 1(1): 11–20.

Calvino, I. (1975), *Invisible Cities*, London: Secker and Warburg.

Cameron, S. (2006), 'From low demand to rising aspirations: housing market renewal within regional and neighbourhood regeneration policy', *Housing Studies* 21(1): 3–16.

Cochrane, A. D. (2000), 'The social construction of urban policy', in Bridge, G. and S. Watson (eds) *A Companion to the City*, Oxford: Blackwell.

Cole, I. and Nevin, B. (2004), *The Road to Renewal: The Early Development of the Housing Market Renewal Programme in England*, York: York Publishing Services.

Colomb, C. (2012), *Staging the New Berlin: Place Marketing and the Politics of Urban Reinvention Post-1989*, London: Routledge.

Crookes, L. (2008), [Field notes collected during observation at public inquiry in Liverpool.] Unpublished raw data.

Crookes, L. (2011), *The making of space and the losing of place: A critical geography of gentrification-by-bulldozer in the north of England*, unpublished dissertation, University of Sheffield.

Davidson, M. and Lees, L., (2005), 'New-build gentrification and London's riverside renaissance', *Environment and Planning A*, 37: 1165–1190.

Dery, D. (2000), 'Agenda setting and problem definition', *Policy Studies*, 21(1): 37–47.

DETR – Department of the Environment, Transport and the Regions (1999), *Towards an Urban Renaissance: The Report of the Urban Task Force*, chaired by Lord Rogers of Riverside. London: DETR.

Dileonardo, M. and Reed, A. (1989), 'Academic poverty-pimping', *Nation*, 249(13): 442.

Dikeç, M. (2005), *Badlands of the republic: Space, politics and urban policy*, Oxford: Blackwell.

Dikeç, M. (2007), 'Space, governmentality, and the geographies of French urban policy', *European Urban and Regional Studies*, 14(4): 277–289.

Fairclough, N. (2000), *New Labour, New Language?* London: Routledge.

Finlay, B. (2007), *Known Fundamental Errors in the Evidence Justifying Housing Market Interventions in Liverpool and Subsequent Conspiracies*, Proof of Evidence, The Urban Regeneration Agency (Edge Lane West, Liverpool) CPO (No 2) Public Inquiry.

Foucault, M. (1977), *Discipline and Punish*, New York: Vintage.

Friedmann, J. (2007), 'Reflections on place and place-making in the cities of China', *International Journal of Urban and Regional Research*, 31(2): 357–79.

Fullilove, M. T., Hernandez-Cordero, L. and Fullilove, R. E. (2010), 'The ghetto game: Apartheid and the developer's imperative in post-industrial American cities', in Hartman, C. and Squires, G. D. (eds) *The Integration Debate: Competing Futures for American Cities*, New York: Routledge.

Goetz, E. G. (2003), *Clearing the Way: Deconcentrating the Poor in Urban America*, Washington: Urban Institute Press.

Goetz, E. G. (2013), 'The audacity of HOPE VI: Discourse and the dismantling of public housing', *Cities*, 35: 342–348.

Graham, S. (2004), 'Cities as strategic sites: Place annihilation and urban geopolitics', in Graham, S. (ed.) *Cities, War and Terrorism: Towards an Urban Geopolitics*, Oxford: Blackwell.

Gray, N. and Mooney, G. (2011), 'Glasgow's new urban frontier: "Civilising" the population of "Glasgow East"'. *City*, 15(1): 4–24.

Hajer, M. (1993), 'Discourse coalitions and the institutionalisation of practice: The case of acid rain in Britain', in Fischer, F. and Forester, J. (eds) *The Argumentative Turn in Policy Analysis and Planning,* London: UCL Press.

Hall, S. (2003), 'New Labour's double-shuffle', *Soundings*, 24: 10–24.

Harley, J. B. (1988), 'Maps, knowledge, and power', in, Cosgrove, D. and Daniels, S. (eds), *The Iconography of Landscape: Essays on Symbolic Representation, Design and Use of Past Environments*. Cambridge: Cambridge University Press, pp. 277–312.

Harley, J. B. (1989), 'Deconstructing the map', *Cartographica*, 26(2): 1–20.

Harvey, D. (1989), 'From managerialism to entrepreneurialism: The transformation in urban governance in late capitalism', *Geografisker Annaler*, 71B(1): 3–17.

Harvey, D. (1996), *Justice, Nature and the Geography of Difference*, Oxford: Blackwell.

Harvey, D. (2008), 'The Right to the City', *New Left Review*, 53: 23–40.

Herb, G. H. (2008), 'The politics of political geography', in Cox, K., Low, M. and J. Robinson (eds) *The Sage Handbook of Political Geography*, London: Sage.

House of Commons (2002), *Housing, Planning, Local Government and Regions Committee. Departmental Annual Report and Estimates 2002, Minutes of evidence: Examination of witnesses (Questions 40–59)*. Accessed 7 November 2008 at: http://www.publications. parliament.uk/pa/cm200203/cmselect/cmodpm/78/2110504.htm

House of Commons Committee of Public Accounts (2008), *Housing Market Renewal: Pathfinders* Thirty–fifth Report of Session 2007–08, Ordered by The House of Commons, London.

Jacobs, K. and Manzi, T. (2013), 'Modernisation, marketisation and housing reform: The use of evidence based policy as a rationality discourse', *People, Place and Policy Online*, 7(7): 1–13.

Jacobs, K. Kemeny, J. and Manzi, T. (2003), 'Power, discursive space and institutional practices in the construction of housing problems', *Housing Studies,* 18(4): 429–446.

Jenkins, S. (2007), 'Once they called it Rachmanism. Now it's being done with taxpayers' money', *The Guardian*, 16 March.

Kallin, H. and Slater, T. (2014), Activating territorial stigma: Gentrifying marginality on Edinburgh's periphery. *Environment and Planning A*, 46(6): 1351–1368.

Klein, N. (2007), *The Shock Doctrine: The Rise of Disaster Capitalism*, London: Allen Lane.

Kretzmann, J. P. and McKnight, J. L. (1993), *Building Community from the Inside Out: A Path for Finding and Mobilizing a Community's Assets,* Chicago: ACTA Pub.

Leather, P., Cole, I. and Ferrari, E. (2007), *National Evaluation of Housing Market Renewal: Baseline Report*, London: Department for Communities and Local Government.

Leeming, C. (2010), *Streetfighters.* In Lees, L. (ed.) (2004) *The Emancipatory City: Paradoxes and Possibilities*? London: Sage. Available online at: http://www.ciara leeming.co.uk/tag/streetfighters/ (accessed 15 August 2015).

Levitas, R. (1998), *The Inclusive Society? Social Exclusion and New Labour*, London: Macmillan.

Lowe, S., Spencer, S. and Keenan, P. (eds) (1998), 'Housing abandonment in Britain: Studies in the causes and effects of low demand housing: Conference papers.' Centre for Housing Policy, University of York.

Macleod, G. and Johnstone, C. (2012), 'Stretching urban renaissance: Privatizing space, civilizing place, summoning "community"', *International Journal of Urban and Regional Research*, 36(1): 1–28.

Massey, D. (2001), 'Geography on the agenda 1', *Progress in Human Geography*, 25(1): 5–17.

Mee, K. (2007), '"I ain't been to heaven yet? Living here, this is heaven to me": public housing and the making of home in inner Newcastle', *Housing Theory and Society*, 24: 207–228.

Mele, C. (2000a), 'The materiality of urban discourse: Rational planning in the restructuring of the early twentieth-century ghetto', *Urban Affairs Review* 35: 628–648.

Mele, C. (2000b), *Selling the Lower East Side: Culture. Real Estate and Resistance in New York City*, Minneapolis: University of Minnesota Press.

National Audit Office (2007), *Department for Communities and Local Government: Housing Market Renewal.* London: The Stationary Office.

NHLP – National Housing Law Project (2002), *False HOPE: A Critical Assessment of the HOPE VI Public Housing Redevelopment Program.* Oakland, CA: NHLP.

Nevin, B. (2001), *Securing Housing Market Renewal: A Submission to the Comprehensive Spending Review.* Produced for the National Housing Federation in collaboration with the Key Cities Housing Group and the Northern Housing Forums, London: National Housing Federation.

Nevin, B. (2007a), *The Urban Regeneration Agency (Edge Lane West, Liverpool) Compulsory Purchase Order (No.2), Proof of Evidence of Brendan Nevin.* Nevin Leather Associates

Nevin, B. (2007b), 'We're more than simply demolition men', *The Guardian,* 27 March.

Nevin, B., Lee, P., Goodson, L., Murie, A. and Phillimore, J. (2001), *Changing Housing Markets and Urban Regeneration in the M62 Corridor,* Birmingham: Centre for Urban and Regional Studies.

New Heartlands (2006), *Housing Market Renewal Initiative: Liverpool Delivery Plan,* Liverpool: New Heartlands and Liverpool City Council.

Nicolaus, M. (1968), '"Fat-Cat Sociology". Remarks at the American Sociological Association', *The American Sociologist*, 4(2): 154–156.

Northern Way (2005), *Moving Forward: The Northern Way Growth Strategy, Section C9, Sustainable Communities,* Newcastle: Northern Way Secretariat. Available online at: http://www.thenorthernway.co.uk/downloaddoc.asp?id=430 (accessed 6 May 2007).

ODPM – Office of the Deputy Prime Minister (2003), *Sustainable communities: Building for the Future,* London: ODPM.

ODPM – Office of the Deputy Prime Minister (2005), *Sustainable Communities: Homes for All*, London: HMSO.

O'Tuathail, G. (1996), *Critical Geopolitics: The Politics of Writing Global Space*, London: Routledge.

Ove Arup and Partners Ltd with Innovacion Ltd for The Northern Way Sustainable Communities Team (2006), *The North's Residential Offer: Policy and Investment Review: Phase 1 Report*, Newcastle: The Northern Way Sustainable Communities Team.

Parker, D. and Long, P. (2004), '"The mistakes of the past"? Visual narratives of urban decline and regeneration', *Visual Culture in Britain*, 5(1): 37–58.

Patillo, M. (2009), 'Revisiting Loïc Wacquant's Urban Outcasts', *International Journal of Urban and Regional Research*, 33(3): 858–864.

Pfeiffer, D. (2006), 'Displacement through discourse: Implementing and contesting public housing redevelopment in Cabrini Green', *Urban Anthropology and Studies of Cultural Systems and World Economic Development*, 39–74.

Pickles, J. (2004), *A History of Spaces: Cartographic Reason, Mapping, and the Geo-coded World*, New York: Routledge.

Pickles, J. (2006), 'On the social lives of maps and the politics of diagrams: A story of power, seduction, and disappearance', *Area*, 38(3), 347–350.

Porteous, J. D. (1986), 'Intimate sensing', *Area*, 18: 250–251.

Porter, L. (2009), 'Struggling against renaissance in Birmingham's Eastside', in Porter, L. and Shaw, K. (eds) *Whose Urban Renaissance: An International Comparison Of Urban Regeneration Strategies*, London and New York: Routledge.

Power, A. and Mumford, K. (1999), *The Slow Death of Great Cities? Urban Abandonment or Urban Renaissance*, York: Joseph Rowntree Foundation.

Slater, T. (2006), 'The eviction of critical perspectives from gentrification research', *International Journal of Urban and Regional Research*, 30: 737–757.

Slater, T., and Anderson, N. (2012), 'The reputational ghetto: Territorial stigmatisation in St Paul's, Bristol', *Transactions of the Institute of British Geographers*, 37(4): 530–546.

Smith, N. (2002), 'New globalism, new urbanism: Gentrification as global urban strategy', *Antipode*, 34: 427–450.

Smith, N. (1996), *The New Urban Frontier: Gentrification and the Revanchist City*, New York: Routledge.

Smith, N. (1979), 'Toward a theory of gentrification: a back to the city movement by capital, not people', *Journal of the American Planning Association*, 45: 538–548.

Stone, D. (1989), 'Causal stories and the formation of policy agendas', *Political Science Quarterly*, 104: 281–300.

Vale, L. J. (1997), 'Empathological places: Residents' ambivalence toward remaining in public housing', *Journal of Planning Education and Research*, 16(3): 159–175.

Wacquant, L. (1997), 'Three Pernicious Premises in the Study of the American Ghetto', *International Journal of Urban and Regional Research*, 21(2): 341–353.

Wacquant, L. (2007), *Urban Outcasts: A Comparative Sociology of Advanced Marginality*, Cambridge: Polity Press.

Wacquant, L., Slater, T. and Pereira, V. B. (2014), 'Territorial stigmatization in action', *Environment and Planning A*, 46(6): 1270–1280.

Webb, D. (2010), 'Rethinking the role of markets in urban renewal: The Housing Market Renewal initiative in England', *Housing, Theory and Society*, first published on 4 January 2010 (iFirst).

Weber, R. (2002), 'Extracting value from the city: Neoliberalism and urban redevelopment', *Antipode*, 34(3): 519–540.

Wildavsky, A. (1979), *Speaking Truth to Power*, Boston, MA: Little Brown.

Zukin, S. (1995), *The Cultures of Cities*, Oxford: Blackwell.

7 Opening the reputational gap

Hamish Kallin

Obsolescence has become a neoliberal alibi for creative destruction, and therefore an important component in contemporary processes of spatialized capital accumulation.

(Weber, 2002, p. 185)

The settled community and those who will choose to come to Craigmillar in the future will expect Craigmillar to be a place, not just a housing estate.

(City of Edinburgh Council, 2005, p. 114)

Introduction

The processes of territorial stigmatisation and gentrification are increasingly seen to be linked (Gray and Mooney, 2011; Kallin and Slater, 2014; Kipfer and Petunia, 2009; Purdy, 2003; Slater and Anderson, 2012; Weber, 2002). The representational devalorisation of space can pave the way for its economic revalorisation. Increasingly, where the state intervenes to promote, enforce and enact the latter, 'strategically *stigmatizing* those properties that are targeted for demolition and redevelopment' (Weber, 2002, p. 173) can work very effectively at the level of justification. But as Slater (in press) suggests, there has been little attempt to integrate rent gap theory into these debates. This is surprising, as the insight of the rent gap model was its emphasis on the necessary *de*valorisation of space as a preface to its eventual *re*valorisation, tying the process of gentrification into a systemic analysis of the uneven development of capitalism. Moreover, through his discussion of the 'frontier', Smith (1986, 1992a, 1996) made it clear that an intensely territorialised form of stigmatisation had a role to play in this process, for it was precisely those areas that suffered from negative media coverage, where 'the contrast with the rest of [the city] could hardly be more stark' (p. 145), that, over time, became susceptible to gentrification. Negative reputation is a corollary to disinvestment, with the two reinforcing each other. Causality in Smith's model lies firmly with the latter, but it is through discursive framing that the emerging geographies of the city's class divisions become known and are reinforced.

This chapter is an attempt to bring the two concepts together by introducing a third. I will call this a *reputational gap*, the contours of which overlay those of the rent gap, but are not synonymous with it. This is not intended as a rival model or an amendment to the rent gap itself, but is deployed as an aid to explain the role that territorial stigmatisation plays in the process of state-led gentrification where it is enacted as 'urban regeneration'. Specifically, I wish to suggest that an emphasis on symbolic metamorphosis – between the denigrated present and the promised future – is necessary to the invocation of rent gaps in situations where the latter may be ineffectual in economic terms. Hence the rent gap model can be increasingly useful even in instances where no market (re)valorisation of the land occurs, precisely because state policy is so fixated on the possibility that it *could* (and, of course, that it *should*) occur. As Harvey (2006) points out, in this era of the 'financialisation of everything' (p. ix), 'capitalism increasingly lives on faith alone' (p. xxvi): thus the wilful articulation of a reputational gap lies at the centre of 'regeneration' policy, because there is precious little else to capitalise on.

To make this argument I will be drawing on three years of research into an ambitious urban regeneration project in Craigmillar, a historically working-class neighbourhood on the south side of Edinburgh, capital city of Scotland. Here 'the role of the state in the formation of stigma, thus the widening of the rent gap and the subsequent facilitation of gentrification' (Kallin and Slater, 2014, p. 1633) can be seen quite clearly. This assertion was drawn from interviews with local residents, highlighting the disjuncture between an at-times fierce sense of local identity on their part and the official discourse of making the new Craigmillar 'a place, not just a housing estate'. At the empirical heart of the present chapter, however, I am drawing on a series of lengthy interviews with those ostensibly responsible for the 'delivery' of the project (the planners, architects, councillors, board members of the regeneration company, and so on; all names have been changed or omitted). A 'blemish of place' – to use Wacquant's (2008) phrase – was invoked every time one of my interviewees sought to justify what they were doing and why. Stigmatisation is articulated at the core of this project in two ways: first, stigmatisation becomes a form of internal justification, morally speaking, by which the aims of the project are rationalised as 'fair'; and second, the fiscal straitjacket of 'austerity urbanism' forces the local state into a double bind: flog off assets to tackle indebtedness whilst gaining the political capital from the stigmatising discourse that legitimises this.

I should clarify before proceeding that I consider this to have been an urban regeneration project based on the logic of state-led gentrification. *State-led* because whilst the project was orchestrated by a 'private company', this was a company owned by the local authority, largely staffed by the local authority, and working on land almost entirely belonging to the local authority. And *gentrification* because this was a project that involved the wilful destruction of a working-class landscape, intentional displacement, the perceived necessity of raising land values and the desire to attract more affluent residents.

I will begin with a brief discussion of the rent gap model itself, identifying the point of departure for my argument, namely that the 'gap' only exists insofar as it is constantly rearticulated. I will then outline Craigmillar's historical trajectory from rural fields to maligned council estate in the space of 60 years, before showing how poverty and tenure are stigmatised, and set to contrast against the 'ideal' place imagined in the council's planning documents. It is here that the idea of a reputational gap comes to the fore, as the trajectory of closing the rent gap only begins once this new future is articulated. I will then conclude by suggesting how we might rethink the role of territorial stigmatisation in the process of gentrification.

Imagining 'potential'

When first detailing what he meant by a 'rent gap', Neil Smith (1979, p. 545) explained it as 'the disparity between the potential ground rent level and the actual ground rent capitalized under the present land use'. At its simplest, then, the model suggests that gentrification is most likely in instances where there is a sizeable 'gap' between the amount of money a landowner is currently extracting from a site, and the amount they could extract if such a site were to be put to a different use. The model is represented below in its simplest form (Figure 7.1). Over time, the value of the building falls (as a site of fixed capital, it will depreciate in value without maintenance or reinvestment). The capitalised ground rent falls with it. The dotted line (potential ground rent) then emerges as a trajectory of 'what could be' if the site were put to a different use. This 'different use' could involve the building being renovated, knocked down and replaced, perhaps simply re-let at a higher price, or the land being deployed for a different function altogether. The red in the model below is the rent gap. At the point where the capitalised ground rent begins its upward trajectory to meet the potential ground

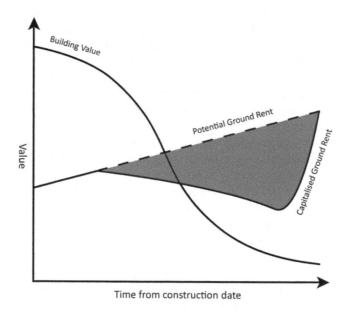

Figure 7.1 The rent gap model.

Source: adapted from Smith (1979, p. 544).

rent, gentrification is occurring. When the two lines meet, the rent gap is closed, though the beauty of Smith's model is its fluidity: once closed, there is no guarantee the gap will remain closed, as it reflects the shifting patterns of investment in the urban land market.

The model is far more than a simple representation of the profit motive, for it alludes to the active devalorisation that opens the gap in the first place. As Eric Clark (1995, p. 1490) later put it:

> In a nutshell, the rent gap constitutes initially an economic pressure to disinvest in the fixed capital on a site, which consequently becomes increasingly inappropriate to the site's 'highest and best' use, and eventually an economic pressure to doom the site to higher intensity and type of use through redevelopment.

Crucially, this was never a model that sought to theorise gentrification as a standalone process. As Smith (1982, 2010) argued, it ties directly into the uneven development of capitalism more broadly, where *dis*investment is a necessary corollary to *in*vestment just as *de*struction is a necessary corollary to *con*struction (Mitchell, 2015). But rent gaps do not, as such, 'exist'. Clark (2015, n. p.) points out that the appearance of the model itself can give 'an image of being rather fixed and determined [as if] it's *out there* and you can find it'. It is important to conceive of the rent gap 'not [as] a mechanistic device, but rather a general structural tendency that follows different paths depending on human agency and context' (Lees *et al.*, 2008, p. 67).

For the purposes of this chapter, I want to emphasise that the dotted line (potential ground rent) is always hypothetical. It only exists insofar as it is *imagined*. Considered as an economistic register, this 'potential' can of course be numerated, but the dominance of fictitious capital means it is increasingly numerated as a fantasy, and one that cannot be detached from the moral questions of what and who land is for. The notion of a 'better' use for land 'contains prior normative assumptions concerning the necessity and desirability of spatial change' (Blomley, 1997, p. 287). In other words, the stigmatisation of that which exists, whether implicitly or explicitly, is fundamental to the legitimacy of 'what might be'.

Existing on the periphery

> [Craigmillar] is an area that has steadily been positioned from the late 1950s as existing on the periphery, both geographically and economically.
> Phil Denning (2004, p. 225)

The land that makes up Craigmillar was not part of the City of Edinburgh until the late 1920s, when the city corporation, seeking space to expand, purchased the estate from an aristocratic family who had owned it since the nineteenth century. It is one of the quixotic oddities of the English language that the word *estate* describes what the land was and what it was set to become, despite the disparity between the landholdings of a rich family ordained by the crown and the tenements of a local authority seeking to house those forcibly displaced from slum clearance schemes. Thus even in its genesis, the modern Craigmillar remains a testament to one of the enduring functions of stigma: the more negative the reputation of an area, the more righteous the light in which those who decree its demolition have been able to bask. The (re)making of nineteenth-century Edinburgh tied the emerging landscape of the modern city into a series of stigmatising and spatially fixated discourses. Once vacated by the city's wealthiest inhabitants, the Old Town came to typify the fearful excesses of the rapidly expanding urban landscape: too densely populated, too poor, too ill, and too unruly. Such evils had to be filtered through the bourgeois imagination (either as fear or moral concern) before the first Improvement Acts in the 1870s began the wholesale demolition of those tracts of land deemed most objectionable (Rodger, 2001; Smith, 1994). The link between social phenomena and a spatial signifier was innate. When Gordon (1979) talks about the way 'status areas' began to emerge alongside the growing inequality, he highlights how pervasively the space of the city was arranged into a social hierarchy. Those at the bottom were inevitably first in line to be demolished, and their demonisation was an essential part of this process.

Niddrie Mains, the first of the Craigmillar council estates, emerged out of this round of slum clearances. Built in the 1930s largely to house those evicted from elsewhere in the city centre, the estate carried with it a displaced stigma that was in evidence almost immediately. Or as Seán Damer (1989, p. 47) puts it, 'the new slum clearance estates and their tenants were pre-stigmatized. They were not meant for 'normal' people. 'Edinburgh's local newspaper declared these 'drab houses' as early as 1934. The scheme was supposedly 'ugly' and 'an abortion' (Craigmillar Festival Society, 1978, p. 2.01). Ugliness, much like beauty, is in the eye of the beholder.

The sustained population growth in Craigmillar was fuelled by industrial expansion. This was an area characterised by coal mines, breweries, factories, and the kind of jobs that are manifestly and routinely written out of the history of the city at large (Crummy, 2004; Denning, 2004). In his meticulously detailed post-war plan for this city, Patrick Abercrombie (Abercrombie and Plumstead, 1949, p. 48) noted that the industrial land in Craigmillar would 'probably remain for such uses for a long time to come,' and advocated the building of further housing nearby to guarantee a steady labour supply. Around the same time, the newly formed National Coal Board (NCB) confidently predicted that several of the pits in the East Lothian coal field would be operational for a century to come (Hutton, 1998). The population in Craigmillar climbed rapidly, reaching almost 25,000 by the mid 1960s. This is despite having been initially planned as an estate to house no more than 8,000 people (Hague, 1982, p. 231), a demographic mismatch that manifested itself in the striking lack of community resources and facilities.

In a story now sadly familiar for so many parts of 'post-industrial' Scotland, Abercrombie and the NCB were wrong. Just 19 years after the Edinburgh plan was published, the coal pits in Craigmillar were closed (Oglethorpe, 2006). The situation was to get far worse as the years wore on. Some 17,000 local jobs disappeared in the decade following 1964 (Craigmillar Festival Society, 1978, p. np.) and by the late 1970s all of the large-scale employers had either shut down or moved elsewhere. By the end of the 1980s mining in the region as a whole was something encountered only in a museum (excepting a brave but short-lived attempt at running the nearby Monktonhall colliery as a workers' cooperative in the mid 1990s). Unemployment in the area skyrocketed and remained stubbornly high. By the late 1970s, residents were keenly aware that 'Craigmillar' had become a shorthand for failure, and routinely the focus of negative discourse in the local media (Craigmillar Festival Society, 1978). Helen Crummy (1992, p. 23), a prominent local activist, recalls that the media increasingly fixated on the most negative aspects of life in the area, and 'so it was that the once proud names of Niddrie and Craigmillar became synonymous with crime and social problems'. The historian Christopher Harvie (1981, p. 167) noted that the 'grim system-built suburbs of East German towns' seemed 'somewhat better endowed' than their equivalents in the Scottish capital.

By the early 2000s this area was by far the most deprived in Edinburgh, and contained the fourth most deprived ward (out of 1,222) in Scotland as a whole (Scottish Government, 2003, n. p.). It is worth noting that the least deprived ward in the country was at this point less than three miles away, also on the south side of Edinburgh. As Taylor (2008, p. 530) notes,

> [T]he capital status of Edinburgh and its recent property boom masks extremes of poverty and affluence: its well planned and central residential, leisure and business complexes discord with peripheral housing estates and involve a geographical exclusion at first out of sight.

At the dawn of the twenty-first century the population in Craigmillar had plummeted by two-thirds since the mid 1960s (Scottish Government, 2006, n. p.). The City of Edinburgh Council, backed by the Scottish Executive (now 'Government'), sought a drastic intervention. Two of the area's largest council estates – Niddrie Mains, built in the 1930s, and Greendykes, built 30 years later – were to be erased from the map. A not-very-private 'private company' called *Promoting and Regenerating Craigmillar* (or PARC) was set up in 2003 to engineer a new urban environment in the spaces that were left. From the void came *potential*, and it is this link I wish to explore.

Planning the rent gap

By May 2007, the few remaining council tenements on the Niddrie Mains estate lay forlornly on death row. The windows were boarded up, the gable ends were awash with graffiti, and the tenants were gone. Local photographer Neil Shaw (2009) aptly called this a ghost town, and future profit was to act as the exorcist. What had been one of the most infamous public housing estates in the 'national demonology' (Jack, 1989, p. 36) was about to become a 'place of choice' (City of Edinburgh Council, 2013, p. 33), or so we were told. The plan sought to erect some 3,000 new homes, a new primary school (finished and opened), a new library (delayed, but now open), a new high school (delayed indefinitely) and a new town centre (delayed; has recently been resubmitted under new plans).

Six years later and Gavin stands in a sales hut clad in shiny corrugated metal, sporting PARC's logo on its sides. The undulating roof is designed to stand out on the roadside, and Gavin is trying to sell houses in the heart of a neighbourhood with some of the lowest rates of homeownership in the city. He remains upbeat.

> Local people laugh at me when I say this, but if you move aside the social problems, this is one of the best sites in Edinburgh. It's between two great shopping centres, it's got great transport, great green space, it's so close to Arthur's Seat, to the city centre.

Arthur's Seat, for those uninitiated in the roll-call of Edinburgh's geological features-cum-tourist attractions, is an extinct volcano whose hulking outline dominates the city's skyline. Proximity to its shadow recalls how close this neighbourhood is to the centre of one of the most affluent cities in the UK. As Gavin (perhaps unwittingly) makes quite clear, these are plans that rely on something being displaced. The future opportunities of 'the site' are expressed as somehow in contrast to the lives of the people who might be (or were) living on it. This is explicit in his words ('moving aside the social problems'). One local resident, replying anonymously to a consultation conducted in the project's early stages, is acutely aware of what this might mean:

> Craigmillar [is] too good a location in Edinburgh. . . chip away until community not here [sic] so property developers make money.
> (Scottish Participatory Initiatives, 2005, p. 20)

Notice how the two quotes – one from a representative of the 'development' company, one from a local resident – are essentially saying the same thing, even if they offer opinions on this process that are at odds with each other. Though not expressed in these terms, both quotes are clear invocations of an imagined rent gap. The potential ground rent is deemed to be much higher than the capitalised ground rent, and this is fundamental to the project. This is not a subversive observation. The plans for the new Craigmillar stop short of deploying the notion of a rent gap in those terms, but they may well have done. PARC's aim, in the words of their chief executive, 'was to move the average house price in Craigmillar from around 60 per cent of the Edinburgh average to 80 per cent of the Edinburgh average' on a 15-year timescale. The founding resolutions of the company state this quite clearly, aiming to 'promot[e] economic development [. . .] on sound commercial profit-making principles so as to maximise the financial return to the Company' (PARC, 2002, p. 2), with an ultimate aim to 'dispose of the Craigmillar Landholdings [. . .] on the best terms reasonably available'. The order of priorities is equally clear, for there are no legal obligations to provide jobs or alleviate poverty. Without rising property prices, this is a 'regeneration' project that cannot work.

This is made starkly clear by the fallout from the crisis of capitalism that began to unfold rapidly in late 2007. As PARC's accountant put it to me sombrely, 'obviously, the world fell apart'. The insolvency of the project's banking partners and the near-complete cessation of an active housing market in the months that followed means that Craigmillar today does not look like a particularly convincing example of 'gentrification'. A significantly higher-than-intended proportion of the properties completed are now owned by housing associations.[1] Much of the land remains in a state of limbo, and less than 30 per cent of the plan is complete (City of Edinburgh Council, 2012). The shifting dynamics of the project are important, and I will reflect on this in more detail below. But the apparent failure of the project on its initial terms should not serve to obfuscate the intentionality of those terms. The fact that the rent gap was *not* closed serves to highlight the interplay between state strategy and gentrification, for the degree to which the local authority was relying on the latter is very much in evidence: stagnant land values engineered a stagnant landscape. In this instance, the *desire* for a rent gap – the insistence that one existed – was of more power than the actual existence of one. There was no market-led revalorisation at work and, as it turned out, the state could not induce one.

Poverty, tenure and potential

If the deprivation in Craigmillar was 'real' (experienced and empirically evident; requiring, few would doubt, amelioration), then the way this deprivation was interpreted, discussed and tackled in PARC's approach is more akin to the political production of reality. To clarify, it is worth recalling Deborah Stone's (1989, p. 282) discussion of causal stories:

> Problem definition is a process of image making, where the images have to do fundamentally with attributing cause, blame and responsibility. Conditions, difficulties, or issues thus do not have inherent properties that make them more or less likely to be seen as problems or to be expanded. Rather, political actors *deliberately portray* them in ways calculated to gain support for their side.

This chimes with the suggestion that 'problems' in the field of housing are engrained through dominance of a discursive space, a dominance that involves policy makers, media representations and public discourse (Jacobs *et al.*, 2003). I want to suggest that the 'causal story' in Craigmillar was very much one of territorialised stigmatisation, grounded in poverty but articulated through tenure, both of which were then held up against the notion of 'untapped potential'.

There is no shortage of work on the myriad ways in which poverty and working-class culture have been heavily stigmatised in both media and political discourses in recent decades. Importantly, the focus on 'problem' *people* has been closely tied to a fixation on 'problem' *places* (Haylett, 2003; Skeggs, 2004). One particularly striking example of this convergence of metaphors was in the run-up to the 2010 General Election, where the Conservative Party's rhetoric of a 'Broken

Britain' was galvanised by a barely restrained indictment of Glasgow's East End and its residents (Gray and Mooney, 2011; Slater, 2014). The place, in this instance, functioned as a shorthand for the lives lived in it. Recent contributions from Tyler (2013) and McKenzie (2015) show how the vilification of working-class women in particular has become central to the legitimisation of welfare 'reforms' and austerity governance more broadly. Since the mid 1980s there has been a concerted shift away from a structural understanding of poverty – where unemployment is the result of systemic job loss – to a pathological understanding, where 'joblessness' is seen as a choice, or simply the result of laziness. This is particularly evident in social and urban policy, where the framing of the problem was increasingly devolved to the scale of the 'local' (Hastings, 2004; Mathews, 2010). Such a discursive shift has an extra-malign effect in areas such as Craigmillar, where the jobs essentially disappeared.

Craigmillar suffered a further form of marginalisation due to its tenure profile being so drastically at odds with the city at large. According to the census data, just over 90 per cent of the residencies in Craigmillar were owned by the local authority in the early 1980s, compared to about 30 per cent for the city at large. The Right to Buy ate away at this homogeneity, but not nearly as much as its architects hoped. Across Scotland, the number of people living in houses owned by their council declined from a high point of 57.6 per cent (on a par with the Soviet bloc) in 1978 down to just 12.8 per cent by 2011 (Adams, 1978, p. 155; The Accounts Commission, 2013, p. 10). In Craigmillar, however, the proportion of local authority housing remained much higher for much longer. Even by 2001, over 50 per cent of the properties in the area were owned by the state. Across the UK, a process of *residualisation* was at work, where (quite predictably) the most desirable properties sold fastest to the wealthiest tenants in the most sought after locations (Forrest and Murie, 1988). On average, the least deprived areas of Scotland retained some 50–60 per cent of their council stock, compared to 81 per cent in the most deprived areas, resulting in a clear pattern of polarisation (Scottish Executive, 2008, p. 39).

As Hancock and Mooney (2013, p. 48) point out, 'social housing estates (or areas where these dominate) and the populations therein are frequently highlighted and represented as being not only vulnerable, but as particular locales where social pathologies and problems flourish'. Following a concerted emphasis on the supposed 'naturalness' and desirability of homeownership over decades of housing policy since the late 1970s, those remaining in the council rented sector were frequently typecast as 'deficient' (Flint, 2003). The council rented sector generally has been increasingly maligned aesthetically, spatially isolated, and politically neglected, confined to a tenancy of 'last resort' with all the attendant perceptions of failure that this brings with it (Kintrea, 2006; McIntyre and McKee, 2012). Reducing the dominance of local authority housing in Craigmillar area was fundamental to PARC's plan. Izzie, one of PARC's key architects, explains:

> I think it's probably one of the earliest council-set objectives for the regeneration. You see that, not just Edinburgh council, you know, everywhere, in the country, in the world, is talking about mixed tenure being a sustainable

community, or one of the key sorts of drivers for creating a sustainable community, and I think one of the failures, not the only failure, of places such as Craigmillar, was the, overriding single tenure sort of environment.

In keeping with this observation, of the 3,000 new homes that PARC promised to build, not one was intended to be a council house.

The kind of media attention that frames an area and its residents as 'bad' is frequently in evidence in Craigmillar. Shortly after the PARC project was initiated, one exasperated resident wrote to the *Evening News* asking why it 'seems to take every opportunity to denigrate the community of Craigmillar' (Wilkinson, 2004, n. p.). This negative perception fuelled PARC's zeal and seemed to offer a convincing moral case for intervention. Danny, who works for PARC at a senior level, tells me that, if the project was successful,

> the rest of Edinburgh would no longer see it as a place to be avoided. When it gets mentioned in the *Evening News* you would no longer find that every second comment at the bottom was pretty much, you know, the people of Craigmillar should be locked up and thrown away, which is currently the comment.

There is a moral commitment in this statement, and it is seemingly a genuine one. It is through notions of fairness and 'helping' that stigma legitimates intervention not only in the sense of external justification (as Weber, 2002, observes), but in terms of an internal justification at the level of the individual and the institution. Danny finds the vitriol 'deeply distressing' because his time spent in the area has attuned him to the unfairness of its reputation. He wants to help, and in this he is not alone. 'You're helping people!' notes Jeremy, a member of the council's planning team, 'that's the best thing about local authorities.' Malcolm, one of his colleagues, grasps his City of Edinburgh Council name-badge as if for emphasis: 'I think', he argues, pausing ever-so-slightly before continuing, '[that] the majority of people involved in being an elected member are doing it because they want to better their area, their community, and the city, you know?' Johnny, a city councillor who sat on the board of PARC, concurs. 'This sounds really silly to me,' he concedes, 'but in a very small way, I'm working towards making the world a better place. But that's what local authority jobs are supposed to be anyway, you know?' These testimonies are not anomalies. Without quoting each individual separately to prove their point, it was overwhelmingly the case that a sense of *helping* appeared to be quite genuine. Clearly this is not unproblematic in and of itself, for the idea of an external authority (be that a governing body or a benevolent individual) 'helping' raises all sorts of issues about power, control and governance, especially when the continued existence of that authority is legitimised by the general perception that it helps. Nevertheless, I believe this is important. Internalising a set of ideas about how 'progress' is achieved and reformulating them into a plan for what should be done is different from a punitive idea being wielded with intentionally regressive effect. This reaffirms that it is through the construction of 'common sense' that ideology works (Hall and O'Shea, 2013).

The deployment of a stigmatising discourse is perhaps clearest in the repeated use of the 'ghetto' label. For Gavin, the story of the regeneration project began when Craigmillar 'became a kind of ghetto area' and this negative reputation meant 'we needed to do something big'. 'Actually', he concedes,

> some local people didn't think we should have knocked down all [those] old council tenements. Personally, I thought it was the right decision, you know? To send out the right message.

One councillor calmly explains that 'we knocked about half of Craigmillar down, and the reason we did that is that we don't want ghettos'. It is important for the reader to realise that Craigmillar is not and has never been a ghetto, for all the reasons that Wacquant (2008) details in the case of the Parisian *banlieues*. My concern is not linguistic impurity (a sociologist with a tick-box list of structural causes cannot say to those who feel *they* live in a 'ghetto' that they are simply wrong). My concern, rather, is that the 'ghetto' is deployed by policy-makers. It becomes part of the process of how to respond to urban deprivation, when this deprivation has not been caused by a process of ghettoisation (Slater and Anderson, 2012). What creates a ghetto is a combination of systemic racism coupled with the need to retain the services of the denigrated 'other'. A neighbourhood such as Craigmillar, however, suffers from a set of problems that are quite distinct. The very disappearance of work in the area recasts the local population as 'surplus', part of an immense reserve army of labour that has little official purpose in an era where consumerism, not production, fuels economic growth (Bauman, 1998). To suggest Craigmillar *is* a ghetto is to equate an area whose population are integral to the functioning of the wider society that effectively imprisons them there (the Jews of seventeenth-century Venice; the African-American communities of twentieth-century Chicago, and so on) with an area whose population are now denigrated as surplus to the function of that wider society, and whose population has gone into freefall from sustained out-migration. This is at very least unhelpful, and at worst a pernicious stigmatising discourse that proscribes the right solution (abolish the ghetto!) to the wrong problem. For once described as a 'ghetto', demolition is imbued with a moral hue that would be much harder to sustain if the project was articulated as the destruction of a working-class community and the mass privatisation of land and the housing on that land. All of this is imbued with bitter irony, of course, because it is the City of Edinburgh Council who have allocated tenants into the housing in Craigmillar for some seventy years.

Another councillor assures me that he does not use the word ghetto 'pejoratively at all, just cos, you know, *Craigmillar. . .*', but rather to advocate the 'need to spread the problems over Edinburgh' because 'if you have a ghetto like that [then] they reinforce each other's behaviour, but if they're spread across Edinburgh they're much easier to handle.' What I find interesting about a formulation like this, aside from the obvious invocation of the flawed idea of 'concentrated poverty', is the way that a territorialised stigma ('the ghetto') is to be dealt with through a literal decoupling of people and place. To liberate the area from its stigma requires, simultaneously, the eradication of the landscape itself (in the most visceral way, through demolition) and the displacement of the 'problem' people. It is in this sense that gentrification becomes an 'anti-stigma' strategy.

The reputational gap

One of the critiques levelled at the rent gap was its inability to predict those neighbourhoods that might gentrify next, despite the fact that Smith (1992b) never intended it to. In particular, those areas where the rent gap seemed largest were often those areas, by virtue of their negative reputation, that seemed most resistant to inward investment. As Hammel (1999, p. 1290) put it:

> Inner-city areas have many sites with a potential for development that could return high levels of rent. That development never occurs, however, because the perception of an impoverished neighbourhood prevents large amounts of capital from being applied to the land. The surrounding uses make high levels of development infeasible, and the property continues to languish. Thus, the potential land rent of a parcel based on metropolitan-wide factors is quite high, but factors at the neighbourhood scale constrain the capitalised land rent to a lower level.

Hammel's point is astute, but the notion of scalar causality needs some attention. Spatial inequality is relational: 'capital grows to a huge mass in a single hand in one place, because it has been lost by many in another place' (Marx, 1990, p. 777). Massey's (1994) discussion of uneven development remains important here. A class analysis, she reminds us, is *relational*. Wealth does not simply accrue in one area because it is not in others, as if a finite resource has been inadequately shared out over space. Rather, the accumulation of wealth in one area is dependent on the accumulation of poverty in another because material affluence in a capitalist society relies on exploitation. In this sense 'factors at the neighbourhood scale' needs to be read more as how factors are spatialised at the neighbourhood scale rather than originating from the neighbourhood scale. 'The poor', Richard Peet (2011, p. 388) reminds us, 'are poor because the rich are rich.'

At the same time, Goffman's (1968) innovation in his classic treatise on stigma was to pose it also as a relational concept. The 'abnormality' of the sub-section within any given population was defined against the 'normality' of the majority, and so 'a language of relationships, not attributes, is really needed' (p. 12). If it is 'metropolitan-wide factors' that create the potential ground rent in a maligned area – proximity to the centre, comparative desirability of neighbouring areas, transport links, population growth, job market, and so on – then it is *also* 'metropolitan-wide factors' that, in large part, create, foster and constantly reaffirm the negative perception of an area, both materially and discursively. And what happens if that negative perception is suddenly heightened, contrasted against a promised future vision of what, to quote Izzie, 'building a good place' is all about? The rent gap is now open.

What the regeneration project tried to do was to create the opportunity for high(er) capitalised ground rents by insisting on their potential. This act of wilful speculation is not abstract. It is, unlike the rent gap itself, very much *out there*, expressed quite clearly in planning documents, promotional leaflets, advertisements, development guidelines and the like. It is 'potential' created through government decree and large-scale privatisation, buoyed by some £60 million of state money, and reinforced continuously through emphasis on the 'new' Craigmillar. This vision undoubtedly involved the denigration of the 'old' Craigmillar, a symbolic and political disinvestment no less violent than the decaying architecture invoked by the downward curves on Smith's model. It is here that territorial stigmatisation functions, conversely, as part of the gentrification process not just as the justification for intervention, but as a necessary component of that intervention.

In some ways, this is a modest observation. The contrast between a denigrated present and a promised future historically lies at the heart of urban planning. Modernism at its most audacious made this contrast more pronounced, or certainly more idealistic (Hall, 2002). But the priorities of urban planning and the rent gap have become far closer in their association, and the act of privatisation reflects this. If the land in Craigmillar was once removed from the market – and in this sense had neither a capitalised nor a potential ground rent of much significance – then this is categorically, *and very intentionally*, no longer the case. This lies at the heart of Smith's (2002) observation that urban policy in the UK is increasingly operating on the basis that gentrification is desirable, and that all other options are unrealistic.

At the heart of this is the wider context of 'austerity urbanism' that has derailed projects such as PARC whilst at the same time imbuing them with a more desperate rationality. As of 2013/14, the City of Edinburgh Council were in debt to the tune of £1.6 billion and rising, representing a rise of 106.7 per cent over ten years (Carrell, 2015 n.p.). This was the same ten years in which the PARC project floundered. If it was the neoliberal cocktail of competitive urbanism, privatisation, and economic growth that made projects such as PARC so predictable in the early years of this millennium, it is the fallout from those presumptions that makes the strategy seem intractable. Behind the straitjacket of neoliberal policy making lies a mountain of debt. The Thatcherite mantra that 'there is no alternative' becomes true, in this sense, because without tackling the debt, all policy decisions have to be made with a view to paying off the interest and keeping the debt levels as low as possible. The indebtedness of the state in this context makes the privatisation of land, and its circulation as fictitious capital seem like good governance. Anything else would be irresponsible. To not strive for the highest possible price, to not, in other words, enact gentrification as an act of accumulation by dispossession, would make no sense for those notionally in control.

Conclusion

I wish to suggest that the reputational gap can help to explain the process of gentrification in areas where the rent gap seems to be open, but ineffectually so.

The difference is subtle, but the reputational gap puts an emphasis on the *articulation of potential*. In this sense, it is the opening of the reputational gap – as a policy – that facilitates the closure of the rent gap. Put simply, the mooted transformation of Craigmillar involved the necessity of rising capitalised ground rents. The way to begin this process was to declare a sudden break in the area's history. This is clearest in the decision to demolish so much of the housing in one go, and was bolstered by declarations that 'Craigmillar has no need or wish to be such a part of the city anymore' (Development Framework Team, 2001, p. 5). The state is not an addendum to the rent gap model, but lies at its heart.

Except, of course, that the project faltered, and in the intervening years, much has changed. At a national level, the ascendance of the Scottish National Party (SNP) in 2007 provoked a noticeable shift in rhetoric. The impact on policy was belated and somewhat erratic, but most noticeable in terms of housing was the scrapping of the Right to Buy (McKee, 2010). The effect on housing policy more generally is uneven and it is difficult at present to judge whether this is more of a set-piece to do with defining the SNP's sense of 'difference' from the UK establishment, or part of a longer-term commitment to an alternative set of priorities. The recent (but vague) commitment to new security of tenure and the (re)introduction of rent controls in 'high-pressure areas' of the private sector offer glimpses of a more radical shift. The role of campaign groups such as Living Rent in the run up to the drafting of this bill should remind us that the framing of policy priorities always involves struggle. In Craigmillar, the tenure profile of the area is more of a genuine 'mix' than its architects ever wanted, specifically because housing associations were, for a while, the only developers interested in building. The council has even shifted some 'units' back into their ownership, and has committed to constructing more. Whether this will be a blip in the trajectory of development or a watershed moment in its changing course remains to be seen.

Note

1 Housing associations are 'not-for-profit' companies who have increasingly taken over the responsibility of providing affordable housing from the local authority.

References

Abercrombie, P. and Plumstead, D. (1949), *A Civic Survey and Plan for the City and Royal Burgh of Edinburgh*, Edinburgh: Oliver and Boyd.

The Accounts Commission (2013), *Housing in Scotland*. Available online at: http://www.audit-scotland.gov.uk/docs/local/2013/nr_130711_housing_overview.pdf (accessed 24 February 2014).

Adams, I. H. (1978), *The Making of Urban Scotland*, Montreal: McGill-Queen's University Press.

Bauman, Z. (1998), *Work, Consumerism and the New Poor*, Buckingham: Open University Press.

Blomley, N. (1997), 'The properties of space: History, geography and gentrification', *Urban Geography*, 18(4): 286–295.

Carrell, S. (2015), 'Why is local government debt so high in Scotland?' *The Guardian*, available at http://www.theguardian.com/news/datablog/2015/mar/05/why-local-government-debt-so-high-scotland (accessed 5 March 2015).

City of Edinburgh Council (2005), *Craigmillar Urban Design Framework*, Edinburgh: City of Edinburgh Council.

City of Edinburgh Council (2012), *Craigmillar Urban Design Framework Review: Options for Consultation*, Edinburgh: City of Edinburgh Council.

City of Edinburgh Council (2013), *Revised Craigmillar Urban Design Project*, Edinburgh: City of Edinburgh Council.

Clark, E. (1995), 'The rent gap re-examined', *Urban Studies*, 32(9): 1489–1503.

Clark, E. (2015), 'Making rent gap theory not true', Paper presented at 'Global Capitalism and Processes of Urban Regeneration: A Tribute to Neil Smith' at MACBA, Barcelona, 15 September 2015.

Craigmillar Festival Society (1978), *The Gentle Giant: Craigmillar Comprehensive Plan for Action*, Edinburgh: Craigmillar Festival Press.

Crummy, H. (1992), *Let The People Sing! A Story of Craigmillar*, Edinburgh: Craigmillar Communiversity Press.

Crummy, H. (2004), *Mine a Rich Vein*, Edinburgh: Craigmillar Communiversity Community Press.

Damer, S. (1989), *From Moorepark to 'Wine Alley': The Rise and Fall of a Glasgow Housing Scheme*, Edinburgh: Edinburgh University Press.

Denning, P. (2004), 'Re-discovering Coketown', in Bell, D. and Jayne, M. (eds), *City of Quarters: Urban Villages in the Contemporary City*, Aldershot: Ashgate.

Development Framework Team (2001), *Craigmillar: Masterplan Design Guide*, Edinburgh: City of Edinburgh Council.

Flint, J. (2003), 'Housing and ethopolitics: Constructing identities of active consumption and responsible community', *Economy and Society*, 32(4): 611–629.

Forrest, R. and Murie, A. (1988), *Selling the Welfare State: The Privatisation of Public Housing*, London: Routledge.

Goffman, E. (1968), *Stigma: Notes on the Management of Spoiled Identity*, Harmondsworth: Pelican Books.

Gordon, G. (1979), 'Status areas of early to mid-Victorian Edinburgh', *Transactions of the Institute of British Geographers, New Series*, 4(2): 168–191.

Gray, N. and Mooney, G. (2011), 'Glasgow's new urban frontier: "Civilising" the population of "Glasgow East"', *City: Analysis of Urban Trends, Culture, Theory, Policy, Action*, 15(1): 4–24.

Hague, C. (1982), 'Reflections on community planning', in: Paris, C. (ed.), *Critical Readings in Planning Theory*, Oxford: Pergamon Press, pp. 227–244.

Hall, P. (2002), *Cities of Tomorrow: An Intellectual History of Urban Planning and Design in the Twentieth Century*, Oxford: Wiley-Blackwell.

Hall, S. and O'Shea, A. (2013), 'Common-sense neoliberalism', *Soundings: A journal of politics and culture*, 55: 8–24.

Hammel, D. (1999), 'Re-establishing the rent gap: An alternative view of capitalised land rent', *Urban Studies*, 36(8): 1283–1293.

Hancock, L. and Mooney, G. (2013), '"Welfare Ghettos" and the "Broken Society": Territorial stigmatization in the contemporary UK', *Housing, Theory and Society*, 30(1): 46–64.

Harvey, D. (2006), *The Limits to Capital*, London: Verso.

Harvie, C. (1981), *No Gods and Precious Few Heroes: Scotland Since 1914*, Edinburgh: Edinburgh University Press.

Hastings, A. (2004), 'Stigma and social housing estates: Beyond pathological explanations', *Journal of Housing and the Built Environment*, 19(3): 233–254.

Haylett, C. (2003), 'Culture, class and urban policy: Reconsidering equality', *Antipode*, 35(1): 55–73.

Hutton, G. (1998), *Mining the Lothians*, Ochiltree: Stenlake Publishing.

Jack, I. (1989), 'Problem Families', *London Review of Books*, 11: 36–37.

Jacobs, K., Kemeny, J. and Manzi, T. (2003), 'Power, discursive space and institutional practices in the construction of housing problems', *Housing Studies*, 18(4): 429–446.

Kallin, H. and Slater, T. (2014), 'Activating territorial stigma: Gentrifying marginality on Edinburgh's "other" fringe', *Environment and Planning A*, 46(6): 1351–1368.

Kintrea, K. (2006), 'Having it all? Housing reform under devolution', *Housing Studies*, 21(2): 197–207.

Kipfer, S. and Petunia, J. (2009), '"Recolonization" and public housing: A Toronto case study', *Studies in Political Economy*, 83: 111–139.

Lees, L., Slater, T. and Wyly, E. (2008), *Gentrification*, London: Routledge.

Marx, K. (1990), *Capital: A Critique of Political Economy, Volume I*, London: Penguin Books.

Massey, D. (1994), *Space, Place and Gender*, Minneapolis, MN: University of Minnesota Press.

Mathews, P. (2010), 'Mind the Gap? The persistence of pathological discourses in urban regeneration policy', *Housing, Theory and Society*, 27(3): 221–240.

McIntyre, Z. and McKee, K. (2012), 'Creating sustainable communities through tenure-mix: The responsibilisation of marginal homeowners in Scotland', *GeoJournal*, 77(2): 235–247.

McKee, K. (2010), 'The end of the right to buy and the future of social housing in Scotland', *Local Economy*, 25(4): 319–327.

McKenzie, L. (2015), *Getting By: Estates, Class and Culture in Austerity Britain*, Bristol: Policy Press.

Mitchell, D. (2015), *Author-Meets-Critics: David Harvey's Seventeen Contradictions and the End of Capitalism*. Available online at https://www.youtube.com/watch?v=kHobOMn_UkQ (accessed 4 May 2015).

Oglethorpe, M. (2006), *Scottish Collieries: An Inventory of the Scottish Coal Industry in the Nationalised Era*, Edinburgh: The Royal Commission on the Ancient and Historical Monuments of Scotland in partnership with The Scottish Mining Museum Trust.

PARC (2002), Written resolutions of Craigmillar Joint Venture Company Limited. Retrieved on 3 May 2015 from https://beta.companieshouse.gov.uk/company/SC234777/filing-history (accessed 15 January 2017).

Peet, R. (2011), 'Inequality, crisis and austerity in finance capitalism', *Cambridge Journal of Regions, Economy and Society*, 4: 383–399.

Purdy, S. (2003), '"Ripped off" by the System: Housing policy, poverty and territorial ttigmatisation in Regent Park Housing Project, 1951-1991', *Labour/Le Travail*, 52: 45–108.

Rodger, R. (2001), *The Transformation of Edinburgh: Land, Property and Trust in the Nineteenth Century*, Cambridge: Cambridge University Press.

Scottish Executive (2008), *The Right to Buy in Scotland: Pulling Together the Evidence*, available at http://www.scotland.gov.uk/Resource/Doc/149718/0039871.pdf (accessed 23 April 2013).

Scottish Government (2003), *Scottish Indices of Multiple Deprivation*. Available online at: http://www.gov.scot/Publications/2003/02/16377/18194 (accessed 3 May 2015).

Scottish Government (2006), *Scottish Indices of Deprivation 2006: General Report*. Available online at: http://www.scotland.gov.uk/Publications/2006/10/13142739/6 (accessed 30 December 2015).

Scottish Participatory Initiatives (2005), *What Do People of Craigmillar Think About the Craigmillar Urban Design Framework?*, Edinburgh: Scottish Participatory Initiatives.

Shaw, N. (2009), *Ghost Town: The Last Days of a Scottish Housing Scheme*, London: Blurb.

Skeggs, B. (2004), *Class, Self, Culture*, London: Routledge.

Slater, T. (2014), 'The myth of "Broken Britain": Welfare reform and the production of ignorance', *Antipode*, 46(4): 948–969.

Slater, T. (in press), 'Planetary rent gaps', *Antipode*.

Slater, T. and Anderson, N. (2012), 'The reputational ghetto: Territorial stigmatisation in St Paul's, Bristol', *Transactions of the Institute of British Geographers, New Series*, 37(4): 530–546.

Smith, N. (1979), 'Toward a theory of gentrification: a back to the city movement by capital, not people', *Journal of the American Planning Association*, 45(4): 538–548.

Smith, N. (1982), 'Gentrification and uneven development', *Economic Geography*, 58(2): 139–155.

Smith, N. (1986), 'Gentrification, the frontier, and the restructuring of urban space', in Smith, N. and Williams, P. (eds), *Gentrification of the City*, Boston, MA: Allen and Unwin, pp. 15–34.

Smith, N. (1992a), 'New city, new frontier: The Lower East Side as Wild, Wild West', in Sorkin, M. (ed.), *Variations on a Theme Park: The New American City and the End of Public Space*, New York: Hill and Wang, pp. 61–93.

Smith, N. (1992b), 'Blind Man's Buff, or Hamnett's philosophical individualism in search of gentrification', *Transactions of the Institute of British Geographers, New Series*, 17(1): pp. 110–115.

Smith, N. (1996), *The New Urban Frontier: Gentrification and the Revanchist City*, New York: Routledge.

Smith, N. (2002), 'New Urbanism, New Globalism: gentrification as global urban strategy', *Antipode*, 34(3): 427–450.

Smith, N. (2010), *Uneven Development: Nature, Capital, and the Production of Space*, London: Verso.

Smith, P. J. (1994), 'Slum clearance as an instrument of sanitary reform: The flawed vision of Edinburgh's first slum clearance scheme', *Planning Perspectives*, 9: 1–27.

Stone, D. (1989), 'Causal Stories and the Formation of Policy Agendas', *Political Science Quarterly*, 104(2): 281–300.

Taylor, Y. (2008), '"That's not really my scene": Working-Class lesbians in (and out of) place', *Sexualities*, 11: 523–546.

Tyler, I. (2013), *Revolting Subjects: Social Abjection and Resistance in Neoliberal Britain*, London: Zed Books.

Wacquant, L. (2008), *Urban Outcasts: A Comparative Sociology of Advanced Marginality*, Cambridge: The Polity Press.

Weber, R. (2002), 'Extracting value from the city: Neoliberalism and urban redevelopment', in Brenner, N. and Theodore, N. (eds), *Spaces of Neoliberalism: Urban Restructuring in North America and Western Europe*, Oxford: Blackwell Publishing, pp. 172–193.

Wilkinson, M. (2004), 'Craigmillar a fine place to be', *Edinburgh Evening News*. Available online at: http://www.scotsman.com/news/craigmillar-a-fine-place-to-be-1-1021094 (accessed 14 July 2015).

8   Voices from the *quartiers populaires*

Belonging to stigmatised French urban neighbourhoods

Paul Kirkness and Andreas Tijé-Dra

Introduction

Ce que l'on est parle tellement fort qu'on en oubliera ce que l'on dit.

[What we are speaks so loudly that what we say will be ignored.]
'Speaker Corner', by Médine (*Démineur*, 2015)

In January 2015, the *Journal du Dimanche*, a weekly newspaper in France, published a list of 64 'ghettos' that had been identified within the country, with approximately 900,000 residents in total.[1] This rather dramatic portrayal of such a large proportion of the territory of the French Republic is not in and of itself particularly novel. In fact, the reality of 'ghettoisation' has been discussed in French academic debates in recent years, with some arguing that it is a phenomenon that exists (Lapeyronnie, 2008), and others stating that 'sensitive urban zones', as they are often referred to, cannot be compared to ghettos in the United States (Vieillard-Baron, 1998; Wacquant, 2008a). As a result of the territorial stigmatisation that affects so many of France's urban neighbourhoods, several urban policy initiatives have been put into place over the past decades in order to deal with the problem of the so-called *quartiers populaires* (working class neighbourhoods). As has been formidably demonstrated by Dikeç (2007), urban policies directed to the so-called *banlieues*,[2] as well as inner city working class districts, have significantly evolved over time, following a change in perception of the stigmatised localities from 'neighbourhoods in danger' to 'dangerous neighbourhoods'. The result of this evolution has been the increased securitisation of such areas, but it has also encouraged the development of a parallel discourse that legitimises the renewal and demolition of significant sections of these neighbourhoods, under policies of renovation that are referred to officially as 'démolition-reconstruction'.

Of course, politicians and the media are not the only ones to use the word 'ghetto' when referring to heavily stigmatised neighbourhoods in France. It is not uncommon to hear references to deprivation, concrete, segregation and ghettoisation in the lyrics of rap artists from such neighbourhoods, as well as in descriptions of some local residents of these stigmatised urban areas. However, it is important to note that such hyperbolic claims – belonging to a 'ghetto', or performing the role of 'gangster' – can also be perceived as small acts of affirmative transgression, appropriation and speaking back to depictions that are made outside the stigmatised neighbourhood (see Kirkness, 2014). In these instances, the term is wielded with a certain amount of pride that could signify belonging as much as it reiterates existing exclusions. Yet, even if we ignore this marker of a sense of attachment, we also need to recognise the existence of rap artists who actively and demonstrably signify their feelings of belonging to a place in which they live, or where they have grown up. In this chapter, we focus on two rap songs in which this is apparent with respect to two heavily stigmatised neighbourhoods in Marseille. We then underline the fact that these attachments do not stand alone and that they are shared by other residents of stigmatised neighbourhoods. To do this we focus on data gathered during ethnographic fieldwork in two other neighbourhoods, both of which are located within the southern city of Nîmes.

The neighbourhoods that we focus on in this chapter have all recently been the targets of a series of renovation programmes intended to deal with 'problem' neighbourhoods. These programmes have *inter alia* been intended to promote social mix, introduce gentrification into traditionally working class neighbourhoods or demolish buildings that were considered to be too dilapidated. As has been the case in other urban contexts (including several that are described in this edited volume; see also Kallin and Slater 2014), they have inevitably resulted in the displacement of people from their local base to other neighbourhoods. We argue that these displacements are ultimately counter-productive and that they result in the dissolution of vital social networks. The argument that we defend is that residents of stigmatised neighbourhoods often develop strong bonds with the people and communities around them, and that they can experience strong territorial attachments to the area where they live, an argument that has also been made in other contexts (see August, 2014).

Unfortunately, territorial stigmatisation leads to these bonds being ignored, thus giving further legitimacy to the policy initiatives that result in the destruction of place, which ultimately deepens territorial stigma instead of eradicating it. We feel that there exists some disregard for the experiences of the residents, as well as a lack of acknowledgement for the performative displays of belonging and the genuine affection for one's place of (albeit stigmatised) residence. The reputation of residents is more widespread than their actual voices and ideas. Our ultimate goal is to present some of these voices, often missed as a result of stereotypical depictions such as that which rapper Médine points to in the epigraph above.

Refuting stigmatising representations

The *quartiers populaires* should not be equated to the famously misrepresented *banlieues* (suburbs), and while some can be found in the surrounding areas of French cities, several of the neighbourhoods that are located within these urban areas are also designated in this way. For the purposes of this chapter, this detail is an important one: the neighbourhoods that we discuss are all inner-city *quartiers populaires*, located within the borders of Marseille (Plan d'Aou and La Joliette) and Nîmes (Valdegour and Pissevin). In large part as a consequence of how the *quartiers populaires* are represented, it is readily assumed that if residents had the possibility to do so, they would try to leave these structures and move to less disreputable neighbourhoods. In the early 1990s, these parts of the urban land-scape were already being referred to as 'neighbourhoods of exile' (Dubet and Lapeyronnie, 1992), implying that residents of these spaces were, in some way or other trapped outside mainstream urban society, at the margins of city life and banned from full participation. Wacquant (2009: 116–117) suggests that 'feelings of disgust and indignity' are internalised by 'those at the bottom of the urban lad-der' as they trickle-down into the neighbourhoods from 'above' or from outside. Evidently, there are enormous material and social consequences to residing in urban areas with negative reputations: address discrimination (from employers, for instance), on the capacity for residents to enact collective action and on the operation of public services and state policy (see also Wacquant *et al.*, 2014). Today, it is well known that unemployment statistics are well above average in a number of France's *quartiers populaires* but there is also intense media covering of miscellaneous news items (such as sporadic unrest or the presence of Islam, for instance; see Sedel, 2009). As such, it is reasonable to assume that a number of residents are (or should be) on the look-out for a means of escape, a form of symbolic self-protection that Wacquant labels a strategy of exit (2008b: 116). However, it is also legitimate to ask whether this is *necessarily* the case.

In a different context, when Hyra (2008) writes about the residents of Chicago's Bronzeville and New York's Harlem, he grants considerable atten-tion to the strong feelings of attachment that his respondents develop for these places, despite the fact that they have both been labelled as 'notorious ghettos'. Notwithstanding the extreme material conditions that result from residing in these stigmatised spaces, Hyra's analysis offers an insight into the fact that the blemish of place is not at once, and always internalised by those who live in stereotyped places. Similar findings can be found in Slater and Anderson (2012), who studied the neighbourhood of St Paul's in Bristol, only to find that residents often displayed forceful feelings of pride for this part of the city, which could be seen to emerge as a direct result of the defamation that is produced outside it. Undoubtedly, there can be severe symbolic consequences to territorial stig-matisation, and the stigmatic imagery may well at times be internalised by the residents of 'tainted' spaces, as Wacquant argues (2009). Yet it is also important to point to the feelings of ambivalence that some inhabitants exhibit in a num-ber of contexts. For instance, Jensen and Christensen's (2012) study of Aalborg East, in Denmark, suggests that residents there exhibited positive or merely ambivalent attitudes towards their neighbourhoods. The respondents quoted in August (2014) were aware of the neglected physical environment and safety issues in Regent's Park, Toronto, they described benefits of living in an area with 'concentrated poverty', including dense social networks of friendship and sup-port, and access to services and agencies that suit their various needs. In a point that she relegated to a footnote of her paper, she states that 'exiting the neighbour-hood was not an option sought by most respondents. They commonly expressed a commitment to remain, viewing RP as a viable place to make their home' (ibid., 1329). Another such example is described in McKenzie's (2012, 2015) study of St Ann's, in Nottingham, in the UK. There she came across the use of the word 'ghetto' to describe the neighbourhood, yet it was used in a way that signified the appropriation of the area by those using it. She notes that the attachment to the neighbourhood is a reality that is not being acknowledged beyond the invisible walls that separate it from the rest of the city. In La Courneuve, the very neighbourhood in which Loïc Wacquant had conducted his own fieldwork in the early 1990s, Garbin and Millington (2012) found that residents did not internalise the language of stigma but responded to it through a series of coping strategies. Sure enough, some of these were submissive strategies (such as retreating to the private sphere of the home), but a number of these were also resistive in nature.

To point to 'resistance' to territorial stigmatisation, is not to be 'glibly celebrating resistance', as Garbin and Millington (2012, p. 2079) have cautioned, but it allows to draw other conclusions than the strictly pessimistic ones that can be found elsewhere. It is to focus on the 'ambiguities of domination/resistance' (ibid.) at work within the very processes of stigmatisation, contestation and spatial appropriations. Kirkness (2013, 2014) has dissected these at length, arguing that, to appropriate a neighbourhood – or a smaller sub-section of this urban area – is already a fundamentally resistive act of 'emplacement' which can has opened the way for residents to perform 'counter-scripts' in which they attempt to repaint the neighbourhood under a more positive light. It is also significant that other forms of stigmatisation, as described in social psychological literature has also focussed increasingly on the modes of contestation to the process, underlining the fact that collective and individual agencies are always at work in producing counter-scripts to stigmatic representations of people and places (e.g. Campbell and Deacon, 2006; Howarth, 2006). This type of collective endeavour is described in Purdy (2003) where some residents of Toronto's Regent Park organised collectively to resolutely battle the 'brutalizing depictions' of the neighbourhood, and 'becoming active players in building a meaningful place' there (p. 50).

Once again, our aim here is to state that a consequence of territorial stigmatisation has been to delegitimize the varied voices of those who live in such neighbourhoods. This is a point that has been made by others in a variety of contexts, but regularly, we see that arguments about mobility and social mix are articulated in reference to 'problem places' in order to supposedly break up the patterns in which residents are said to have been 'trapped into' (see Wacquant, 2007; Gray and Mooney, 2011; August, 2014). Such arguments are important inasmuch as they apply strictly to urban quarters because they are never really deployed in relation to the country. People who spend their entire lives in a small village are never described as the prisoners of spatial constraints yet they often undoubtedly lack the comforts and amenities that urban life is said to offer its residents. This point is an important one, since it helps to underline the fact that the relationship between stigma and 'renovation' is constructed upon other bases: a process that is presented as in the interests of residents, the so-called renovations serve other purposes of neoliberal urbanism and the interests of developers and the urban competitiveness. For Steinberg (2010, p. 222) 'a policy predicated on the claim that the demolition of their homes will advance the interests of the very people whose homes are being destroyed is a preposterous sham'. Elsewhere, this has been clearly argued with relation to Edinburgh's state lead gentrification of Craigmillar by Kallin and Slater (2014). In effect, more often than not, it is in the interests of the state or a city government to strategically stigmatise a neighbourhood in order to better justify future demolition, redevelopment and gentrification of an area (see Weber, 2002).

Arguing that demolitions tend to erode the stock of affordable housing, Pierre Gilbert (2009) adds that displacement harms social capital, a resource that is otherwise very potent in working class neighbourhoods. The relocation of populations, following demolition of their homes and displacement, is described by the psychiatrist Mindy Fullilove (2004) as 'root shock', leaving important psychological and emotional scars. She has not been alone in demonstrating that these consequences exist, since Marc Fried (1969) had already described the phenomenon of 'grieving for a lost home' in his study of trauma that residents of West Boston endured as a result of being forcibly removed from the neighbourhood. Porteous and Smith (2001) have also related the feelings of stress, anguish and hopelessness that the victims of 'everyday domicide' can expect to experience. Notwithstanding, they note that these have become acceptable since it is so commonly acceptable that the residence will have a better life elsewhere, away from the defamed neighbourhood. The voices of residence, on the other hand, are rarely included in such conclusions and, as we try to show here, they might signal very different needs.

It is our contention that, observable within the immense range of rap lyrics (which can, themselves, be read as a dialogue between different rappers, audiences and the targets of rap's sometimes confrontational style), is a common desire to shed light on 'difficult neighbourhoods' in a way that also recasts these spaces as 'home' and places of 'belonging'. Indeed, as we show in the following section, this sense of belonging and the pride that rap artists display for the areas where they have grown up has become one of French rap's trademarks cultural criteria of credibility: a rapper that leaves his or her neighbourhood can often expect to lose the attentive audience from France's *quartiers* that 'speaking for' marginalised spaces can at times generate. Rappers are residents of such spaces and, much like many other residents, they reclaim them through representations of their 'lived experiences' that are then sanctioned and appreciated by an audience in a process that generates credibility as well as cultural visibility and presence.

Displaying place attachment through rap music

French rap lyrics commonly shed light on 'difficult neighbourhoods' in a way that can also recast these spaces as 'home' and as places of 'belonging'. French rappers often originate from *quartiers populaires* and many still inhabit these neighbourhoods. In addition to the importance of 'representing' in rap music, they can reclaim these areas through articulations of their lived experiences (Forman, 2000;

Hammou, 2014, 70–90). Of course, they are not oblivious to the often harsh living conditions of those living in the *quartiers populaires* and they also create critical discourses[3] about their lived space, so it is common to assume that rappers speak of decay, concrete, violence and a feeling of being trapped. However, being aware of the state sanctioned flaws that one's neighbourhood exhibits, is not to have internalised the negative representations of such places. As the two examples from Marseille that we discuss here demonstrate, rappers refer back to local networks and strengthened neighbourhood feelings that can be said to be the very consequence of the language of stigma that is spoken outside the neighbourhood.

Before tracing the manifestations of place attachment in French rap music, it is important to clarify the way in which we analyse it here. Meeting the demands of the assumed constitutive openness of stigmatising discourses, we refer to methods of enunciative pragmatics that derive from French discourse analysis (Williams, 1999; Maingueneau and Angermüller, 2007; Angermüller, 2011). According to its concepts, linguistic markers configure each act of enunciation and its executed utterances on a formal level. They help to mobilise the contexts to which texts relate and they point to a range of possible interpretations. This means that functional words, and notably those carrying indexical information, instruct a readers' possible interpretations of enunciative contexts. The interplay of the concepts of *deixis* and of *polyphony*, which we apply below, help to highlight these discursive mechanisms.

Deictica (deriving from Ancient Greek *deixis*, 'to display') are linguistic particles that reveal the personal, spatial and temporal context of an utterance and forge one speaker's subjectivity (i.e. indexical words like 'I', 'here', 'now') (Angermüller, 2011: 2994). Contextual spatial references display the unconscious positioning mechanisms of language that contribute to the spatial definition of the social: the differentiation between a 'we', which is located here, and a 'they' that is to be found elsewhere, is crucial in constituting space, place and social relations (Glasze and Mattissek, 2009).

On the other hand, the concept of *polyphony* describes the dialogistic dimension of enunciative acts. In order to position themselves, speaking subjects often relate to preceding discourses and points of view, whose authors are 'absent' within the very situation of enunciation. Thus, the speaking subject is made to conduct a (potentially dissenting) multitude of their own statements, as well as those made by others – establishing hierarchical and relational orders. Signs for polyphony are argumentative connectors (i.e. 'no' or 'maybe'), as well as adjectives, verbs and adverbs of doubt and negotiation that establish relationships between different utterances (Roulet, 2011). Drawing on one's contextual, social background knowledge the reader/listener attempts to fill the identified point of views within the polyphonic play.

> To utter (*énoncer*), therefore, means setting a zero point from which a communicative space unfolds whose contextual parameters (time, space, person) need to be determined by the reader.
>
> (Angermuller, 2011: 2994)

This communicative space notably unfolds antagonistically within coping or resistive practices and subjectivities, as they always must relate to their stigmatising 'Other' in order to gain sense and identity.

The first example that we wish to discuss is a track entitled 'Les Cités d'Or' and it was written by Psy4 de la Rime, a rap group from Marseille. It illustrates the discursive traces that point to the feelings of ambivalence that some residents experience in relation to their *cité*. The song expresses the attachment to a stigmatised neighbourhood that many residents come to embody, even as they remain keenly aware of negative reputations and the structural consequences that these have. The track includes vocal participation from two rappers whose stage names are Soprano and Alonzo.[4]

In thie song, spatialised stigma is vehemently negotiated. At first, the *cité* of Plan d'Aou, in Marseille's 15th district and 'notorious' *quartiers nord*[5], receives an alternative toponym: the negatively loaded place name is substituted with the neologism 'Cosmopolitania'. This culturalistic re-labelling moves beyond a narrow sense of place, by suggesting a spatially defined identity beyond national belongings. Through the use of polyphonic markers, the rapper Soprano debates the frequent conjunction of his neighbourhood with issues of economic scarcity in the following passages. Such shortages are debated alternatively by putting a number of things into perspective and through the rejection of economistic and (spatially) external speakers. He uses the word 'despite' when he refers to the limits imposed by the economic situation of many residents, and he then opposes that to the his experience that inhabitants 'always' find solutions to problems that emanate from this situation. In doing so, the author acknowledges the taken for granted deficiency, but modifies the widespread bleak representations that suggest that the needs of *cité*-dwellers will never be entirely satisfied (i.e. food supply) as a result of where they live. In fact, according to the rapper, such problems are always solved due to the strong social networks that are identified.

The speaker reproduces this utterance at some other point in the song in an altered polyphonic way using the words 'even if (on two occasions)', 'too much' and 'not', as well as through the constitution of a collective identity located within the *cité*. In a series of verses, he stresses that despite the burdens of structural poverty - such as the condescending treatment that residents receive in financial institutions - the frugality of the 'we' is put forward and possible economistic objections are distanced. The point is that, contrary to collective representations, there actually exists a certain type of 'normality' in the neighbourhood, expressing itself through mutual sympathy, within the present as well as the deictically signalled future.

Alonzo, another member of the group, starts the second verse with what is an essentially exclusionary gesture, using the negation 'cannot' when referring to the possibility that the locally non-initiated and 'remote listener' might ever really understand his proffered points of view. The attachment to Plan d'Aou cannot be seized from a spatial, political, medial or societal distance. Instead, *it can only be grasped through the lived space*. Most importantly, this points to a re-appropriation of the inhabitant's self-signification, which is often marginalised through hegemonic discourses and their internalisation. These include a number of ready-made scripts, such as the notorious but pitiable *banlieue/cité*-youth. A few verses down, the rapper uses the words 'I', 'we' and 'our', all in association to the neighbourhood itself. By deictically integrating himself into to the collective place-bound identity through the use of pronouns, the speaker presents himself as a 'correspondent' for his neighbourhood, zigzagging between negative representations in order to articulate a more positive sense of belonging.

While the neighbourhood is seen to be falling apart, the rappers insist that this does not necessarily apply to the existing social tissue: an enriching familiarity compensates for the darker sides of everyday life. The future of Plan d'Aou is not depicted in the same quintessentially depressing way as when it is discussed by non-residents of the neighbourhood. In contrast to another stereotypical but anonymous assumption that *cité*-youths are prone to deviant behaviour, the rapper asserts the existence of moral integrity in the 'here and the now', manifested through mutual respect between residents. Another re-toponymisation of Plan d'Aou can be read in the imagery of 'cities of gold', the title of the song and of the album on which it appears, consolidating the strong expressions of attachment. This primarily eschews stereotypes and constitutes a certain degree of belonging and of a strength that emanates directly from social richness. In this sense, inhabitants (of these or similar places) are interpellated as integral parts of the 'cities of gold'.

Rappers also stand up for inhabitants through interventions and active engagement against urban renewal policies that aim to 'revitalise' stigmatised city centres. In Marseille, this is a process that is taking place in former working class *quartiers*, surrounding the Old Port and neighbouring the former docks. These areas are marked by poverty and a high percentage of (post)migrant inhabitants. Together with the *quartiers nords*, they strongly contribute to Marseille's nationwide image of a poor, post-industrial and post-migrant harbour town. Since 1995, the *quartiers* near the former industrial centre have been targeted by a large-scale urban renewal program (Euroméditérranée I and II), with the intended objective of turning Marseille into a renowned cultural and economic Mediterranean centre (endorsed by local, regional and national planning agencies). By granting subsidies for developers and investors, town and state officials have paved the way for publicly subsidised large-scale operations, including demolitions, renovations (into luxury apartments), resettlements and the construction of modern offices and museums. All of these are being constructed in very densely populated, but 'socially weak' areas (Jourdan, 2008). The proclaimed goal has been to attract middle class residents and to strengthen the city's tertiary sector. This top-down induced gentrification, camouflaged by cultural and economic labels, never in fact led to the desired economic and urban growth. Symbolically, this was a failed intervention. The high profile target group has not yet been attracted to the newly rebuilt city centre, which is witnessing a steady population drop, the rise of vacant apartments that in turn become subject to speculation, and the somewhat 'sterile' environment has continued to breed stigmatisation (Rescan, 2015).

In 2013, when Marseille became the European Capital of Culture, the municipality tried to enhance the new cultural image of the remodelled downtown. Local activists, which included the female rapper Keny Arkana, believed that this was yet another ruthless attempt to gentrify the old *quartier populaire* with no consideration for those who still lived there. Of course, this population does not correspond to the new image that the city has been trying to create, neither materially, nor in terms of identity. Teaming up with other activists, Arkana set to work on a song and produced a video documentary,[6] both of which attempt to re-label Marseille as the capital, not of culture, but of '[social] rupture'. The song itself expresses how the adherence to the *quartiers populaires* of downtown Marseille also forges strong feelings of attachment and place-based identity.

In her lyrics, Keny Arkana starts by excluding possible positive interpretation of downtown development of La Joliette[7] with the use of the strong marker 'certainly not' when stating that this is not what the residents of the area would have wanted. By using the collective 'we' when she discusses the rejection for these policies, Arkana associates herself to the frustration and she then goes on to reject the propositions that these redevelopments have led to (even partial) improvements to the lived experiences of local inhabitants. For long-term residents, La Joliette has not changed for the better and what has been done to date has not been in the interest of the local collective identity ('we', 'our'), which she claims to represent. The large-scale evictions that she implies (she states that there have been 'hundreds of' these) have also had strong impacts on the social structure and daily life of the area, but they have also demonstrated the strong emotional bonds to the city. As a large percentage of Marseille's downtown population is of North African descent and has been resettled elsewhere today, this identity aspect of 'her' (in the song, appearing as the collective 'our') Marseille seems to have disappeared. The polyphonic 'seems like' in combination with this disappearance demonstrates Arkana's weakening sense of hope and the uncertainty that she feels with regards to the possible undoing of what has already been undertaken for the sake of city redevelopment.

The rapper furthers her point of view by explicitly critiquing property-led city development, which she describes as being to the disadvantage of current residents. In fact, the economic interests of non-residents bring about a diminished (or 'demolished') quality of life to those who would otherwise have been happy enough to stay put. This is clear when one considers the example of one of Marseille's shopping streets, the Rue de la République and its Haussmannian buildings that in part belonged to la Joliette. There, demolition and displacement has led to the withdrawal of social and economic life, as investors followed different strategies intended to close the rent gap and to valorise former social housing and traditional shops, selling them to developers. In many cases, social housing operators ended up buying up the very stock of housing that had been redeveloped by the developers when these were unable to sell it. As a result, social life is now heavily fragmented: social and physical space is marked by a strange mosaic of vacant and degraded homes, as well as (often empty) luxury apartments and renewed social housing units (Un Centre Ville Pour Tous, 2015).

This is also why the streets that she vocally claims as 'ours' have changed for the worst in the present. In a drastic comparison, Keny Arkana compares the segregation stemming from these processes of urban capitalism to a certain kind of urban 'Apartheid', pointing to a profound social fractures. In the same context she tackles hegemonic representations of the *quartiers populaires*, re-labelling them as 'zones' in which a sense of community has clearly existed according to her (unlike the term zone would otherwise suggest). Before the said 'Apartheid' could take root, the 'zone' was also the place of 'true' belonging, given all the 'real' joy and friendship of the inhabitants. This image differs strongly from the (partially acknowledged, 'not just') hegemonic stigma that attaches itself to Marseille's city-centre as a place where inhabitants are unrestrained and are engaged in violent criminal activities (in the song, she denies the presence of guns in the neighbourhood). Keny Arkana concludes that the ongoing renovation of her city only displaces the inhabitants along with their poverty (due to the collapse of the local working class), while maintaining social inequality. This urban regime, in combination with a punitive state, leads to a bourgeois 'cleansing' ('disinfect') of the city centre for 'culture's sake', which is marketed to tourists as a new city brand, at the expense of undoing the long established identity of the city as a working-class town.

Everyday displays of attachment to stigmatised place

Similar findings can be read into research conducted in two French housing estates that border one another in the southern French city of Nîmes. Valdegour and Pissevin are regarded by non-residents as *banlieues*, which as has already been said, is a term that has taken on a host of stigmatising connotations. The first discursive move of residents has at times been a counter-constitution of the researcher as non-resident and therefore as someone who is unable to truly understand life in the housing estates. Jean, a long-term resident and retired school-teacher explains that:

> Journalists, and maybe researchers too, they always somehow assume that we'd be trying to get out of our neighbourhoods. I never hear the question about what is positive here even as I'd be able to tell you a lot. Then again, I'm not sure you'd understand. That is because there are many things that exist here and that just don't exist in the rest of the town [Nîmes].[8]

In stating that 'I'm not sure you'd understand', Jean is effectively re-inscribing the difference between the spatially non-initiated and those who have come to know the estates through everyday experiences of the lived space. This differentiation is at work in the words of Psy4 de la Rime, in the first of the two set of lyrics that are discussed in the previous section, but it is also at work in the voices of other housing estate residents. This keenness to separate the journalistic/research knowledge produced *about* housing estates and the knowledge generated through the everyday lived experience that develop within these spaces is not intended as a glorification. The recognition of what can be thought of as negative neighbourhood aspects are present in the above rap lyrics just as they are present within the narratives of other residents.

Bilal, another respondent who is very engaged in helping neighbourhood youths, demonstrates this understanding clearly while emphasising how attached he is to the neighbourhood and its residents:

> It would be a massive fib to say that everything you read in the [local daily paper] is nonsense because we have huge problems here. You've got the guys who hang out instead of going to school. You've got the mums who've just immigrated here and who don't dare to go out into the neighbourhood because of its reputation. And they are sometimes right, because you look at some of the kids wrongly and they will launch into a show of force. All that is true. I think you get that elsewhere though and the only big difference here is that a lot of the people have to make do on an everyday basis. My feeling is that it is because the difficulties of that we have here, that we have this kind of solidarity happening. There are maybe twelve thousand people in Pissevin, and I know pretty much all the youths by name.

Petty crimes such as the dealing of marijuana or the noisy occupation of public spaces are often associated with *banlieue* youths. Bilal recognises the anxieties and potential antagonisms that take hold within housing estate spaces but he rightly asks whether these are restricted to such neighbourhoods. In an attempt to explain the 'incivilities' of some youths in the *cité*, he blames the unequal access to goods and services which leads residents to 'make do'. In the words of Psy4 de la Rime, this corresponds to the 'lack of an economy' which does not stand in the way of the ability of residents to cope with some of the structural effects of neighbourhood stigmatisation. What is more, this in turn becomes one of the reasons that justify Bilal's most important point which he describes as neighbourhood solidarity. He associates himself to the adolescents and young adults in one sweeping discursive gesture by stating that he can name them (implying that they too know who he is). There is an assumption that Bilal is talking about mutual respect between residents here, much as the issue is raised in the song 'Les Cités d'Or'.

Amina, an unemployed mother of four children who are schooled in the neighbourhood feels similarly that:

> Some kids get up to no-good but the neighbourhood itself isn't the problem. It isn't fair to say such things. If there was more money coming into for the school or for us to have training and if there were more jobs out there . . . Then you wouldn't hear anything saying this place is a bad place to live. The only time we ever see the colour of investment is when they come here to destroy an apartment building. [. . .] My kids somehow hang out in the neighbourhood and they can get up to no good, like every kid anywhere. The only difference is that they might feel like it is healthier to be outside when it is 30 degrees than locked into our apartment. And why wouldn't they be outside anyway? It is their neighbourhood too.

Pissevin is 'their' neighbourhood as well as much as it is hers and, as observers, we should consider the socio-economic reasons that have projected this urban space's representation towards stigmatisation instead of focussing on the every-day deficiencies. Like Keny Arkana, there is a recognition that the population of *quartiers populaires* are thought of as secondary and that investments are made only to see them moved away. Asked if she would ever consider leaving her neighbourhood for a different part of Nîmes, Amina's answer is explicit:

> Some people would gladly swap their apartment because they dream of a gar-den. Of course, that is appealing. But you live with what you have, and I hap-pen to very much like what I have. Our garden, well the kids, you see them outside. It is there, outside, everywhere in the neighbourhood. [. . .] You go to [local public square] and watch. You see old men who talk together. You will see me chatting with my friends. . . There are all sorts of people. The bad boys, they find unnoticeable places to do their nonsense.

Amina is consciously insisting on her perception of the neighbourhood as different to the ways in which she expects non-residents to understand its spaces. The pub-lic squares that local and national media paint as problematic meeting spaces for delinquent youths are repainted as 'collective gardens' and counter-discursively evaluated in what they signify as positive for residents. Her attempt to repaint Pissevin as an 'ordinary' place takes a definite direction when she explains that 'nonsense' (ie. illegal or threatening activities and people) are mostly hidden from view, as they are elsewhere in the city and beyond. She erases the divide between the delinquent youths that have been constituted into the objects of journalistic fascination and academic study by referring to 'bad boys', a marginal group that are to be found 'everywhere'. Like Arkana, she refers to the activities that are also part of the everyday: friendships, and the fact that it is not all about criminal activity (Arkana's reference to 'rifles').

Normalising the neighbourhood is common in the words of residents. Nacira, a resident of Valdegour who won an award that enabled her to set up a recycling company, chose her neighbourhood as a place in which to set up *Le Coin Ecolo* ('the green corner') and stated that she would 'never have set it up anywhere else'. She explains that 'The people who work with me here, I know we have relations of respect that don't exist outside here. These are relations that we have because we grew up here.' Much like the mutual respect described by Psy4 de la Rime, Nacira believes that those who make Valdegour (instead of the *cité d'or*) are reliable and trustworthy and work well together, unlike what non-residents are told to expect.

All these examples, as well as the many more that were recorded in the 39 interviews conducted with estate residents in Nîmes, can lead us to conclude that there are clear attachments that residents deploy over the spaces that they have grown up in, that they reside in or that they have established networks of friends. It would shock no-one to hear that such relations existed outside stigmatised neighbourhoods. And this once again points to the pervasiveness of the discursive gap that is perpetuated nationally between the *quartiers* and the rest of French territory. The stigmatisation that has affixed itself to the image of contemporary French housing estates would suggest that residents might not be so attached to places that have often severe consequences when one resides in them. Partly these common representations are at the heart of the vast programmes of demolition that have been enabled through the mass injection of funds for 'urban renovation' and dispensed by the ANRU since 2003. Certainly there are other reasons that are given for proceeding to dismantle or raze to the ground a series of building blocks (see Belmessous *et al.*, 2005), but the territorial stigmatisation surrounding specific neighbourhoods is then used as a further reason, to then displace residents in the name of social and spatial re-engineering.

A short article entitled 'Démolition' appeared in the *Journal de Valdegour*, a local associative newspaper written by and for locals. Written Atman K., it mourns the destruction of Galilée, one of the residential tower blocks in the housing estate. Here is a revealing excerpt:

> It has started, the Galilée block is set to be demolished, after the Newton block, the Jean Perrin tower, the Archimède block, the bridge to Archimède street [. . .]. [T]he falling stones are as many memories from my childhood that crumble and break like those sad bits of concrete. Soon Galilée will be gone and a large emptiness will take its place, aerating a neighbourhood that the city wishes to be less dense, and this emptiness will drown my heart for many years to come. It is between the warm and comforting walls of a spacious apartment in Galilée that my life began [. . .]. The image of this building remains one of the most beautiful in my childhood memories [. . .]. I now understand better the expression: 'things were better before.'

The author interlinks his personal narrative - indeed, his very birth - with the falling concrete that once represented home, and we can see echoes of Arakana's concern with minds being demolished, along with the walls (2). The Galilée tower is perceived as a symbolic victim of the desire of the municipality to impose certain urban aesthetics on Pissevin and Valdegour without truly consulting the very people who, it is imagined, will be the beneficiaries of such amendments. When Atman K. writes that he understands how one could say 'it was better before', this is dissonant to the usual understanding of this sentence when it is uttered in relation to French *cités*. 'It was better before', is usually read in relation to the golden age of France's housing estates, when the housing blocks represented real improvements in living conditions for so many, whether working class or immigrant populations. Sonia, another resident explains that she was sad when the authorities tore down the Archimède block in Valdegour. Her cousins lived there in the 1980s and in light of these recent demolitions, she recalls all her fond memories of the residential building:

> They say it is nice because without Archimède you can see more of the sky and stuff like that. But all those memories, all those families with all those stories. Basically, it has all gone. I don't think they were worth anything beyond Valdegour. So no, I don't think it is nicer or prettier now.

Sonia is another example of someone whose choice has been to remain active in the estate of Valdegour even though her financial situation and that of her husband would enable them to leave. Her statement, aligned with those of others mentioned above, suggests once again that some residents enjoy and desire a residence in what are thought of as stigmatised neighbourhoods. In all these statements, as in many of the interviews conducted for this research, a deep perception of the estates as home is testament to the fact that residents develop attachments to their place of residence against all odds. They are well aware of negative representations but these are secondary to beliefs in togetherness, the social networks that are enabled in 'living together' and the security that it provides. Of course, this is not to say like 'all is for the best in the best of possible worlds'. Nonetheless, it is to say that when French rappers express an often ambiguous affection for stigmatised estates, they do not do this in ways that are disconnected from the everyday sensitivities and perceptions of other *cité* residents. Together, they re-negotiate negative representations, accentuate others, insist on their collective agency while underlying the structural relations of power that also lead to incapacity. They also stress how much their place of residence means to them, something which is too often merely ignored in media coverage and political commentary, as well as institutional decision making bodies and urban planning officials.

Those wishing to discredit the lyrics of rappers, whether 'hardcore' or not, as speaking from a different position to that of the supposed 'ordinary' resident may find that rappers re-articulate common themes that echo through large housing estates throughout France. In effect, the strong attachment to places that rap lyrics redraw in a language quite different to that or media depictions is felt by residents at the level of everyday life as they situate themselves, their histories and social networks. Not only do residents embody the neighbourhood with pride, much as rap lyrics encourage them to do, but they offer counter-discursive strategies that should push planners to perceive estates not just as meagre run-down and failed projects from the past. These stories should help to underscore the fact that these are also, for several people, a home. Demolition of this entity without very concrete and participative relocation plans can only be seen as ineffective and futile in the long run.

Certainly, there are several residents who hope that one day, the possibility will present itself for them to move out, and several respondents in Pissevin and Valdegour stressed that they often had this desire. In particular, the elderly and newly arrived immigrants suffered terribly from the anxieties spawned by the neighbourhood's reputation. Even though the fact that these neighbourhoods are so multicultural is at times one of the reasons that lead residents to develop anxious relations to the Other, this is something that has been praised by the rap group Psy4 and their vision of the estate in which they live as 'Cosmopolitania'. Encounters with strangers can be both threatening, leading to a desire for mobility, but in the right setting, they can be liberating and at the heart of the reproduction of the public realm. This is a vital aspect of life in the *quartiers populaires*, including its public spaces and the affective conduct of its residents. This potential is a facet that has been consistently ignored by officials in charge of urban policy who tend instead to visualise neighbourhoods as mere laboratories for their attempts at spatial restructuring and take their decisions by working around the desires of residents, something which is deeply counterproductive (Deboulet, 2006).

Conclusion

Perhaps we should return to the epigraph by Médine at the start of this chapter: *'ce que l'on est'*, what these residents are is defined by where they live. They are the residents of stigmatised places and this results in their being tainted to the point where what they say or feel about their home-space becomes valueless. Yet, when the links that people establish with their stigmatised neighbourhoods are pushed to one side, there is a danger of unintentionally providing legitimacy to the destructive politics of 'renovation' projects. In France, 'urban renewal' has tended to promote the demolition of stigmatised building blocks *in the very name of the residents that have been displaced as a consequence of these demolitions.* As we show in this paper, this is deeply problematic because when apartment blocks are torn down, or when tenants are evicted in order to renovate buildings in supposedly run down city centres, the strong affective territorial attachments that we have highlighted are also destroyed in the process. Importantly, and as many of the essays in this book also argue, we wish to hypothesise that *one of the consequences of territorial stigmatisation is the development of a profound attachment to place* for a number of residents. It is our hope that the *quartiers populaires* can be redefined and understood as meaningful places for those who reside within them, beyond the discourses that constitute them as places to leave, spaces that are thought of as insecure and 'tainting'. As such, we aim to defend a vision of the *quartiers populaires* that is grounded in a right that we perceive as fundamental: the right for people to live in their *quartiers*, which can be better formulated as the 'right to stay put'. Programmes of urban renewal seem to imply the renewal of local populations, and this is something that is deeply harmful to the communities that are displaced. The right to stay put, so famously coined by Chester Hartman and his colleagues in 1984, is of necessity intertwined with that of the right to the city. Here, a reminder of David Harvey's oft-quoted definition is useful:

> [t]he right to the city is far more than the individual liberty to access urban resources: it is a right to change ourselves by changing the city. It is, moreover, a common rather than an individual right since this transformation inevitably depends upon the exercise of a collective power to reshape the processes of urbanization. The freedom to make and remake our cities and ourselves is, I want to argue, one of the most precious yet most neglected of our human rights.
>
> (Harvey, 2008: 23)

It seems that the rap artists whose lyrics we have discussed here, as well as the residents that were interviewed in Nîmes, would aspire to this idealised vision of the city in which they live. It is evident that the right to remake ourselves and the neighbourhoods in which we reside is linked to positions of power that residents of *quartiers populaires* are often lacking. The right to the city as we understand it is intended to empower residents, and, in addition the focus on the intersections of social and spatial mobility – in the sense of equal access to urban space – that emerges in many of the texts on the 'right to the city', we point to a *right not to be mobile*. It grants citizens the right to be present in public (stigmatised) space and it is a right that some believe should be reclaimed when these public spaces are overly controlled, policed or privatised. It comes attached, we argue, with the right for residents to stay put, even as this right is denied by the novel forms of neoliberal urbanism that are increasingly enacted in French cities. As Mitchell (2003) has astutely argued in his major contribution to the understanding of the concept, the right to the city cannot exist without some level of uncertainty and the loosening of 'controlled urbanism'. We would argue that along with the right to stay put comes a right to partly help shape the future of one's neighbourhood. This has long existed in neighbourhood associations and committees that were not always associated with the middle and wealthier classes. A real right to the city, in the French *quartiers populaires*, would be one where residents choose whether they wish to move or not and are further able to co-determine the social, cultural and physical future of their neighbourhood is ignored in urban policy decisions that label residents as deviant.

Our concern is that scholars who focus only upon the very real inadequacies of life in a stigmatised neighbourhood should also remind their readers, as well as planners and political decision-makers that these hardships do not mean that the neighbourhood is value-less for its residents. The right to the *quartier*, understood as we would hope, offers the possibility of resisting current threats to urban (dis) enfranchisement. As Howarth has stated, '[s]tigma is as much about the resistance of identities as the reduction of identities; it is a dialectical process of contestation and creativity that is simultaneously anchored in and limited by the structures of history, economics and power' (2006: 450).

Notes

1 The map is available here: http://www.lejdd.fr/Societe/Cette-carte-des-64-ghettos-de-France-qui-n-existe-pas-714524 (accessed 16 June 2016).
2 Although the term *banlieue* has taken on a loaded meaning whereby it is associated to the most stigmatised neighbourhoods in France, it literally translates as 'suburb'. But not all suburbs are stigmatised, and not all stigmatised neighbourhoods are located within the suburban areas of cities.
3 This section focuses solely on rap lyrics for pragmatic reasons. It is important to stress that French rap that originates in and describes stigmatised neighbourhoods, does not necessarily stay locked within a small subcultural sphere. It seeps into other social fields. For instance, certain critical rap artists such as *La Rumeur* (Tijé-Dra, 2014) intervene in the political field publicly and in affirmative ways. Others introduce rap workshops to youths from their own neighbourhoods in order to promote 'emancipation' and 'self-consciousness' through the practice of rap music (Baumann *et al.*, 2015).
4 Sadly, the record companies that own the rights to the lyrics would not let us publish them in this chapter unless we paid them a large fee. We were unable to speak to the artists directly – Keny Arkana and Psy4 de la Rime – and we are sure they would have

disapproved of the ways in which the record companies prevent the advancement of social scientific enquiry that corresponds with artistic intentions by putting up financial barriers. Therefore, we would particularly like to thank Médine and Julien Thollard, from Din Records, for being so enthusiastic about us using one of Médine's lines in our text. Nonetheless, we call francophone readers' attention to the easily and legally researchable content on video platforms such as YouTube and rap lyrics web pages on the internet.

5 The *quartiers nords* – a collective signification for the northern inner city districts of Marseille – are well known nationwide and popularised in the media for their desolate urbanistic condition, high unemployment rates, and ongoing rivalries between narcotic-dealing gangs that have escalated to the point of using arms of war.

6 http://www.dailymotion.com/video/x2fwi5r_mini-docu-marseille-capitale-de-la-rup ture-20-13-min_music.

7 An old working-class neighbourhood near Marseille's former harbour.

8 Interviews were conducted in French and the translations are our own. The names of the respondents have been altered in this text.

References

Angermüller, J. (2011), 'From the many voices to the subject positions in anti-globalization discourse: Enunciative pragmatics and the polyphonic organization of subjectivity', *Journal of Pragmatics*, 43(12): 2992–3000.

August, M. (2014), 'Challenging the rhetoric of stigmatization: The benefits of concentrated poverty in Toronto's Regent Park', *Environment and Planning A*, 46(6): 1317–1333.

Baumann, C., Tijé-Dra, A. and Winkler, J. (2015), 'Geographien zwischen Diskurs und Praxis: Mit Wittgenstein Anknüpfungspunkte von Diskurs- und Praxistheorie denken', *Geographica Helvetica,* 70(3): 225–237.

Belmessous, F., Chignier-Riboulon, F., Commercon, N. and Zepf, M. (2005), 'Demolition of large housing estates: An overview'. In R. Van Kempen, K. Dekker, S. Hall and I. Tosics (eds), *Restructuring Large Housing Estates in Europe*, pp. 193–210, Bristol: The Policy Press.

Campbell, C. and Deacon, H. (2006), 'Unravelling the contexts of stigma: From internalisation to resistance to change', *Journal of Community and Applied Social Psychology*, 16(6): 411–17.

Deboulet, A. (2006), 'Le résident vulnérable: questions autour de la démolition', *Mouvements*, 47–48(5): 174–181.

Dikeç, M. (2007), *Badlands of the Republic: Space, Politics, and Urban Policy*, Oxford: Blackwell.

Dubet, F. and Lapeyronnie, D. (1992), *Les quartiers d'exil*, Paris: Seuil.

Forman, Murray (2000), '"Represent": Race, space and place in rap music', *Popular Music*, 19(1): 65–90.

Fried, M. (1969), 'Grieving for a lost home: Psychological costs of relocation', in Wilson, J. (ed.) *Urban Renewal: The Record of Controversy*, Cambridge, MA: MIT Press.

Fullilove, M. (2004), *Root Shock: How Tearing Up City Neighborhoods Hurt America*, New York: Ballantine Books.

Garbin, D. and Millington, G. (2012), 'Territorial stigma and the politics of resistance in a Parisian *banlieue*: La Courneuve and beyond', *Urban Studies*, 49(10): 2067–2083.

Glasze, G. and Mattissek, A. (2009), 'Diskursforschung in der Humangeographie: Konzeptionelle Grundlagen und empirische Operationalisierungen', in Glasze, G. and Mattissek, A. (eds), *Handbuch Diskurs und Raum,* Bielefeld: Transcript.

Gray, N. and Mooney, G. (2011), 'Glasgow's new urban frontier: "Civilizing" the population of Glasgow East', *City*, 15(1): 4–24.

Hammou, Karim (2014), *Une histoire du rap en France*, Paris: La Découverte.

Hartman, C. (1984), 'The right to stay put', in Geisler, C. and Popper, F. (eds), *Land Reform, American Style*, Totowa, NJ: Rowman and Allanheld.

Harvey, D. (2008), 'The right to the city', *New Left Review*, 53: 23–40.

Howarth, C. (2006), 'Race as stigma: Positioning the stigmatized as agents, not objects', *Journal of Community and Applied Social Psychology*, 16(6): 442–451.

Hyra, D. (2008), *The New Urban Renewal: The Economic Transformation of Bronzeville and Harlem*, Chicago, IL: The University of Chicago Press.

Jensen, S. Q. and Christensen, A.-D. (2012), 'Territorial stigmatization and local belonging', *City*, 16(1–2): 74–92.

Jourdan, S. (2008), 'Un cas aporétique de gentrification: la ville de Marseille', *Méditerranée*, 111: 85–90.

Kallin, H. and Slater, T. (2014), 'Activating territorial stigma: Gentrifying marginality on Edinburgh's periphery', *Environment and Planning A*, 46(6): 1351–1368.

Kirkness, P. (2013), *The Territorial Stigmatisation of French Housing Estates: From Internalisation to Coping with Stigma*, unpublished PhD thesis, The University of Edinburgh.

Kirkness, P. (2014), 'The *cités* strike back: Restive responses to territorial taint in the French *banlieues*', *Environment and Planning A*, 46(6): 1281–1296.

Lapeyronnie, D. (2008), *Ghetto urbain: Ségrégation, violence, pauvreté en France aujourd'hui*, Paris: Robert Laffont.

Maingueneau, D. and Angermüller, J. (2007), 'Discourse analysis in France: A conversation', *Qualitative Social Research* 8(2). Available online at: http://www.qualitative-/r/esearch.net/index.php/fqs/article/view/254 (accessed 17 June 2016).

McKenzie, L. (2012), 'A narrative from the inside, studying St Ann's in Nottingham: Belonging, continuity and change', *The Sociological Review*, 60: 457–475.

McKenzie, L. (2015), *Getting By: Estates, Class and Culture in Austerity Britain*, Bristol: Policy Press.

Mitchell, D. (2003), *The Right to the City: Social Justice and the Fight for Public Space*, London: The Guilford Press.

Porteous, J. D. and Smith, S. (2001), *Domicide: The Global Destruction of Home*, London: McGill-Queens' University Press.

Purdy, S. (2003), '"Ripped off" by the system: Housing policy, poverty, and territorial stigmatization in Regent Park housing project, 1951–1991', *Labour/Le Travail*, 52: 45–108.

Rescan, Manon (2015): 'A Marseille, le centre-ville résiste toujours à la gentrification', *Le Monde*. Available online at: http://www.lemonde.fr/logement/article/2015/06/11/a-marseille-le-centre-ville-resiste-toujours-a-la-gentrification_4652108_1653445.html (accessed 29 May 2016).

Roulet, Eddy (2011), 'Polyphony', in Zienkowski, J. and Verschueren, J. (eds), *Discursive Pragmatics*, Amsterdam: Benjamins.

Sedel, J. (2009), *Les médias et la banlieue*, Lormont: Le Bord de l'Eau.

Slater, T. and Anderson, N. (2012), 'The reputational ghetto: Territorial Stigmatization in St Paul's, Bristol', *Transactions of the Institute of British Geographers*, 37(4): 530–546.

Steinberg, S. (2010), 'The myth of concentrated poverty', in Hartman C. and Squires, G. (eds), *The Integration Debate: Competing Futures for American Cities*, New York: Routledge.

Tijé-Dra, Andreas (2014), ,Rapper vs. Polizei? Zu einem französischen Verhältnis', *Sub\ Urban, Zeitschrift für kritische Stadtforschung*, 2(2): 131–136.

Un Centre Ville Pour Tous (2015), 'Rue de la République. Onze ans après...et toujours inachevée'. Available online at: https://www.google.de/url?sa=t&rct=j&q=&es rc=s&source=web&cd=1&cad=rja&uact=8&ved=0ahUKEwjiuOb4x7jMAhXC0x QKHWuzDgsQFggkMAA&url=http%3A%2F%2Fwww.centrevillepourtous.asso. fr%2FIMG%2FFiles%2F20151012-CVPT-Republique.pdf&usg=AFQjCNG8z- RUQRoXrTVUWXqp_rrHp2Ww2A (accessed 29 May 2016).

Vieillard-Baron, H. (1998), *Banlieue, ghetto impossible?*, Paris: Éditions de l'Aube.

Wacquant, L. (2007), 'Territorial stigmatization in the age of advanced marginality', *Thesis Eleven*, 91(1): 66–77.

Wacquant, L. (2008a), *Urban Outcasts: A Comparative Sociology of Advanced Marginality*, Cambridge: Polity.

Wacquant, L. (2008b) 'Ghettos and anti-ghettos: An anatomy of the new urban poverty', *Thesis Eleven*, 94: 113–118.

Wacquant, L. (2009), 'The body, the ghetto and the penal state', *Qualitative Sociology*, 32(1): 101–129.

Wacquant, L., Slater, T. and Pereira, V. (2014), 'Territorial stigmatization in action', *Environment and Planning A*, 46(6): 1270–1280.

Weber, R. (2002), 'Extracting value from the city: Neoliberalism and urban redevelopment', in Brenner, N. and Theodore, N. (eds), *Spaces of Neoliberalism: Urban Restructuring in North America and Western Europe*, Oxford: Blackwell.

Williams, G. (1999), *French Discourse Analysis: The Method of Post-Structuralism*, London: Routledge.

9 'This is my "Wo[1]"'

Making home in Shanghai's Lower Quarter

Yunpeng Zhang

Introduction: the Expo-induced *chaiqian* as a *fanshen* opportunity

The World Expo is an internationally recognised exhibition which, in essence, is an enormous transient spectacle in which nation-states participate in order to display their national prowess, levels of technological progress and to educate the public. The privilege of hosting the event has been pursued by local growth regimes worldwide in the belief that the event attracts footloose capital and boosts local economies. Residential displacement – a dark side of the story – is usually buried deep by the event-state/market-media complex.

Globally speaking, mega-events such as World Expos and the Olympics have a notorious track record of unleashing dire consequences upon disadvantaged populations living locally, including the obliteration of the neighbourhoods in which they live (COHRE, 2007; Greene, 2003; Olds, 1998). In preparation for Shanghai to hold the World Expo 2010, more than 18,000 registered households were uprooted from the city's waterfront neighbourhood in order to clear the land for pavilions and event facilities. However, this number does not reflect the real scale of total displacement as it excludes the migrants, those who were renting private properties and those people who lost their homes to event-related infrastructures outside the parameters of the Expo Park. With these figures in mind, what is perhaps most astounding is the speed with which these processes of displacement have occurred. In one earmarked neighbourhood, the demolitions took no more than three months.

In Chinese, the process of 'demolition and relocation' is named *chaiqian*. The word is used interchangeably with *dongqian*, a term that focuses on the spatial event of relocation, thereby downplaying the violence involved in property dispossession and the destruction of home. Indeed, the Chinese media and official accounts almost always portray the Expo-led *chaiqian* in a positive light. This is in part a consequence of the fact that *chaiqian* usually allows the displaced to purchase apartments at a discounted rate from newly built, modernist, planned communities in remote locations. *Chaiqian* is thus celebrated as a social intervention, lifting those who are otherwise perceived as incapable of changing their living conditions, out of poverty. The reasoning is nicely captured in the local saying, 'the poor rely on *chaiqian* to turn over their bodies (*fanshen*)'.

It is noteworthy that *fanshen* is a word stemming from the Chinese revolution. It is currently deployed by the Communist Party of China (CPC) to manufacture legitimacy and cultivate loyal subjects (Hinton, 1967). *Fanshen* relates to class struggles and is only achievable through class struggles. However, the objective is more than liberation and empowerment. It aims to transform the established orders, class relations and power structures, and more specifically, to turn them upside down. So what does *fanshen* mean in the context of *chaiqian*? What exactly are the sources of oppression that are denoted, and does *chaiqian* truly deliver a form of emancipation or empowerment and transform class relations and power structures?

To answer these questions, this chapter relies on a collection of official accounts, literary texts, newspaper clips, artistic products and interviews with the displaced. Most material was gathered during six months of fieldwork in Shanghai in 2012. Interviews with the displaced were semi-structured and conducted either in Mandarin or Shanghainese. Due to the difficulty of obtaining access, most interviews took place in public spaces, such as hallways, community gardens, food stalls and streets. For several reasons, many of the displaced who participated voluntarily in one-to-one interviews quit before I had the chance to fully introduce my project and obtain their details. Others, who had heard about my project from their neighbours and friends, came to find me directly. To comply with research ethics, 121 interviews were retained for final analysis. Younger generations (under 30s) are under-represented because many did not live in resettlement compounds. However, this is not a significant limit for the present paper considering that the displaced with whom I managed to speak had longer housing trajectories and were more knowledgeable about family histories. They had also been more involved in negotiating resettlement deals.

The chapter argues that the '*fanshen* by *chaiqian*' discourse is underpinned by a cultural logic that stigmatises and devalues the demolished neighbourhoods and their residents. It obscures diverse forces and processes that consistently perpetuate structural inequalities and lock the displaced into vulnerable positions. It also disrespects the numerous efforts of the displaced to make a decent home for their families under conditions that are sometimes complicated. Whilst resettlement apartments in planned communities have kept some of the displaced sheltered, those who rebelled against the demolitions in any way were evicted and never rehoused. In any case, the displaced all had to invest heavily in efforts to rebuild their social networks and their communities from scratch. Therefore, what *chaiqian* seems to really entail is not empowerment or emancipation, but exploitation and devaluation as well as the domination of capital over the needs of the displaced. To counter the *fanshen* rhetoric, this paper investigates the formation of the pre-Expo site and the political construction of the vulnerabilities of the displaced. Against this backdrop, it then examines the ambivalent and sometimes contradictory lived experiences of the displaced in order to understand the ways in which they value their neighbourhoods and to understand how their political agency may have been affected.

Sit(e)ing stigma

The concept of territorial stigmatisation was developed by Wacquant (2007, 2008a, 2008b). It explores the production and circulation of negative representations of places and their denizens and the ways in which these profoundly undermine the well-being and self-image of the inhabitants. Meanwhile, Wacquant suggests that the dominant elites benefit both materially and symbolically from the incorporation of negative imagery by residents of stigmatised places.

The notion owes a lot to Bourdieu (1999 [1993]) and Goffman (1974[1963]) but Bourdieu's influence is more significant in the development of Wacquant's theorisation (see Jensen and Christensen, 2012). It is therefore useful to have some understanding of Bourdieu's political sociology, particularly his spatial triad. In a short book chapter entitled 'Site Effects', Bourdieu (1999 [1993]) explains the intricate relations between social space, physical space and mental space. The key to his elaboration is the double inscription of social space and the naturalisation effect. He maintains that social space is translated into physical space (albeit in a blurred way), which in turn is reproduced in mental space through thought and language. The underlying mechanism is that social space structures the distribution of and the access to goods, services and agents. Due to different combinations and volumes of capitals (economic, social, cultural and symbolic), agents have different capacities to appropriate goods and services, and therefore create social and spatial distances. Through the mediation of durable physical space, social space is translated into conceptual categories, boundaries and divisions, which over time cultivates one's sense of being in/out of place. In Bourdieu's words, 'there is no space in a hierarchized society that is not itself hierarchized and that does not express hierarchies and social distances, in a form that is more or less distorted and above all, disguised by the *naturalization effect* [emphasis original] produced by the long-term inscription of social realities in the natural world' (ibid., p. 124).

The naturalisation effect is a result of incessant social struggles that take place on material and symbolic levels. Bourdieu (1977, p. 164) holds that 'every established order tends to produce (to very different degrees and with very different means) the naturalization of its own arbitrariness'. Its mechanism occupies a central theme in Bourdieu's political sociology. For him, material violence plays a less important role than legitimatisation in securing the positions of the dominant groups who are rich in legitimate capitals. Those social groups are successful in producing and universalising their point of view and making it the legitimate order, the truth and the rule of the game. When the authority and legitimacy of such views are unquestioned and when the views are internalised by the subordinated, structural inequalities and injustices are reproduced with the complicity of those who are the primary victims. This is what Bourdieu calls 'symbolic violence', an insidious effect of which is to make the dominated group believe that 'their actions are the cause of their own predicament and that their subordination is the logical outcome of the natural order of things' (Bourgois, 2002, p. 223).

Territorial stigmatisation is one example of such struggles and Wacquant's analysis sheds light on its political functions in contemporary society. Territorial stigma is a relational construct through dominant beliefs in norms and social values. It reinforces the ways in which 'Others' are imagined and polices the boundaries between 'us' and 'them'. It can conjure up aversive feelings – disgust, contempt and anger (Tyler, 2013) – and in turn, these abject feelings can undermine potential feelings of sympathy and solidarity. When activated in political discourses, it can be used to manufacture a consensus that enables violent classification practices and increased coercive measures (Kallin and Slater, 2014). It offers powerful ideological justifications for the deployment of exceptional means, often outside legal confines, to secure the accumulation of capital whilst producing massive insecurities for the victims in the process (Gray and Porter, 2015). More perversely, as in the case of *chaiqian*, it creates ripe conditions for the moral inversion that defines certain types of intervention as good, not only for the greater society, but also for the victims themselves.

Wacquant also looks into the corrosive effects of territorial stigmatisation through journalistic and statistical representations. In his early analysis, he is more concerned with Bourdieu's symbolic violence, paying attention to the internalisation of territorial stigma that negatively affect the identities and self-worth of residents, triggering their disidentification with their neighbourhoods through 'symbolic self-protection practices' (Wacquant, 2008a, p. 116). His observation shows that territorial stigmatisation can induce feelings of shame and guilt and that these lead to the development of a number of coping practices, such as concealing one's address or avoiding visits from friends and relatives (Wacquant, 2008b). In many cases, residents withdraw from their neighbourhoods to the private spheres of the family, an expression of mutual distanciation and lateral denigration (Wacquant, 2008a: 116). Escape, when possible, is a shared pursuit amongst the residents according to Wacquant.

Wacquant's work only partially captures the lived experiences of residents of defamed places. His early theoretical framework underestimates the complexity in the process of constructing meanings of, and developing bonds with places and thereby excludes a wide range of emotional connections with places (see Manzo *et al.*, 2008; Manzo, 2005; Jensen and Christensen, 2012). It gives insufficient attention to the functions of different places – tainted or not – for inhabitants. Indeed, for victims of territorial stigma, who are most likely to have limited legitimate capital, their neighbourhoods allow them to secure a foothold and lay down the roots from which they can derive resources that are denied elsewhere. They can also lead to the development of a sense of community, security and belonging. Through many place-making practices, stigma can be strategically appropriated, contested or managed (Kirkness, 2014). Therefore, flight away from stigmatised neighbourhoods is not necessarily – at least not by default – a shared desire. Empirical studies are required to map out the diverse ways in which people make sense of place and tease out the tension between escape and captivity.

Siting the Expo: 'The place in need of surgery'

Initially, the area on which the Expo would take place had not been shortlisted by Chinese planners because it was occupied by industrial operations, warehouses, self-built housing and planned residential compounds, whose tenure and property relations were very complicated. Many companies (e.g. Shanghai No.3 Steel Company and Jiangnan Shipyard) were deemed of strategic importance to the national economy and their relocation would require approval from the central government. This was also why this area was largely left untouched by previous waves of gentrification taking place elsewhere.

The ultimate choice of the Expo site was inspired by an Italian and her fellow students who presented the idea to the technocrats at the Bureau International des Expositions (BIE), the intergovernmental organisation in charge of overseeing World Expos, which is based in Paris (Wang and Zhu, 2007). The response was very positive and the message was quickly relayed to the bidding committee in Shanghai which was expected to secure the bid at any costs (Chen, 2010). Thus, the possibility of siting the Expo on the waterfront was rapidly studied and rationalised. This was soon backed by a group of cultural elites, comprised of professors, planners, journalists and artists. The shared goal was to justify the new siting choice and mobilise society to the hosting of the event. In doing so, they also legitimatised displacement and forced evictions as a necessity under exceptional circumstances.

Legitimatising discourses centre around several themes. First, they laid greater emphasis on distressed housing conditions, disordered built form and poorly developed public facilities. In a book commissioned by the local authority to celebrate the uprooting of thousands of families within a short time, renowned journalist Wu Jichun recalled his experience of visiting Shanggang area on the east side of the Expo Park:

> Many small wooden shacks in cheap tilted roofing weighed down by some bottles stood along the stagnant filthy water. The windows were broken. Smells of decomposed vegetables were everywhere around the shacks. This was the description of the miserable living conditions of the Shanghainese twenty years ago written by a French sociologist. This was hardly any exaggeration. When I arrived in Pudong in 2005, what I saw were rows of run-down houses, with many small simple shacks at the back. Five people lived inside this same room, dining, pooping, and sleeping, all inside this small room'
>
> (Shanghai Pudong New District Expo-Induced Chaiqian
> Command Centre *et al.*, 2006, p. 78)

In photo albums produced by local middle-class artists (Lu, 2010; Wei, 2010; Chen, 2011), a clear choice of colours was used to contrast the past with the present. The extravagant new Expo pavilions are portrayed in their actual colours, whilst the demolished neighbourhoods are predominantly printed in black and white. Although black-and-white is a common technique in journalistic photography and documentaries to claim neutrality and authority, it also filters the distractions and forces the audience to concentrate on the objects, the themes and the allegory that the photographers are attempting to convey through their way of seeing. Representations of those neighbourhoods focused on disordered physical space, the lack of modern conveniences and 'uncivilised' habits such as residents emptying their night stools, visibly urinating in public space, dining in narrow alleys, dining in narrow alleys and migrant children playing next to piles of garbage.

The disordered built form was also connected to the legality of tenures of those who were displaced. Renovated buildings or shacks, constructed to accommodate changed housing needs or increase household income, were referred to as *'weizhang dajain'* (illegal construction) (see Sun and Cen, 2005). Not only did this suggest that the Expo-induced *chaiqian* followed the letter of the law, but it accentuated the impression that those to be displaced were poaching on public land, with little respect for law and order. This generated an opportunity to disregard the displaced who insisted on compensation for the *weizhang dajian* as greedy and selfish. Ultimately, the discourse gave legitimacy to the pre-emptive demolition of 'illegal' structures, thus reducing the overall costs of *chaiqian*. Moreover, it insinuated that the displaced were the cause of poor public facilities and services, as their building practices had worsened the communal environment and reduced access to the area for ambulances, fire engines, buses and taxies.

Second, the neighbourhoods that were listed for demolition were seen as containers for large concentrations of marginalised and stigmatised social groups, such as laid-off workers, welfare dependent groups, the disabled, the sick, lower-income families or people who had returned to Shanghai after being forcefully sent to the countryside (known as *zhiqing*). Insistence on the fact that several people had criminal convictions or that they were migrants without local citizenship was also common. In Pei's report on Bailianjing, on the east side of the Expo site, distressed housing conditions and poverty suffered by residents there was considered by the head of the grassroots organisation as a cause of social conflicts ranging from small quarrels to armed attacks (Pei, 2006, p. 179). In an interview with the director of a government agency in charge of demolition, residents who fell under those categories were once again framed as 'problems' and 'challenges' because, with nothing to lose, they were presumed to be more likely to take advantage of the state and hold out until the eviction crew gave in to their 'unreasonable' demands (06/2012).

In the bidding report submitted to the BIE, people from the Expo site were represented as inferior to younger and better-educated groups, as well as foreign expats (Shanghai Expo Bidding Committee, 2002). They were depicted as part of the urban maladies to be dealt with and relocating them elsewhere was seen as a solution. Such medical language is echoed by Gui Xinghua, a local poet, in his book *City Heartbeats* (2008). Praised by Fu Liang (an important figure in the Fudan Poetry Club) in the afterword of the book for his 'acute sense of political responsibility' (p. 255), Gui dedicated a canto to the Italian who had inspired Chinese planners and brought about the changes to Shanghai.

> You collected the data of stone gates housing
>
> You connected the accumulated glory and past sorrows
>
> You mapped out the golden waterfront line
>
> You wanted to patch up a fissure
>
> The turning point of the Huangpu River
>
> Each inch of the land there is gold to glitter
>
> Your choice was firmly backed by the warm wind from Pudong!
>
> You have never been to an operation room
>
> You did not realise the 'surgery' on the city would require so many tailor made scalpels
>
> (Gui, 2008, pp. 26–27)

The first metaphor of gold mine admits that siting the Expo on the waterfront is also about reaping the rent gap (Smith, 1987). In fact, relocation of the residents and businesses on the Expo site was partially funded with the expected land revenues from post-Expo redevelopment in mind (Cui, 2008). The metaphoric reference to operations and surgical tools is more intriguing. Framing the city as a human body that follows the natural process of growth and decline has the power of making people believe that the removal of malignant tissue is the right choice in order to return to a healthy, functioning city. It testifies that urban planning (and political governance) is a 'pseudo-science', pretending to function as a division of medical science but which serves the interests of the private developers instead (Goodman, 1971, p. 67).

Third, the legitimatising discourses activated collective memories associated with *penghuqu* (slums or shanty towns) and *xiazhijiao* (lower quarters) by casually assigning the demolished neighbourhoods to this category, based on their positions in the spatial hierarchy, both historically and according to their built forms. The latter was naturally opposed to the 'standard' exemplified by newer modernist planned apartments. Local understandings of *Xiazhijiao* and *penghuqu* are loaded with pejorative connotations that refer to 'troubled' places with distressing landscapes in comparison to *shangzhijiao* (upper quarters) (Pan, 2005). In colonial Shanghai, inhabitants of *penghuqu* were believed to originate from a region known as Subei, an obscure and contested area without coherent boundaries (Honig, 1992, p. 20). Even though they were officially recognised as belonging to the Han ethnicity, just like the rich migrants from Ningbo and Canton in upper quarters, those thought to come from Subei were generalised as a social category defined by poverty, tastelessness, filth and crudeness (ibid). The term *subei* became used in an adjectival form. Lacking political, social and economic capitals, *subei* people relied on native-place based networks and association as a source of security (Perry, 1993). Many had relations with organised crime based on cronyism. As a consequence of periodic violence and because of these connections with gangsters, the *penghuqu* was defamed as a place of evil, for outlaws and was perceived as a no-go area for the *shangzhijiao* residents (Honig, 1992).

To this day, *penghuqu*, *xiazhijiao* and *subei* remain and shape the social geography of Shanghai and the cognitive structure of Shanghai citizens (Chen, 2012). Previous generations of *subei* may have laid down their roots and become Shanghainese, and some have moved out (for work, education, etc.) and distanced themselves from their old neighbourhoods. However, their *subei* origin is still remembered by residents with strong native-place identities. *Subei* people are still believed to be different from '*zhengzong*' (authentic) Shanghainese for their calculative mentality, manners and accent. They are joined by new generations of migrants, constituted as a unified group of *waidiren*, the new victims of ethnic domination in contemporary Shanghai. In a similar way to *subei* people in colonial Shanghai, their languages, accents, dress codes and behaviours are frowned upon by local citizens for their lack of taste. Denied access to affordable housing, many seek cheap accommodation in places such as the demolished neighbourhoods in the pre-Expo site. In both official and popular discourses, this new generation of migrants is seen as a latent threat to social stability and public security and must therefore be monitored and controlled. *Penghuqu* therefore became a durable container for the uncivilised, poor and deviant throughout the history of Shanghai. According to a local artist who surveyed local taxi drivers' views on Shanghai's best and worst places, the pre-Expo site on the east side topped the list of no-go areas for the filth and disorder (interview, 07/2012).

Legitimatising discourses conflate people entrapped in places that have issues with problematic populations in troubled places. Built upon this cultural logic, *chaiqian* for the pursuit of property-led development and private gain, is cunningly conflated with the logic of state welfare. Instead of undermining the ruling regime, it paints it in a positive light and allows it to derive legitimacy from participating in welfare provision. Against this background, *chaiqian* is normalised as a good thing, a long overdue attempt towards redistributive justice and a social intervention that somehow addresses structural inequalities.

Yokes chained the body: the real legacy of history

Justifications for the siting choice are right about housing distress, distinctive built forms and poverty. However, they failed to pursue a vitally important question – what causes and entraps the displaced in those neighbourhoods in the first place? If *chaiqian* is about transforming oppressive relations, it is then important to track down the roots of oppression. This section sets out to examine the mechanisms that consolidate the positions of those who deprived and stigmatised in the socio-spatial hierarchies under the governance of CPC.

Instead of addressing Shanghai's uneven geographies, when the CPC took over Shanghai it remained insensitive to established residential patterns and housing structures (Lu, 1999). To accumulate political capital rapidly, the CPC did improve the living conditions of *penghuqu* dwellers with the offer of state employment and state-sponsored neighbourhood improvements in the formative years of the new regime (Huanle Renjian Editorial Committee, 1971). However, such efforts were short-lived as a result of the defects in political ideology, economic organisation and social policies – particularly, housing policy.

Under Mao, Shanghai quickly fell out of the political elites' favour due to its 'shameful' connections with the bourgeoisie. Its position was also undermined by the national prioritisation of industrial development and military defence. Shanghai became an industrial powerhouse and did not receive corresponding financial transfers and investments from the central government as a consequence (Howe, 1968). This was in turn reflected in the fiscal restraints on investing and improving collective consumption facilities and infrastructures. As for housing, it was defined as a non-productive sector, investment into which was further restrained. The consequence of severe housing shortage and structural inequality is still felt today (ibid).

To showcase the superiority of socialism over capitalism, the CPC committed itself to reducing, if not actually eliminating, both exploitation and inequality. Housing provision during this time was organised primarily through the *danwei* system (work units), which tied the fate of the workers to the prosperity and power of their employers and fostered greater dependence of workers on their line managers and work units (Walder, 1986). Access to housing was decided according to one's rank, family size, current housing conditions, housing history and, importantly, through political connections. In this way, the Party created a stratified housing system under actually existing socialism (Wu, 1996). However, the new organised consumption was out of the reach to *penghuqu* dwellers because of severe housing shortage. During this time, most residents took on the renovation work themselves partly because they could now afford to do so, with stable employment, but more importantly because their dwellings were no longer habitable or adequate given changed family structures. This was a hidden form of exploitation of the *penghuqu* inhabitants under socialism, because salaries were deliberately low given that most consumption was organised collectively (Chen, 2012).

The housing history of Mrs Wu (in her 80s) is representative of many families that I worked with. She migrated to Shanghai during wartime and lived in a boat before moving into a hut built in Pudong. She renovated her house twice. The first time was in the 1960s when the roof was blown away in a storm. She and her husband then decided to build a small room with bricks. With no access to building materials, she scouted all over the city and collected discarded building materials. They eventually built their first decent home with great effort – her husband almost died of exhaustion. In the early 1980s, they started to renovate again because of her son's persistent begging to improve his chances of getting married. Once again, the renovation cost her family a fortune and they accumulated large debts, owed to their relatives.

The living conditions of *penghuqu* dwellers were also worsened by repressive socialist urbanism. Under the prevailing accumulation strategy and the ideology of building cities for production, spaces of reproduction were deliberately planned and constructed in close proximity to industrial operations. Between 1959 and 1984, industrial development rapidly expanded in the pre-Expo site as a result of the appropriation of land and through attracting workers to settle in the vicinity. To control the sprawl of *penghuqu*, existing inhabitants were encouraged to increase land use efficiency by adding additional stories rather than building further structures on vacant land (Chen, 2012). This was effectively controlled by the nationalisation of land in the city, as well as strict planning licensing. Against this backdrop, it is not difficult to understand the unique shapes, forms and styles of renovated *penghuqu* dwellings and the mixed land use of the pre-Expo site – both were considered nightmarish to modernist planners and architects.

The neoliberal revolution in China led to the most destructive attacks on *penghuqu* dwellers. To begin with, the restructuring of the state economy destroyed the iron rice bowls[2] of many workers and subjected them to the forces of the market. Job insecurity became widespread and this was followed by systematic privatisation, from land and state-owned enterprises, to medical care and education. Mostly employed as industrial workers in the state sector, *penghuqu* dwellers suffered severe losses. Those who were retired partially escaped the direct effects but there have been obvious consequences on the inter-generational transfer of resources in a place that is otherwise known for Confucian familial ethics. Familialism is a built-in value in social policy design, which binds the family as a welfare unit responsible for the well-being of its members (Chen, 2012). For these reasons, pain, depression and insecurity caused by one reform easily transcends generational gaps and permeates the entire family.

On the other hand, housing reforms, reinforced and solidified structural inequalities amongst workers. The privatisation of existing housing stock mostly benefited workers with higher ranks, wider political connections and more financial resources (Logan *et al.*, 2010). Most public housing tenants from the Expo-site did not benefit from this reform because of a lack of financial means and political connections. Since 1998, the cancellation of in-kind housing benefits has disadvantaged latecomers to the neighbourhood, people on 'waiting lists' for public housing, and those who were formerly excluded by the socialist housing system. The latter now have to rely on the market-based housing provision system, but the difficulties for all have been dramatically exacerbated.

The development of the housing market has given rise to a state-led real estate/finance complex. Owing to the incentive structures (e.g. tax-sharing system, cadre evaluation and promotion systems, land use quota systems) created for local state agents, they are encouraged to capitalise on land within their jurisdictions and pursue property-led development. Through the monopoly of land available for construction, local states have created an entry barrier to housing development which promotes developers with pockets deep enough to acquire land parcels and carry out projects with high yields. The consequences are inflation of property bubbles (Haila, 1999), inadequately developed public housing and the rise in unaffordability of housing. Apartments with modernist façade in planned communities are naturalised as the standard built form. As flawed consumers, the low purchasing power of *penghuqu* residents foreshadowed their fate in the emerging housing market. Although they are usually located in prime neighbourhoods, their homes have been discredited and their value has depreciated in the market for deviating from the standard.

'Your living hell is my Wo': serial displacement, home-making and ambivalence

In this section, I seek to give voice to the displaced and to their feelings about their homes and communities as an attempt to challenge dominant narratives. The chapter aims to unravel their ambivalent lived experiences and capture the tension between dependency and independency within the group of the displaced, both on an individual and a collective level. It is clear that negative representations of *penghuqu* finds some resonance amongst some of the displaced, especially those who are stuck in poverty and unable to upgrade their housing. They have internalised the stigma when describing the neighbourhood, they echo dominant representations (ie. lack of modern conveniences, small and dark rooms, unregulated building patterns).

A few of the respondents believed that urban environments provided by places such as their neighbourhood would be damaging to the prospects of future generations. Mr Ding was one of those concerned parents who feared that poor living conditions before *chaiqian* were responsible for his son's several failed relationships (interview, 05/2012). Mrs Yuan, on the other hand, was concerned that crowded living conditions, poverty and the lifestyle might have undermined her son's masculinity. Her view was shared by Mr Huang (late 20s) who commented on his next-door neighbour (male, mid 30s) as being stingy, calculative, shallow and a freeloader, 'a typical *xiaoshimin* [little Shanghai citizen]' (interview, 02/2012). He suggested that it had something to do with this person sharing a living space with his grandparents in a crowded home. These kinds of views are examples of mutual distanciation and the corrosion of the sense of self (Wacquant, 2007; Wacquant *et al.*, 2014). They reflect the negative links that many of the displaced had developed to the neighbourhoods.

Two other sources of negative bonds were a worsening community environment and the concentration of polluting factories. As a result of rapid commercial housing development and industrial expansion, many neighbourhoods were increasingly under the threat of flood and fire. Industrial smoke and dust from chemical plants and manufacturing factories in the pre-Expo area were seen by many as an inevitable part of their life. They did not pursue any political demands to clean up the urban environment as a result of a sense of powerlessness and a somewhat rational sense pushing them to privilege everyday subsistence. As Mrs Li (40s), a long-term resident from the Expo site, remarked, 'for the poor who are struggling to have their basic needs met, a better physical environment is a luxury to pursue in life. Unless you die instantly from the pollution, chronic contamination is a part of your fate' (interview, 06/2012).

Such negative bonds with place were also present amongst those who refuted the stigmatising rhetoric. But they also emphasised that dominant representations had underplayed their continuous efforts to make a home and create a functioning community in difficult times, and had narrowed the meanings of homes and communities for them. To understand their point fully, it is vital to get an idea of their life histories and housing trajectories. Many had often endured the threat and experience of displacement in the course of their lives. Here, one should stretch the meaning of displacement to include forms of displacement that capture the loss of one's place in society without being physically dislocated (Feldman *et al.*, 2011, p. 10). A broader definition allows us to better understand the complicated experience of serial forced displacement that many of the displaced have gone through, their reinforced attachment to their homes and communities, and their mixed feelings about the state.

Many of the displaced had personal experiences or family histories that included forced displacement during their formative years as a result of various political campaigns during the Maoist era. Some went through the loss of home under socialist transformation and the Cultural Revolution. For instance, Mr Deng (interview, 06/2012) and Mr Ding (interview, 04/2012) had seen their family properties confiscated by the CPC for their 'suspicious' bourgeois background. Many others had been forcefully sent to the countryside (*zhiqing*) or forced to relocate to support the development of inland regions (*zhinei*) or frontier regions (*zhibian*) under various campaigns. Due to strict mobility controls, they were then entrapped in situations resembling incarceration. This meant more than the loss of home and community. It came with a loss of status, as well as Shanghai citizenship and a rupture of their identities. The upheaval was extremely painful for the *zhiqing*, who were forced to leave their homes as teenagers to engage in agricultural labour in the countryside and compete against skilled farmers for subsistence. Negative experiences of displacement motivated their attempts to return to Shanghai as soon as the policy relaxed in the era of reform. However, due to the interruption of their education and their inadequate skills, most of them encountered enormous difficulties in surviving the harsh labour regime. They have mostly been stuck in poverty ever since.

Mrs Yang, a returned *zhiqing*, refrained from visiting doctors for her ankle, injured during her time in the countryside, because her income did not allow her to spend much on medical bills in the post-Mao era (interview, 04/2012). Mr Wu, sent to Xinjiang, spoke about this episode of his life with anger and frustration. His return to Shanghai was only made possible by marrying a woman with local citizenship status but with whom he did not get along well. He provided for his family and paid for his son's education by setting up a small grocery store (interview, 05/2012). Mrs Wang, whose husband was sent to Jiangxi, encountered enormous difficulties in finding jobs upon their return to Shanghai. Her family of four were crowded into one room. She worked as a cleaner to make ends meet (interview, 05/2012).

For those who were fortunate enough not to be targeted by the political campaigns, this did not mean an escape from the possibility of displacement. Indeed, as I have stated above, they often encountered physical displacement as a result of a rampant property-led accumulation regime that began in the 1990s. For many, the Expo-induced displacement was the second time that their homes were taken from them and their communities destroyed. Remarkably, this had occurred in a timespan of less than two decades. Mrs Guan (interview, 06/2012) and Mrs Xie (interview, 03/2012) had both been displaced once before the Expo-induced *chaiqian*. For Xie, with whom I had several follow-up interviews, both episodes of displacement were experienced as 'robberies'. She complained, 'you have to move no matter if you agree or not. It is not like you have a choice. They come and tell you, give me your home or we'll evict you' (interview, 05/2012).

A more widely shared experience was the loss of place as industrial workers. Under the post-Mao labour regime, the working class was once again relegated to the bottom of society. Many of the displaced were laid off and although some were able to find new poorly paid jobs, several others became unemployed permanently due to their age, education, skills or health problems. This last group has since been made to live off meagre unemployment benefits. In fact, loss of working-class pride, status and privilege is widely shared amongst the displaced. Such experiences aggravate their sense of injustice and sensitise them to the pitfalls of the reforms such as glaring inequalities, decaying social morality and many other social ills.

> The CPC is doomed. Chen Liangyu [former mayor of Shanghai] was imprisoned. Why? Internal fights inside the Party. Chairman Mao once said there was another Party within the CPC, another faction within the faction. He was talking exactly about this kind of things. Back to Chen Liangyu's case, how bad is he? How corrupted is he? It's all about the factional conflict. What prospect does this kind of CPC have?
>
> (Mr Qian, laid-off from his factory before retirement, in his 60s, interview 05/2012)

Against this backdrop, the displaced have often looked for protection inside their communities and relied on communal ties for support and security. They had a functional attachment to their neighbourhoods. Economically, they benefited from living close to their work places. Some ran small businesses within their neighbourhoods (such as Mr Wu, mentioned above). Instead of withdrawing from their neighbourhoods (Wacquant, 2008a), they developed intense and intimate interpersonal relations with each other. Partly, this was because of their low mobility. It was also a result of the peculiar spatial form of their communities. Due to land supply shortage and overcrowded living conditions, much of their everyday life took place on narrow alleys or communal space outside their homes. This included dining, cooking or washing clothes, for instance. Domestication of public space encouraged face-to-face interaction on a daily basis. Visiting their neighbours or chatting on their doorsteps was the most common leisure activity prior to displacement.

Such intensive contact also fostered the development of supportive networks and altruistic behaviour. When I brought up the issue of community life, Mrs Zhu (40s) told me how grateful she had been to her former neighbours for babysitting her child and for taking care of her when she was sick at home. She recalled, 'you didn't need to ask [for their favours]. They would just come, telling you, 'come back when you're done at work and I will pick up your son from school' (interview, 03/2012). Miss Du (30s) told me about her surprise when an ex-neighbour of her grandfather came to find her mother and insisted on attending her grandfather's funeral. Du's mother, who joined later, recalled the stressful moments when Du's grandmother passed away because of a sudden heart attack. Du's mother was a *zhiqing* sent down to Anhui Province and was still working there at that time. She travelled back to Shanghai as soon as she could but as she told me, 'we were overwhelmed and shocked by it. We did not know about the rituals. In the end, it was those neighbours who helped us to get through it all' (interview, 04/2012).

By modern standards, it seems that the crowded and dense space of the neigh-bourhood raises issues of privacy. The displaced were aware of this. Mr Ma joked, 'if you are cooking something nice at home, I can smell it at mine. You cannot hide it. Everyone will know about it. You have no secrets there. If there is a visitor to your home or there is anything going on at your home, everyone will know about it. People will ask about it and get to the very bottom of it' (interview, 04/2012). Residents at the time, like the West-enders in Herbert Gans' (1962, p. 21) ethnography, 'knew so much about each other that there is no need for prying' after many years of close contact. According to Mrs Zhu, her neighbours even recognised her relatives, friends and colleagues. As she commented, 'it made you feel safe. Strangers dare not enter' (interview, 03/2012). Mr Dong (60s) shared Zhu's view, 'the old place is very safe. No crimes. Why? Those seniors who often chatted by doorsteps knew too well who were outsiders. They would recognise strangers and ask. Here [at the resettlement compound], I don't even know the names of my next-door neighbours!' (interview, 06/2012). Because of the relative stability of their group, social control was also strong. Such public gaze, organic and bottom-up helped to maintain some form of social order and morality.

> On my way to the food market, if I bought a lot on that day, neighbours would ask what the occasion was. I would tell them that my brother-in-law and my sister-in-law were visiting their parents on that day. They would then know, the son and daughter of my parents-in-law were performing their filial obli-gations and caring about their aging parents. Otherwise, they would remind them to visit their parents more often next time.
>
> (Mrs Zheng, 05/2012)

Their homes, however unforgiving and substandard these might have appeared to outsiders, offered an additional source of pride as many of the displaced com-mented that they were their *wo*. Speaking of their old homes, they rarely hid their pride at having built a home at all, especially considering all the hardship they had experienced. Mr Jiang (60s), a father of two, bought a small 20-square-metre room with a tiny balcony on the second floor of a self-built housing unit from a local farmer after he returned to Shanghai as a *zhiqing*. To accommodate changes in his family of eight after both his sons were married, they modified the eave and converted it into a small loft inside his home, thus doubling the living space. Later, he altered the structure of the balcony and used it as an independent kitchen.

Likewise, Mrs Zhou (50s) managed to build a detached and well-decorated house before it was demolished and she was forcibly evicted. Whenever her family had any savings, they invested in the upgrading of their housing. The latest one was when her son was married. She asked her brother to redecorate the interiors. She recalled, 'I had this pattern of clouds on the wall. My house was the best in my neighbourhood'. She then explained to me the personal value of her home in tears,

> when you were in bad situations, you needed to find a way to get by. You had to think and work, earn money with your bare hands. No one dared to criticise you if you worked hard enough. . . after all those hardships, I built such a decent home. Now, my lifelong work is gone.
>
> (interview, 03/2012)

Her account highlights the meaning of the materiality of home for many of the displaced. In the course of their life journeys, it was these homes that had enabled the creation of roots, making the places sanctuaries to their memories, their hard work and to the ingenious improvisations that resulted in the built structures. By keeping their families sheltered, their home-making practices also served to demonstrate their moral and social standing. As the most crucial component of their family wealth, the homes were a symbol of what was a family legacy in need of careful preservation. The maintenance and accumulation of family assets established their moral positions in the lineage of their families and kinships. It is also important to underline that the practical understanding of a home as family wealth also has the effect of stabilising family relations and encouraging mutual support (Tucker, 2010). The loss of family legacies thus tended to induce guilt and shame, especially to the senior generations. Such feelings are made worse by a housing market that made it difficult for displaced families to provide for the housing needs of coming generations.

Economic and socio-psychological bonds that the displaced had for the place in which they lived state-led city-wide gentrification and land expropriation. This is particularly strong for those whose housing conditions were improved to be compatible with their changed family structures and circumstances. They contested the defamatory label of *penghuqu*, as well as the other ways through which their neighbourhoods were devalued. Residents of stigmatised neighbourhoods acknowledged that there were significant differences between their own homes and the now prevalent modern high-rise apartments in the city. They nonetheless stressed that there was no norm to which they should conform: not in building design or spatial organisation, nor in consumption patterns or lifestyle. They laid great emphasis on their own efforts to improve their neighbourhoods and pointed out the overgeneralisations and the distortions that were present in the dominant representations. To them, *penghuqu* is an obsolete word from the past.

Our neighbourhood is definitely not *penghuqu*. *Penghuqu* describes huts made of straws. Ours had tiles. . . they were just not like this kind of *gong-fang* [apartments in planned communities].

(male, 60s, displaced from Puxi, interview, 05/2012)

They said my home was part of *penghuqu*. But, on what grounds? My old home was well built and maintained. The difference was that we built it ourselves. . . we had a lot of migrants who rented in our neighbourhood. But if *penghuqu* is defined by a large number of migrants, there are countless *penghuqu* in Shanghai

(female, 40s, displaced from Pudong, interview, 03/2012)

Although they claimed that my home was part of *penghuqu*, it was definitely not. If, as they argued, houses without an independent kitchen, shower and toilets were the defining characteristics of *penghuqu*, we had all of those. When they delivered this apartment [her resettlement apartment], it was in raw state – no toilet, no shower, no kitchen. I had to install them all by myself. Would this apartment not count as *penghuqu*?

(female, 50s, displaced from Pudong, 03/2012)

Although they contested the label of *penghuqu*, like those who internalised the stigmatising rhetoric, they also welcomed the idea of external intervention to improve their conditions. These desires were abused by the eviction crews. The offer of discounted apartments in modernist planned communities enabled access to facilities, services and status symbols that were absent in their neighbourhood. But this offer was conditional on them giving away their homes and accepting a relocation elsewhere, exactly as the compound word *chaiqian* suggests. *Qian* (relocation) to a better living environment is preconditioned on *chai* (demolition) of homes and communities. Those who were stuck in extreme poverty were more receptive to this idea since they were unable to improve their conditions. Once they accepted the offer, demolition followed and made their neighbourhood difficult to live in for those who hoped to stay put. This division effectively drained away territorial resources that might have bonded the residents to organise collective resistance. The exploitative logic of *chaiqian* ignores the repeated oppressions that had helped to chain the displaced to their neighbourhoods. In accepting *chaiqian* agreements, the displaced legitimatised the appropriation of their own properties and submitted to an aggressive form of neoliberal housing regime.

Conclusion

This paper has critiqued the cultural production of territorial stigmatisation and its circulation in legitimatising the siting decision of the Expo that resulted in the loss of homes and communities for thousands of families. I argued that the stigmatising representations failed to account for the layered structural forces shaping the formation of the defamed neighbourhoods, leading observers to turn a blind eye to the lived experiences of those occupants who, through constant struggles, made this place their sanctuary. The chapter also engages with the symbolic struggles over representational space and delegitimises the dominant discourse by offering a counter-narrative of the lived space from the displaced.

I have teased out the mixed feelings of the displaced toward their neighbourhoods and the state, as a result to the experiences of serial displacement, under-investment and their inability to fundamentally alter their subordinate positions. Dependence on external forces – primarily the state – to improve their living conditions surrenders their power to control the production of space in a neoliberalising China. As the chapter shows, some of the displaced – especially the poorest – welcomed relocation arrangements in order to have improved living environments which they would otherwise be unable to obtain alone. According to Kearns and Mason (2013), relocation in this case should not be called displacement because people choose and desire to move. This slightly insensitive argument overlooks the fact that choosing where one wishes to live is frequently denied to some social groups as a result of structural inequalities. Willing movers are often victims of symbolic violence, and people whose desire for a better life has been manipulated. They experience the same constraining forces as those who fight to stay put.

In exploring the ways in which the displaced cope with stigma and structural inequalities, the chapter has paid significant attention to the place-making practices that have allowed people to derive a sense of belonging and security, thus enabling the existence of social support networks and a functioning community. These place-making practices have little to do with settling for the lesser evil or accepting the inevitable. We must not overlook the fact that stigmatised neighbourhoods have offered the inhabitants an anchor that allowed them to gain access to valuable material and symbolic resources, as well as some psychological well-being. At times, this was helpful in undoing some of the hardships that they would endure within and outside the workplace as a result of a now largely unregulated labour market. For some commentators, this could be interpreted as a symptom of resignation, yet such place-making practices, through which inhabitants seek security and respect inside their stigmatised neighbourhood, can also be read as a rejection of wider societal norms and dominant values (McKenzie, 2012, 2015). These practices may even counter the eroding and destructive effects of territorial stigma, allowing the denizens from 'tainted places' to negotiate and define their identities in different ways.

With this in mind, we can redefine *chaiqian* under the property-led development in China. It is not about allowing the displaced to capture and share the value created from land development as the politicians, planners and developers would have us believe. Rather, it is about extracting values – often unrecognised under the dominant economic system – from the victims of urban land development process. Those investments in emotion, labour and money must be respected when it comes to arranging displacement. The challenge for critical scholarship and activists is to fight for the recognition of those place-making practices and for its growth into a counter-hegemonic value system.

Notes

1 *Wo* means home here. There is a Chinese saying, *jin wo yin wo, buru ziji de caowo*, which can be translated as 'Other people's mansions, made of gold or silver, are inferior to a hut of one's own.' With an emphasis on the materiality of the dwelling space, the saying expresses strong sentiments of self-respect and dignity espoused by those living in precarious living conditions, in addition to the emotional experiences of belonging, familiarity, comfort, security, rootedness and place attachment that are typically associated with home (see Young, 1997; Blunt and Dowling, 2006).
2 A Chinese expression referring to stable employment, regular income and guaranteed benefits.

References

Blunt, A. and Dowling, R. M. (2006), *Home,* London: Routledge.
Bourdieu, P. (1977), *Outline of a Theory of Practice,* Cambridge: Cambridge University Press.
Bourdieu, P. (1999 [1993]), 'Site effects' in Bourdieu, P. (ed.), *The Weight of the World: Social Suffering in Contemporary Society*, Stanford, CA: Stanford University Press.
Bourgois, P. (2002), 'The violence of moral binaries: Response to Leigh Binford', *Ethnography,* 3(2): 221–231.
Chen, Y. (2011), *Shanghai Shibo Yingxiang Jiyi [Memory of the Expo in Photos: A Photographer Witness Lived in the Former Site of the Expo Park]*, Shanghai: Shanghai Fine Art Publishing House.
Chen, Y. (2012), *Chengshi Zhongguo de Luoji [City: Chinese Logic]*, Beijing: SDX Joint Publishing Company.
Chen, Z. (2010), *Shenbo Jiyi: Dashijian Xiaoxijie [Memoir of Bidding for the Expo: Mega-event and Trivial details]*, Shanghai: Oriental Publishing Center, Shanghai.
COHRE (2007), *Fair Play for Housing Rights: Mega-Events, Olympic Games and Housing Rights*. COHRE.
Cui, N. (2008), *Zhongda Chengshi Shijian xia Chengshi Kongjian Zaigou: Yi Shanghai Shibohui Weili [Mega-event Led Urban Spatial Reconstructuring: Case Study of the Expo]*, Nanjing: Southeast China University Press.
Feldman, S., Geisler, C. C. and Menon, G. A. (2011), *Accumulating Insecurity: Violence and Dispossession in the Making of Everyday Life,* Athens, GA: University of Georgia Press.
Gans, H. J. (1962), *The Urban Villagers: Group and Class in the Life of Italian-Americans*, New York: Free Press.
Garbin, D. and Millington, G. (2012), 'Territorial stigma and the politics of resistance in a Parisian banlieue: La Courneuve and beyond', *Urban Studies,* 49(10): 2067–2083.
Goffman, E. (1974[1963]), *Stigma: Notes on the Management of Spoiled Identity*, New York: J. Aronson.
Goodman, R. (1971), *After the Planners*, Harmondsworth: Penguin.
Gray, N. and Mooney, G. (2011), 'Glasgow's new urban frontier: "Civilising" the population of "Glasgow East"', *City,* 15(1): 4–24.
Gray, N. and Porter, L. (2015), 'By any means necessary: Urban regeneration and the "state of exception" in Glasgow's Commonwealth Games 2014', *Antipode,* 47(2): 380–400.
Greene, S. J. (2003), 'Staged cities: Mega-events, slum Clearance, and global capital', *Yale Human Rights and Development Journal*, 6(1): 161–187.

Gui, X. (2008), *Chengshi de Xintiao [The Heartbeat of the City, in Chinese]*, Shanghai: Shanghai People's Publishing House.

Haila, A. (1999), 'Why is Shanghai building a giant speculative property bubble?', *International Journal of Urban and Regional Research*, 23(3): 583–588.

Hinton, W. (1967), *Fanshen: a Documentary of Revolution in a Chinese Village*, New York: Monthly Review Press.

Honig, E. (1992), *Creating Chinese Ethnicity: Subei People in Shanghai, 1850–1980*, New Haven, CT: Yale University Press.

Howe, C. (1968), 'Supply and Administration of Urban Housing in Mainland China – Case of Shanghai', *China Quarterly*, (33): 73–97.

Huanle Renjian Editorial Committee (1971), *Huanle Renjian: Shanghai Penghuqu de Bianqian [World Transformed: The Changes of Penghuqu]*, Shanghai: Shanghai People's Publishing House.

Jensen, S. Q. and Christensen, A.-D. (2012), 'Territorial stigmatization and local belonging', *City*, 16(1–2): 74–92.

Kallin, H. and Slater, T. (2014), 'Activating territorial stigma: Gentrifying marginality on Edinburgh's periphery', *Environment and Planning A*, 46(6): 1351–1368.

Kearns, A. and Mason, P. (2013), 'Defining and measuring displacement: Is relocation from restructured neighbourhoods always unwelcome and disruptive, *Housing Studies*, 28(2): 177–204.

Kirkness, P. (2014), 'The *cités* strike back: Restive responses to territorial taint in the French *banlieues*', *Environment and Planning A*, 46(6): 1281–1296.

Logan, J. R., Fang, Y. P. and Zhang, Z. X. (2010), 'The winners in China's urban housing reform', *Housing Studies*, 25(1): 101–117.

Lu, H. (1999), *Beyond the Neon Lights: Everyday Shanghai in the Early Twentieth Century*, Berkeley, CA: University of California Press.

Lu, J. (2010), *Rebirth: Eight Years' Photography on the Creation of the Expo Site*, Shanghai: Shanghai literature and Art Publishing Group and Shanghai Jingxiu Wenzhang Publishing Group.

Manzo, L. C. (2005), 'For better or worse: Exploring multiple dimensions of place meaning', *Journal of Environmental Psychology*, 25(1): 67–86.

Manzo, L. C., Kleit, R. G. and Couch, D. (2008), '"Moving three times is like having your house on fire once": The experience of place and impending displacement among public housing residents', *Urban Studies*, 45(9): 1855–1878.

McKenzie, L. (2012), 'A narrative from the inside, studying St Anns in Nottingham: Belonging, continuity and change', *Sociological Review*, 60(3): 457–475.

McKenzie, L. (2015), *Getting By: Estates, Class and Culture in Austerity Britain*, Bristol: Policy Press.

Olds, K. (1998), 'Urban mega-events, evictions and housing rights: The Canadian case', *Current Issues in Tourism*, 1(1): 2–46.

Pan, T. (2005), 'Historical memory, community-building and place-making in neighborhood Shanghai', in Ma. L. and Wu, F. (eds) *Restructuring the Chinese city: Changing society, economy and space*, Oxford: Routledge.

Pei, G. (2006), 'Qishi' [Enlightenment], in Shanghai Pudong New District Expo-Induced Chaiqian Command Centre, Shanghai Pudong Construction Bureau and Shanghai Expo Affairs Coordination Office in Pudong (eds) *Pudong Shibo Dadongqian [The Grand Displacement in Pudong]*. Shanghai: Shanghai Literature and Arts Publishing House.

Perry, E. J. (1993), *Shanghai on Strike: The Politics of Chinese Labor*, Stanford, CA: Stanford University Press.

Shanghai Expo Bidding Committee (2002), *The Bidding Report for 2010 Shanghai World Expo China*, Shanghai.

Shanghai Pudong New District Expo-Induced Chaiqian Command Centre, Shanghai Pudong Construction Bureau and Shanghai Expo Affairs Coordination Office in Pudong (2006), *Pudong Shibo Dadongqian [The Grand Displacement in Pudong, in Chinese]*, Shanghai: Shanghai Literature and Arts Publishing House.

Slater, T. and Anderson, N. (2012), 'The reputational ghetto: Territorial stigmatisation in St Paul's, Bristol', *Transactions of the Institute of British Geographers*, 37(4): 530–546.

Smith, N. (1987), 'Gentrification and the rent gap', *Annals of the Association of American Geographers*, 77(3): 462–465.

Sun, X. and Cen, Y. (2005), Pudong shibo dongqian: Yangguangxia de chengxin [Pudong Expo-induced dongqian: Integrity and trust under sunshine]. *People's Daily*, 12 August 2005.

Tucker, I. M. (2010), 'Everyday spaces of mental distress: The spatial habituation of home', *Environment and Planning D*, 28: 526–538.

Tyler, I. (2013), *Revolting Subjects: Social Abjection and Resistance in Neoliberal Britain*, London: Zed Books.

Wacquant, L. (2007), 'Territorial stigmatization in the age of advanced marginality', *Thesis Eleven*, 91(1): 66–77.

Wacquant, L. (2008a), 'Ghettos and anti-ghettos: An anatomy of the new urban poverty', *Thesis Eleven*, 94(1): 113–118.

Wacquant, L. (2008b), *Urban Outcasts: A Comparative Sociology of Advanced Marginality*, Cambridge: Polity Press.

Wacquant, L., Slater, T. and Pereira, V. B. (2014), 'Territorial stigmatization in action', *Environment and Planning A*, 46(6): 1270–1280.

Walder, A. G. (1986), *Communist Neo-traditionalism: Work and Authority in Chinese Industry*, Berkeley, CA: University of California Press.

Wang, K. and Zhu, W. (2007), Shibohui, Chengshi Gengxin de Keneng [Expo necessitates Shanghai's Renewal, in Chinese]. *Sanlian Shenghuo*, Shanghai.

Wei, M. (2010), *Turning Point: The Story of Zhoujiadu 2004–2010*, Shanghai: People's Photography.

Wu, F. (1996), 'Changes in the structure of public housing provision in urban China', *Urban Studies*, 33(9): 1601–1627.

Young, I. M. (1997), *Intersecting Voices: Dilemmas of Gender, Political philosophy and Policy*, Princeton, NJ: Princeton University Press.

10 From Social Hell to Heaven?

The intermingling processes of territorial stigmatisation, agency from below and gentrification in the Varjão, Brazil

Shadia Husseini de Araújo and Everaldo Batista da Costa

Introduction

When asked what she thinks about her neighbourhood, the Varjão, a small suburb of Brasília and an independent administrative region of the Brazil's Federal District, Maria[1] responds with conviction: 'Here it is heaven! It is mother's lap.'[2] Maria uses this typical Brazilian expression to signify just how comfortable she feels in this otherwise disreputable neighbourhood. She was the first person that we encountered at the start of our fieldwork. At first, the statement partly surprised us seeing as, only a decade ago, the Varjão had achieved a national status as one of 'the worst places to live' and as a 'social hell on earth'. The name of this neighbourhood had long circulated in media discourses as a signifier for misery, violence and despair. In everyday conversations, the Varjão functioned as a synonym for a place where 'people live beyond law and order' (Lúcio, 2009, p. 58). Traces of this representation still prevail in contemporary discourses and are still reproduced in the local media coverage. A look at the daily *Correio Braziliense* illustrates this: in nearly 80 per cent of the articles about the 'Varjão' published within the past five years, it is directly connected with terms such as homicide, assassination, rape, stabbing and drug trafficking.[3] Yet, all the other residents of the Varjão that we interviewed confirmed Maria's description and expressed a strong feeling of belonging and affection for their neighbourhood:

> Well, I can't leave the Varjão, you know? [. . .] It's good to live here. I like it. Sometimes I go out at night, open the gate, look at the children in the street [. . .]. Here, it's not dangerous, as people say it is. [. . .] It's a neighbourhood that I like a lot.
>
> (IP4,[4] see also IP1, IP2, IP3, IP5, IP6, IP7)

Behind these opposing depictions – Heaven versus Social Hell – lies the Varjão's development from its origin as a squatter settlement in the 1960s, until the consolidated neighbourhood and administrative region of Brazil's Federal District that it is today. In this chapter, we aim to shed light on this development, arguing that its underlying force has been constituted by the close entanglement between territorial stigmatisation and governmental (re)actions, gentrification, as well as resistance and agency from below. Without the latter, the Varjão would neither exist nor would it have been able to evolve into what it is now. We suggest that the outcome of these intermingling (but conflicting) processes may be thought of in terms of an urbanised[5] and gentrifying periphery which is on the rise in valued locations in Latin America's large cities.

We approach the Varjão with a theoretical framework built in a complementary manner to Loïc Wacquant's conceptualisations of territorial stigma (Wacquant 2007, 2008a; Wacquant *et al.*, 2014), on Raúl Zibechi's notion of 'territories in resistance' (2012), and on the idea of gentrification as a generalised urban strategy (Smith 2002; Lees *et al.*, 2015). By contrasting these different theoretical thoughts, we hope to uncover and (to a certain degree) to overcome their respective blind spots in order to understand the development of the Varjão. This approach shall contribute to the conceptual debates on the urbanised and gentrifying periphery in Latin American cities. The empirical analysis is mainly based on seven in-depth interviews with inhabitants of the Varjão; five of them are or were active members in the local resident organisations (all conducted between December 2014 and July 2015). Additionally, we refer to several informal conversations that we held with elderly residents at the Centre for Senior Citizens (*Casa do Idoso*) and a short complementary document and media analysis, including newspaper articles, TV reports, and documents published by resident organizations from both the Varjão and adjacent neighbourhoods.

Territorial stigmatisation, agency from below and gentrification

As has already been demonstrated in the introduction to this book, resistance and agency from those who dwell in stigmatised districts is still a relatively neglected topic (exceptions are Slater and Anderson, 2012; Garbin and Millington, 2012; Kirkness, 2014). This may not at first seem surprising when leading scholars observe that '[t]erritorial stigmatisation [. . .] undermines the capacity for collective identification and action of lower-class families' (Wacquant 2008a, p. 116). According to Wacquant, this holds true for the *banlieues* of Paris, the ghettos of Chicago, as well as for 'the ill-reputed *favelas* of Brazil, the *poblaciones* of Chile, and the *villas miserias* of Argentina' (Wacquant 2008a, p. 117; see also Wacquant *et al.*, p. 6). His theoretical thoughts have not only been cited and discussed in urban research in the so-called Global North, but they have also been applied to the South. Many of the existing studies about denigrated urban districts in Latin American cities have tended to confirm Wacquant's findings, pointing out, for example, that residents seemingly internalise the stigma that is attached to their neighbourhood (e.g. Rodrigues and Rodrigues, 2014), or how stigmatisation results in the urban poor being blamed for the situation in which they find themselves (e.g. Bayón, 2012). Other recent investigations have focused on territorial stigmatisation as a form of symbolic violence (e.g. Cornejo, 2012), on the role of mass media in the reproduction of territorial stigmata (e.g. Sibrian, 2015), or on how territorial stigma gives legitimacy to military action and strengthens the penal state (e.g. Wacquant, 2008b). What all of these studies have in common is that they approach those who live in these 'damned districts' primarily as victims, and thus, they tend to emphasise a viewpoint that underestimates resistance and agency, as well as the power that the stigmatised may develop.

The Uruguayan journalist and political theorist Raúl Zibechi has made an important intervention that is helpful in confronting this tendency. In a recent critique of Wacquant and other 'analysts committed to the First World poor' – he also draws on the work of Pierre Bourdieu, Manuel Castells and Antonio Negri – who 'are unable to see the [urban] peripheries as anything but a problem, defined in negative terms', he points out that '[t]he inhabitants are never regarded as subjects' (Zibechi, 2012, p. 203). Based on a large variety of different case studies in Latin American cities, he shows how residents are able to conquer and construct their own territories, how they act and resist, and how they 'give shape to counterhegemonic social relations that create the new territorialities' (ibid., p. 209). The author uses the word counterhegemonic in order to stress the potential of 'the marginalised' in challenging neo-colonialism, neoliberalism as well as modern state logics.

Taking Zibechi's intervention seriously, we aim to scrutinise the interrelation between the Varjão's stigma and the different forms of agency and resistance of its residents. At the same time, we do not wish to 'romanticise' the Varjão's resistance movements, since they have neither been consistently united nor acting in the interests of the poorest residents, nor have they always operated in ways that one could describe as counterhegemonic. It is also important to note that once the residents had achieved one of their priority objectives – the regularisation of the Varjão (or at least parts of it) – resistance became increasingly fragmented and the government was not only able to co-opt it and to slow it down, but it was also able to sweep aside parts of the resisting movements, in order to pave the way for neoliberal development and gentrification. Many inhabitants themselves have welcomed the emergence of gentrification, not least because it seems to have led, at least to a certain degree, to a resignification of the Varjão, and with this, to less discrimination of those who (still) live there.

Against this background, we also seek to provide new inputs for the debates about territorial stigma and gentrification. In 2002, the urban geographer Neil Smith had already argued that

> the process of gentrification, which initially emerged as a sporadic, quaint, and local anomaly in the housing markets of some command-center cities, is now thoroughly generalized as an urban strategy that takes over from liberal urban policy. No longer isolated or restricted to Europe, North America, or Oceania, the impulse behind gentrification is now generalized.
>
> (Smith, 2002, p. 427)

According to Smith (2002, p. 439), gentrification has, on the one hand, 'descended the urban hierarchy' from big cities to smaller cities, and from city centres to the margins. On the other hand, it has also 'diffused geographically' with reports of gentrification in many different parts of the world. This certainly holds true for

Latin American cities, as evidenced by the increasing scholarly output on this topic (e.g. Betancur, 2014; Herzer *et al.*, 2015; Cummings, 2015; Jones, 2015; Janoschka *et al.*, 2014; Jones and Varley, 1999). Furthermore, there is also a small but growing number of contributions that focus on so-called 'slum gentrification' (Asensão, 2015) or gentrification of (former) irregular and informal settlements (Cummings, 2015). With the case study on the Varjão, we hope to contribute to a deeper understanding of the specific gentrifying processes in (former) irregular and only partly regularised urban settlements, as well as of their relations with territorial stigmatisation and agency from below. While most studies about territorial stigma focus on gentrification as neoliberal and state-led processes that 'overrun' the residents (e.g. Sakizlioglu and Uitermark, 2014; Lees *et al.*, 2008; Slater, 2006), we highlight the fact that residents sometimes do not resist gentrification because they actually welcome its initial effects.

Third, and finally, we seek to widen the perspectives on Brazilian cities in the Anglophone academic literature. Much attention has been given to the territorial stigmatisation of favelas in Rio de Janeiro, São Paulo, or, to a lesser degree, Salvador da Bahia. Much less has been written about Brazil's capital and the problem of territorial stigmatisation that operates in several of its suburbs. Yet, Brasília and the Federal District are examples of particular importance, as territorial stigmatisation has been central to their urban dynamics since the capital's foundation in 1960.

Brasília was planned and developed in order to transfer the capital from Rio de Janeiro to a more central location in the country. As the new national capital, it was intended to be 'different' from the other cities, reflecting modernity, progress and social equality. Yet, the plan was contradictory, above all because it did not consider room for the working class, who *de facto* built the city. These workers came predominantly from the poorest regions of Brazil, and as such were stereotyped as socially inferior and backward. Their early informal settlements, located within Brasília, were systematically removed to very distant locations where they would not disrupt the image of Brazil's capital as modern and progressive.[6] Notwithstanding, several of these settlements somehow resisted the relocation programs and in some way remained within or attached to Brasília. One of these settlements is the Varjão, whose history has been marked by the stigmatisation of its residents from the very start.

The Varjão and its stigma

The name Varjão is a derivation of the words *vargem* or *varzéa*, synonyms that, when literally translated, mean lowland or floodplain. This is a reference to the Varjão's location in the watershed of the Ribeirão do Torto ('the Large Crooked Stream') close to the point where the stream flows into Brasília's artificial Paranoá Lake. By the end of the 1960s, the first migrant workers began to settle in this area, unwilling to move to locations that were further away from the place where they worked and the city which they were *de facto* constructing, as desired by the government. The pioneers built their shanties with wood from the area, leftover material from Brasília's constructions sites and used the water supply from the Ribeirão do Torto.[7] After the construction of Brasília, many employees were dismissed and began to work in other low-paid sectors, such as landscaping, cleaning and domestic services (Calmon, 2010).

Over time, the population grew rapidly and reached 3,200 residents in 1988 (Lúcio, 2009), the year in which the government of the Federal District under José Aparecido de Oliveira (1985–88) attempted to forcibly remove this irregular settlement. The arguments for the removal lay not only in the stigma attached to the informal settlement which was seen as incompatible with the image of Brasília as a modern and progressive city. A second aspect, which is related to the first, was that the surrounding areas were (and still are) wealthy neighbourhoods. From the start, these spaces had been intended for the construction of housing districts that would cater to the more affluent residents of the newly found city (the Península of the Lago Norte, the Setor de Mansões do Lago Norte and, later, Taquari, see Map 10.1). The co-existence with an area like Varjão was seen as deeply problematic. Third, the area of the Varjão was located in an environmental protection area, where any kind of urban expansion was prohibited. Notwithstanding, the population of the Varjão continued to grow, the inhabitants organised themselves and resisted every attempt to remove their homes entirely and have them relocated. In 1991, they achieved the regularisation of parts of their neighbourhood. After its consolidation the

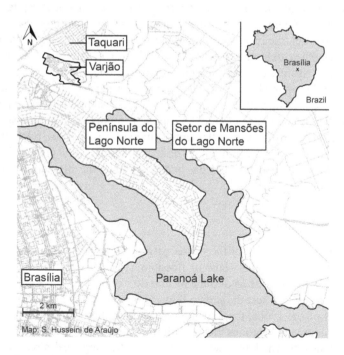

Figure 10.1 The Varjão, its surrounding wealthy neighbourhoods and Brasília.

Varjão underwent various urbanisation and development projects and, in 2003, became an independent administrative region of the Federal District of Brazil.

As a result of this development, most of the residents that we talked to agreed that the image of the Varjão has become far less negative. 'Prejudice against the Varjão still exists' (IP3), but its stigma seems to have become weaker and does not dominate all discourses anymore (IP3, see also IP2, IP4 and IP5). Besides this, the location of the Varjão is extremely valued today: 'People [from outside] come here to live because of the good bus connection, transportation to the Lago Norte, the [other] areas [are] close. . . so, people want to live here' (IP4). The Varjão may certainly not be understood as a destigmatised neighbourhood, but at least as a territory which has become increasingly associated with new, alternative and more positive meanings. In what follows, we first elaborate on the Varjão's historical stigma from the perspective of the interviewed residents and, second, shed light on the (re)actions by the government and the adjacent neighbourhoods.

The historical stigma

The historical stigma attached to the Varjão was above all structured by the notion of 'invasion' (*invasão*). In Brazilian urban discourses, this term refers to 'illegal occupations' of public territory. The Varjão was typical of these so-called 'invasions', here conducted by predominantly non-white and stigmatised migrants from poor rural regions of Brazil in search of work and of a better life. In many Brazilian cities such 'invasions' are also referred to as *favelas* and are usually associated with illegality, violence, criminality, and poverty. When asked where the stigma was originating from, all of the respondents suggested that they realised that the stigma was being reproduced at a number of different levels. However, they all identified the adjacent wealthy neighbourhoods as one of the main sources for this. As proof, they cited certain documents distributed by the associations in these districts, such as the manifest entitled 'Varjão: unwanted *favela* in the Lago Norte', which was written in 1991 by the Residents' Commission of the Block I in Lago Norte. This describes the Varjão with terms such as 'disorganized growth, homicide, assault, theft, and drugs' (cited in Medeiros, 2004, p. 69). Furthermore, the interviewees mentioned school, working place, or the media as places and discourses where stigmatisation is reproduced: 'I once watched a TV report about the Lago Norte. . . and the reporter said: 'next door lives the misery', you know? And this would be the Varjão. This made me very angry' (IP4).

On the one hand, all of the interviewees seemed to be outraged and tried to oppose the stigma when describing the Varjão. On the other hand, all of them reproduced its key elements, especially when referring to the past. According to Wacquant (2007), such a contradiction is not an exception. Rather, this is commonplace among those who dwell in stigmatised places. Examples of this reproduction are exemplified in the following quotes: 'The Varjão was a *favela*, a poor city, [. . .] the Varjão was only shacks, untreated sewage and waste ran through open air ditches, brought diseases. . .' (IP3). 'There were years in which you couldn't leave any clothes on the laundry line, [because] the next day, you would not find them there. [. . .] All of the Varjão was invasion' (IP5). 'The city was very violent. (. . .) We were already afraid of the weekends' (IP4), because murder and violent assaults tended to occur mostly at night during the weekends (a fear that was also raised by IP3 and IP2).

Territorial stigmata are directly connected with various forms of discrimination exercised in many different spheres of life (Wacquant, 2007). Therefore, it is not surprising that all of our interviewees had personal stories about discrimination to tell: 'I studied in the Asa Norte [a district of Brasília]. Fellow students asked me "Where do you live?" "I live in the Varjão!" And all the other students moved their chairs away from me. This was really bad' (IP3, also referred to by IP2, IP5). 'We felt ashamed to say: "I live in the Varjão." There was a company, and if you were from the Varjão, they didn't hire you. [. . .] We had to say that we live in the Setor de Chácaras in order to be employed. They thought that everyone [from the Varjão] is a thief, you know?' (IP4).

The stigma and the (re)actions

As stated by Wacquant, Slater and Pereira (2014), stigmatised territories 'elicit overwhelmingly negative emotions and stern corrective reactions driven by fright, revulsion and condemnation, which in turn foster the growth and glorification of the penal wing of the state in order to penalize urban marginality' (p. 5; see also Wacquant, 2008b). This was clearly the case with the Varjão. For a long time, the Varjão was a serious problem for the government of the Federal District. It was felt that it was somehow contrary to what Brasília should have represented, and it seemed to threaten the adjacent wealthy neighbourhoods. 'From the beginning onwards, no one wanted the people to live and to settle down in the Varjão. The people had to face police, dogs, horses. . . I witnessed all of this' (IP2). The government's politics concerning the Varjão oscillated until the end of the 1980s between the aim of total eradication on the one side, and half-hearted acceptance on the other.

The governor José Aparecido de Oliveira (1985–88) implemented two programs aiming at the eradication of all the so-called 'invasions' within the Federal District. The programs were known as 'Return with Dignity (*Retorno com Dignidade*)' – this was designed to send the workers of Brazil's Northeast to their region of origin –, and 'Surroundings with Dignity (*Entorno com Dignidade*)' – this aimed to 'at least' send the residents of 'invasions' out of the Federal District to be settled in the adjacent federal states Goiás and Minas Gerais. When the residents of the Varjão refused to leave, the government tried to remove them by force. Yet, residents always came back and rebuilt their homes, as we show in the next section of this chapter.

Considering the limited success of the eradication politics under Governor José Aparecido de Oliveira and the increasing pressure exercised by social movements, the subsequent governments under Joaquim Domingos Roriz (1988–90, 1991–95) and Wanderley Vallim da Silva (1990–91) followed a radically different route. They launched new housing programs which resulted in the fixation of the Varjão in 1991 and various urbanisation projects from the 1990s onwards. These projects basically aimed at the transformation of the Varjão from an area of so-called 'invasion' to a legalised modern neighbourhood – one that would neither contradict Brasília's image as the modern, beautiful and progressive capital nor disturb the rich neighbours. The projects aimed at the implementation of infrastructure, social services, and modern housing, the application of security measures, including the establishment of a security council, and finally, they involved the selective and often violent removal of families. Due to limited financial resources, changes of governments, bureaucratic and environmental problems, the projects have often been interrupted and could not be completed to the extent of the original plans.[8] Whereas resident associations of the surrounding wealthy districts had generally supported the attempts to remove the Varjão entirely and had mobilised against its fixation at the beginning of the 1990s (Lúcio, 2009, p. 48), when it became apparent that the entire removal was practically impossible, most of the Lago Norte's residents began to support the urbanisation projects of the government (see Medeiros, 2004, p. 69).

While the Varjão's stigma has resulted in many supposedly corrective reactions by the government, the residents of the Varjão have *both internalised the stigma and fought against it*. They have fought for the Varjão as their territory, as their place to live, and they have always sought to improve their living conditions. They have successfully fought against many 'corrective actions', while at the same time, they have accepted and supported some. That is to say, in the terms of Zibechi (2012, p. 198), that the marginalised are subjects and that they *can* achieve something.

Resistance and agency from below

Zibechi's notion of 'territories in resistance' (2012) is particularly useful to understand and conceptualise the Varjão's social movement and its territoriality until its fixation in 1991. In this section, I explain why this is before shedding light on the period beginning with the Varjão's regularisation and incorporation into the formal urban system, a period that saw agency and resistance becoming increasingly fragmented and absorbed by state led urbanisation projects.

The Varjão until 1991: an urban territory in resistance

The plan to remove the migrant workers from Brazil's Northeast to localities situated far outside the city had already been made when Brasília was still under construction. Yet, many workers had preferred to live close to their workplace and to avoid spending their little money and time on transportation (Medeiros, 2004, p. 10), so they started to occupy places in Brasília, building their homes and organising their lives together with others in the same situation. This was not only a huge act of territorialised resistance. With Zibechi (2012), we may understand the territoriality that the movement introduced as new and counterhegemonic:

The social relations within them and the subjects who form them are what make up the territories [. . .]. Those who produce space embody differentiated social relations rooted in territories. This is not limited to the possession (or ownership) of land, but is rather about the organization of a territory that has different social relations and that embodies the subject. It is the subject's need to give shape to counterhegemonic social relations that create the new territorialities.

(Zibechi, 2012, p. 208)

The author identifies five characteristics of such urban territories in resistance (p. 227), all of which apply to the Varjão. First is the movement of '*rural migrants* arriving in cities that are the centers of power for the dominant classes. The mass influx of rural population into the cities changes social, economic, and cultural relations [. . .]. This is a form of resistance to elite power and an affirmation of the popular world' (ibid; emphasis in the original). Brasília has been constructed as *the* centre of power. Recruited as work force to build the city and attracted to Brasília's new labour market, large parts of the poor rural population moved to the city – and stayed there, although the migrants were originally expected to return from where they came once the new capital was constructed. As a consequence, they challenged and changed the social, the economic, and the cultural relations of the envisioned capital right from the start.

Second, '[t]he spaces they construct (settlements, shantytowns, popular neighborhoods) are *different* from the traditional city of the middle and upper classes. They have been constructed differently (that is, collectively) and urban space is occupied and distributed differently within them (based on solidarity, reciprocity, and egalitarianism)' (ibid). Directly connected with this is the third characteristic of urban territories in resistance: 'Economic initiatives for survival arise in the popular territories that often take the form of a different, counterhegemonic economy' (ibid). Both aspects hold true for the Varjão. Though the area of the Varjão was relatively near to the migrant's work-place, it was also absolutely disconnected from the city, without any basic infrastructure or services provided by the government (IP2; see also Medeiros, 2004, p. 10). The social relations created by the pioneers where built upon solidarity and friendship, otherwise it would have been impossible to make a living in this place (IP2; Medeiros, 2004, p. 17). What Garcés has said about the occupation La Victoria (Chile), also applies here: the settlers had to 'work together and innovate, drawing on every bit of knowledge and all of their skills', which was 'an enormous exercise' (Garcés, 2002, p. 138).

Fourth, '[e]xplicit or implicit forms of popular power are born in these self-constructed spaces', where '[s]tate logic appears to be subordinate to popular/community logic' (Zibechi, 2012, p. 227). In the case of the Varjão, people organised forms of direct control of the territory and regulated the relations between them in their own ways. Although within the territory, state logic was apparently not hegemonic, the residents learned to act within state logic to develop and defend their territory. One of the most important steps in this direction was the foundation of the Resident Association (*Associação dos Moradores*) in 1982.

Through this representative group, the community was increasingly heard when demanding basic infrastructure that the government should provide. 'We fought for infrastructure, then came asphalt, sewage [. . .]. We founded the newspaper *Varjão em Ação* in order to receive special attention of the authorities [. . .]. We began to distribute this paper in the Senate, in the Congress, in the Legislative Chamber [. . .]' (IP3). As all of the interviewees stated, the situation became gradually better, but all stress that it was a difficult fight to lead:

> The improvement is the result of the residents' struggle [. . .]. Because of the discrimination our children suffered in the schools of the Lago Norte, we dedicated [ourselves] to education and achieved the construction of [our own] school. The nursery, it was [constructed of] wood, [it was then] extended, this is the community nursery today. Then we received electricity, street lighting [. . .]. Then came drinking water from the CAESB[9] that we didn't have before. Look, when we didn't have [something], we met and elaborated with the residents the best way to get it, to mobilise, to go to the Buriti Palace[10], we went there very often [. . .]. Everything we achieved was because of our struggle.
>
> (IP2, see also IP3, IP4, IP5)

Finally, Zibechi concludes that '[c]ontrol of these territories enables the urban popular sectors to resist, stay put, and survive even as the powers that be seek to break them, whether by disguising their differences, through co-optation, or by neutralizing initiatives' (Zibechi, 2012). Besides the control, we argue that in the case of the Varjão it was above all the common struggle for a place to live and the fight against the governmental attempts to remove the Varjão with the long-term objective to achieve legalisation that bounded the residents together.

When the settlement began to grow in the 1960s and 1970s, governmental surveillance was increasingly present in the Varjão. Medeiros (2004, p. 16) states that they 'threatened the residents with the destruction of their shacks and prohibited to construct new ones' (see also IP2). However, inhabitants did not disappear but stayed put, and, when necessary, reconstructed their homes. It is only when social pressure heightened as a result of the crisis and that the Federal District feared that it was losing control of urban dynamics in the neighbourhood, that the government began to attend to the population's demand for housing and better infrastructure. This was nonetheless performed in ways that involved attending to individual concerns as 'emergency' measures, rather than paving the way for the legalisation of the settlements (ibid., p. 17; see the following section). That is to say that, until the mid 1980s, the relation between the Varjão's residents and the government was more ambiguous and instrumental than consistent.

This relation changed with the government under José Aparecido de Oliveira (1985–88) and his attempt to eradicate the Varjão. At the time, the measures applied by TERRACAP and the police resorted to using violence, but once they had destroyed the shacks, they also removed all of the potential raw material that would be necessary for residents to rebuild. Against this background, resistance became much more difficult and required a much higher level of organisation and participation. Residents had to find food and accommodation for the residents who had lost their homes; they had to gather material to rebuild the houses; they had to organise the protection of belongings, especially during daytime, when many of them had to leave the Varjão for work; and as a consequence, a solution had to be found for those remaining in the Varjão for this purpose, and who could no longer go to work (Medeiros, 2004, pp. 55–67). Every resident participated in the resistance movement, because all of them depended on the group's support and help. The local resident associations (the *Associação dos Moradores* and the *Grupo União*, founded in 1988) played a key role in the organisation of the movement (Medeiros, 2004, p. 57). When three months later the new government under Joaquim Domingos Roriz (1988–90, 1991–95) came to power, the situation changed once again. His new housing policies opened up realistic perspectives for the legalisation of the Varjão. What followed were several years of meetings and negotiations with the government, as well as with the resident organisations of the adjacent rich neighbourhoods prior to the fixation and legalisation of parts of the settlement that were achieved in 1991.

The Varjão since 1991: urbanisation and fragmentation of the resistance movement

Zibechi (2012, pp. 189–263) shows that urban territories in resistance change over time and that they may break as counterhegemonic movements and territories, due to changing political, economic and social conditions. While the author does not analyse the evolution of such territories after they have been (re)conquered by the state, we want to show with our example of the Varjão which path such former urban territories in resistance may follow. Once (partly) regularised and as such incorporated into state logic, the state began to dissolve as the 'antagonistic Other' which had been threatening the Varjão's existence. At first glance, the objectives of the residents seemed to have been in line with the government's urbanisation projects and public policies in the Varjão. In fact, the residents achieved infrastructure and housing to a large extent through these projects. The problem has been that the way in which the projects have been carried out and implemented did and still does not always correspond to the residents' ideas, interests and social dynamics. However, the residents' critique could not be translated into unified resistance anymore. After the area was officially recognised, solidarity became gradually more fragile. People have resisted, but in a rather fragmented manner, and key actors of the movement were often co-opted by the government.

Over the years, the population of the Varjão grew rapidly. According to one of our interviewees it reached up to 12,000 inhabitants in its most crowded period (IP1). Many considered the 'thinning out' of the Varjão's population as necessary. This included the government and the residents of surrounding neighbourhoods, but this was also defended by inhabitants of the Varjão itself (IP2, IP4). From the government's perspective, this was the first essential step to transform the territory into an 'organised modern space'. The resident associations as well as many other inhabitants of the Varjão supported this measure, arguing that the Varjão was too small for so many people and that violence and criminality was far too high. Therefore, the government established together with the associations, criteria for those who should leave:

> So, what was done: families living in the Varjão for a long time, with many children, could stay. Those with less children and [members] indicted for criminal offences had to leave. This was decided by all [of us], and we suffered with this, because there were those who didn't want to leave [. . .]. There were residents, good people, who didn't harm anyone, and there were those damned people stuck in the middle of it all. [. . .] Then, [these] people went to many different [and distant] places.
>
> (IP2)

Many of those who had to leave according to the established criteria tried to resist and to stay, but they could no longer count on the vivacity of the Varjão's resistance movement as it was prior to the recognition and urbanisation of the settlements. Although people were pained at having had to take difficult decisions, all of our interviewees regarded the selective removal as necessary and, in the end, as good (IP2, IP3, IP4).

Additionally, the government used another measure to 'thin out' the Varjão's population and to 'clean' the neighbourhood from its so-called 'invasions' (IP5). At the beginning of the 2000s, it relocated those who lived in irregular housing within the entire Varjão to the neighbourhood's last block (the so-called Transition Area), promising to construct social housing for the families in this area. However, the government has not kept its promise meaning that the people of the Transition Area continued to live in self-constructed homes until 2013. During this period, the only measure that the government carried out in this area was to regularly destroy shacks when the area became too crowded because people from outside moved in (Lis, 2009). One year before the FIFA World Cup in Brazil had started, the government decided to 'clean up the Varjão again' (IP5), and the Transition Area in particular. The government promised to construct proper houses for the families in the Transition Area within nine months. It therefore relocated the majority of the inhabitants from there to distant satellite cities in the Federal District and provided a small rental subsidy (*Correio Braziliense*, 2013).

To date the government has not built the promised houses. As we write these words, the relocated families have been waiting for them; 'they are waiting, many families are suffering, with little children' (IP2, see also IP3, IP4, IP5). Initial attempts to take back the area were successfully stopped by the government. It then implemented further bureaucratic measures, extra fees, and new criteria that have weakened the former residents of the Transition Area. In turn, these residents founded the Association of the Transition Area to defend their demands. However, 'the government is complicating bureaucracy, [. . .] exacerbating things for us. So, what does the government do: 'Let's charge these people in order to make them give up.' And when we give up, they will sell [the land]' (IP5). On top of this strategy, the government has tried to exhaust the association by

playing the waiting game: 'The people of the government say 'not today, tomorrow. . . later. . .'' (IP5). And much as the residents who were removed within the
urbanisation projects of the Varjão, the Association of the Transition Area cannot
count on the solidarity of all other residents. As one of its members said, 'in the
Varjão, people are divided. Many want that these residents [from the Transition
Area] stay; many others want them to go away, because it's small here' (IP5).

Besides the selective removal of residents, another criticised aspect has been
the quality of government housing constructed within the urbanisation projects.
The outer appearance of the constructions was prioritised, as were low costs,
ahead of practicality, functionality and sustainability. As Calmon (2010, p. 90)
has shown, '[. . .] the urban projects for [the] Varjão were inspired by the modernist ideals of functionalism, particularly in term[s] of housing. The results were that
the infrastructure improvements and the pleasant aesthetics of the new constructions [has been] disconnected from the most relevant socio-economic dynamics
of the population' (see also Lúcio, 2007, 2009). One of the main issues has been
the small size of housing units, with their mere 30 m² and only one bedroom for
families, regardless of the number of members. The residents had accepted the
housing units because this was one of the conditions for being allowed to stay in
the Varjão. Yet, many residents resisted by building extensions to their homes
only one month after the handing over of the units. This was conducted without any order and was said to 'resemble once again the *favelas*' characteristics'
(Calmon, 2010, p. 91; see also Lúcio, 2007).

When we asked the Varjão's inhabitants about current public policies, many
of them stated that the government, these days, would not do anything (IP2, IP3,
IP4, IP5); 'it is the resident who does!' (IP4), especially when it comes to recrea-
tional activities (IP4, IP5). Besides infrastructural improvements, the government
would above all neglect public services, and education (IP2, IP3) as well as social
programs for youth in particular (IP2). The current neglect by the government has
been explained through a lack of willingness to truly engage with the area as well
as governmental changes. Of course, an additional explanation can be found in the
recent economic crisis and its consequences in Brazil. One might also assume that
the government's lack of recent involvement is due to the fact that it has fulfilled
its principal aims: the Varjão has received a minor 'facelift' and it is (thought to
be) safer than it was ever since the removal of residents and the establishment
of a successful Security Council (IP1, IP2, IP3, IP4). Last but not least, large
parts of the Varjão have been regularised. In other words, the neighbourhood has
been transformed from the counterhegemonic territory to a (more or less) modern
administrative region of the Federal District – integrated within state logic. This
way, the government has made the Varjão accessible for neoliberal development
and gentrification.

The Varjão: an urbanised and gentrifying neighbourhood

Neil Smith's (2002) idea of gentrification as a generalised urban strategy is
observable in many different places around the world:

A vast number of cities around the world, from Mumbai to Rio de Janeiro, from Santiago to Cape Town, from Buenos Aires to Taipei, are simultaneously experiencing intensive and uneven processes of capital-led restructuring with significant influxes of upper- and middle-income people and large doses of class-led displacement from deprived urban areas.

(Lees *et al.*, 2015, p. 441)

Though generalised, there is evidently no simple trajectory of gentrifying areas, but rather 'multiple gentrifications in a pluralistic sense' (Lees *et al.*, 2015, p. 442). In Brazil, gentrification processes are happening in urban centres as well as in valued locations in the periphery of large cities where an increasing demand puts pressure on the housing supply in urban areas. This is also a process that has begun to occur in the so-called invasions and *favelas*. While the existent bibliography addresses gentrification mainly in the *favelas* of Rio de Janeiro,[11] it is also observable in other cities, such as São Paulo or Brasília, where housing supply is equally a problem.

The gentrification process that the Varjão shares with other (former) irregular settlements in valued locations in the Brazilian capital almost always begins with governmental attempts at removing irregular settlements in their entirety. If such a removal turns out to be impossible, these settlements are subjected to state-led urbanisation programs, which do include regularisation, the implementation of infrastructure and social services, housing programs as well as security and pacification measures. However, these also include selective removals and the dispossession of residents. These public policies are often not based on democratic processes and the participation of affected residents, but on the cooptation of key leaders and movements, on non-transparent decision-making, and on measures that slow down and exhaust residential resistance movements, as has been the case in the Varjão. Typical of such urbanisation processes is the relative neglect of additional social programs and an emphasis on measures that enable property speculation (Cummings, 2015) – this has also been felt strongly by our respondents. The opportunities that the Varjão's urbanisation has created for speculation were described by one of the residents as follows:

Today, lots of residents have sold their land [. . .], and people from outside with greater purchasing power are buying, turning it into buildings, shops, and bedsits, to make money. There is speculation here. Those who formerly fought for their land, they sell it today. Where you have a large building, the owner isn't one of the Varjão's old residents, because the working class has difficulties. Look at my house here, with improvised roof tiles! The worker, the former resident, does not have financial resources to construct a building. These are people from outside who construct the buildings in order to rent.

(IP2, see also IP3, IP5)

Besides real estate speculation and commercial development, all the other key elements of gentrification – increasing numbers of middle-class newcomers, rising property values, rising costs of living, displacement of poorer residents – were mentioned in the interviews. 'There are lots of people from outside. There is a building here, and we even don't know any of its residents' (IP4, see also IP2). 'There are almost more newcomers than original residents here. The Varjão became attractive' (IP2). With the growing demand for land and housing by new-comers, property values and housing costs are increasing dramatically, affecting those who live in rented accommodations in particular. 'The rental rates have increased a lot, now, it's expensive, it's not cheap! A bedsit costs R$ 700 [per month] now, it seems. This is expensive!' (IP2, see also IP1, IP3, IP4, IP5). As a consequence, poor residents who cannot afford the growing costs are mov-ing away, in most cases to distant and cheaper satellite cities within the Federal District: 'They move away, to Itapoã, when rental costs became too high for them, you know?' (IP4).

The interviews showed that there is so far no significant movement against the gentrification processes. This can be explained, on the one hand, with the fact that some resistance movements died with the removal of large parts of the popu-lation and others have been weakened by specific governmental strategies. On the other hand, the interviews also revealed that many (old and new) inhabitants of the Varjão have welcomed the gentrification process. All of our interviewees regarded the Varjão's development in sum as very positive, even though some stated that it was sad to see people being removed (by force) and others moving away because of the rising costs. But from their perspective, the Varjão's story is a great success story, based to a large extent upon their own struggle, and has led to a much easier life, a much better image of the Varjão and to less discrimination. The effects of gentrification are perceived, but (still?) not regarded as *the* princi-pal problem, at least not by those who (still) live there.

> Today, I am a member of the Commercial Association [. . .], and as an activ-ist (*lutadora*) and resident for more than 35 years, I think, in a few years, there won't be many of the old residents left. This may be negative, but it may be positive, too. If you look: today, I have a small store, close to the residential buildings, this is good. . . Everything became nice.
>
> (IP2)

The word 'value', the idea that the Varjão is being valued and is growing eco-nomically are seen as something positive: 'It is valued [. . .] because there are many newcomers. [. . .] I think this is good, [because the Varjão] has the chance to grow.' (IP5).

> [The recent development] is good, because the city is growing [economically]. It's changing. . . the scenery, you can see all the nice buildings that were con-structed, you know? Those who are from here don't have the means to construct buildings, do they? So, they sell, buy a better house [elsewhere], and the person [from outside] constructs a building and brings income to this place.
>
> (IP4, see also IP5)

Ultimately, the respondents measured the positive effects of the resignification of Varjão and felt that they suffered less discrimination as a result of the gentrification process. According to them, the spatial stigma haunting the Varjão still exists, but it has been severely weakened. Using a typical expression, one respondent said: 'You don't feel it under your skin anymore (*Não se sente mais tanto na carne*)' (IP2).

Conclusion

The development of the Varjão may be understood as the outcome of the intermingling process of territorial stigmatisation and governmental (re)actions, resistance and agency from below, as well as gentrification. While existing studies on stigmatised, urbanised and/or gentrifying neighbourhoods in large Latin American cities often neglect the role of agency from below, we have argued that it has been fundamental for the Varjão's development.

As an irregular settlement until 1991, the Varjão may be understood as a 'territory in resistance' (Zibechi, 2012) whose inhabitants had to confront brutal governmental attempts to destroy the neighbourhood and remove the population. The social relations among residents of the Varjão were built upon solidarity and friendship in order to survive. The state as the 'antagonistic Other' strengthened these bonds among the inhabitants who fought together for the fixation of the Varjão, aiming for both a secure housing perspective and an improvement of their living conditions. Once the Varjão had begun to be regularised, urbanised and incorporated into state logic, the resistance movement evolved. The state ceased to be considered an 'antagonistic Other', and as a result, the bonds of solidarity among residents weakened, and the resistance movements gradually fragmented. Nevertheless, some movements continue to resist against specific governmental policies and resident agency is still fundamental for the Varjão's development today.

Although the effects of gentrification have become increasingly visible – including the fact that a growing number of the poorest members of the community have been expelled or chosen to leave due to the rising costs of living in the Varjão – there is so far no significant movement against it. In contrast, many residents welcome some of the key consequences of gentrification: whereas current public policies neglect their neighbourhood, newcomers and investors from outside bring economic development to the Varjão. One might ask why the interviewed did not reflect more critically on the gentrification process, while seeming somewhat blind about their own potential futures. To find an explanation, we have to consider that life in the Varjão when it was an irregular settlement has been described as extremely difficult. Whereas the concept 'territory in resistance' and social relations based upon solidarity and friendship sound beautiful, the other side of the coin was that 'we had nothing' (IP2) – in other words: extreme poverty as well as precarious and insecure housing conditions – and 'we were much more discriminated before than we are now' (IP2), being stigmatised as inhabitants of 'social hell'. The interviewees are probably well aware that urbanisation and gentrification are *not* 'heaven', even if they have described the contemporary Varjão using such attributes. Yet, for those who (still) live in the Varjão, urbanisation and gentrification (as processes in which many residents have played an active role in bringing about) have led to much easier living conditions, as well as to a resignification of their neighbourhood (at least to a certain degree), and with that, to less discrimination. The visible negative effects of gentrification seem to be more acceptable than the problems of the past. Obviously, those residents who had to leave the neighbourhood because of higher rents or those who have been forcibly removed will tell a different story (which will be addressed in future research).

In sum, the interviewed residents see the development of the urbanised and gentrifying Varjão as positive and to a large degree as the outcome of their own fight and struggle in which the government has taken on an ambiguous role between support, neglect and confrontation. The follow-up question is, how far are these findings generalisable and applicable to other urbanised and gentrifying neighbourhoods in Latin American cities. Lees, Shin and Lopéz-Morales (2015) observe 'that urban spaces around the world are increasingly subject to global and domestic capital (re)investment to be transformed into new uses that cater the needs of wealthier inhabitants' (p. 442). With respect to irregular settlements in valued urban locations in many Latin American cities, this can either entail the entire removal of these constructions, or their regularisation and urbanisation, with an emphasis on measures enabling land and property speculation as well as the displacement of the poorest residents (Freeman, 2012; Cummings, 2015). Our hypothesis is that in such neighborhoods resistance and agency from below is always fundamental. Even after the start of processes of regularisation and urbanisation, residents' agency and resistance continue to play a key role in the development of the neighbourhood. Yet their relationship to the state also evolves when understandings that it is some antagonistic institution dissolve. The extent to which this altered perception results in the fragmentation of resistance, and a change in *how far* residents are willing to resist, are questions which call for further empirical research. As are questions pertaining to how far residents welcome and even support gentrification in areas like the Varjão. Such an agenda would not only pay tribute to Zibechi's call to approach those who dwell in stigmatised districts as subjects, but also contribute to the development of urban theory that decentres knowledge production from the so-called Global North.

Notes

1 The name has been altered in order to guarantee anonymity.
2 All translations from Portuguese into English have been made by Shadia Husseini de Araújo.
3 See http://www.correiobraziliense.com.br/busca/?c=busca&q=Varj%C3%A3o, accessed 28 October 2015.
4 To guarantee anonymity of the interviewees, we use the abbreviation IP for interview partner and the numbers 1–7 according to the chronological order of the interviews we conducted with different residents of the Varjão.

5 Here, the terms 'urbanised' and 'urbanisation' are literal translations from the Portuguese words *urbanizado* and *urbanização*, which refer to the regularisation of urban squatter settlements, to the incorporation of their territories into state logic, and to the opening up of access for neoliberal urban development.
6 For an anthropological critique of Brasília as a modernist city see Holston (1989); see also Paviani (1996, 2010a, 2010b), Costa (2012) and Costa *et al.* (2013).
7 For detailed accounts of the Varjão's history see Medeiros (2004) and Lúcio (2007).
8 See Lúcio 2007, 2009; and Calmon 2010 for an overview and a critical discussion of the programs.
9 *Companhia de Saneamento Ambiental do Distrito Federal*, the Environmental Sanitation Company of the Federal District.
10 Seat of government of the Federal District.
11 This is especially the case in the context of the FIFA World Cup 2014 and the Olympic Games 2016; see e.g. Freeman, 2012; Tadini, 2013; Cummings, 2015; Steinbrink *et al.*, 2015.

References

Asensão, E. (2015), 'Slum gentrification in Lisbon, Portugal: Displacement and the imagined futures of an informal settlement', in Lees, H., Shin, L. and López-Morales, E., (eds), *Global Gentrifications: Uneven Development and Displacement*, Bristol: Policy Press, pp. 37–58.

Bayón, M. C. (2012), 'El "lugar" de los pobres: Espacio, representaciones sociales y estigmas en la ciudad de México', *Revista Mexicana de Sociología*, 74(1): 133–166.

Betancur, J. J. (2014), 'Gentrification in Latin America: Overview and critical analysis', *Urban Studies Research*, 2014: 1–14.

Calmon, T. C. D. P. (2010), *Advancing Social Sustainability in Brazil: Planning Strategies for Socio-Spatial Integration in Varjão, Brasília*, Master's Thesis: The Pennsylvania State University.

Cornejo, C. A. (2012), 'Estigma territorial como forma de violência barrial. El caso del sector El Castillo', *INVI*, 27(76). Available online at: http://www.scielo.cl (accessed 1 November 2015).

Correio Braziliense (2013), 'GDF inicia operação de desocupação de casas de 370 famílias no Varjão', 21 February. Available online at: http://www.correioBraziliense.com.br/ (accessed 1 November 2015).

Costa, E. B. d. (2012), 'Intervenções em centros urbanos no período da globalização', *Cidades*, 9(16). Available online at: http://revista.fct.unesp.br/ (accessed 1 November 2015).

Costa, E. B. d., Silveira, B, Severo, D., Araújo, E., Beserra, F. and Carmo, T. (2013), 'Metropolização, patrimonialização e potenciais de conflitos socioterritoriais em Brasília (DF)', *Espaço & Geografia*, 16(1): 325–367.

Cummings, J. (2015), 'Confronting favela chic: The gentrification of informal settlements in Rio de Janeiro, Brazil', in L. Lees, H., Shin, L. and López-Morales, E. (eds), *Global Gentrifications: Uneven Development and Displacement*, Bristol: Policy Press, pp. 81–100.

Freeman, J. (2012), 'Neoliberal accumulation strategies and the visible hand of police pacification in Rio de Janeiro', *Revista de Estudos Universitários*, 38(1): 95–126.

Garbin, D. and Millington, G. (2012), Territorial stigma and the politics of resistance in a Parisian banlieue: La Courneuve and beyond', *Urban Studies*, 49(10): 2067–2083.

Garcés M. (2002), *Tomando su Sitio: El Movimiento de Pobladores de Santiago. 1957–1970*, Santiago: LOM.

176   *Shadia Husseini de Araújo and Everaldo Batista da Costa*

Herzer, H., Di Virgilio, M. M. and Rodríguez, M. C. (2015), 'Gentrification in Buenos Aires: Global trends and local features', in Lees, H., Shin, L. and López-Morales, E. (eds), *Global Gentrifications: Uneven Development and Displacement,* Bristol: Policy Press: 199–222.

Janoschka, M., Sequera, J. and Salinas, L. (2014), 'Gentrification in Spain and Latin America: A critical dialogue', *International Journal of Urban and Regional Research,* 38(4): 1234–1265.

Jones, G. A. and Varley, A. (1999), 'The reconquest of the historic centre: Urban conservation and gentrification in Puebla, Mexico', *Environment and Planning A,* 31(9): 1547–1566.

Jones, G. (2015), 'Gentrification, neoliberalism and Loss in Puebla, Mexico', in L. Lees, H. Shin, L. and López-Morales, E. (eds), *Global Gentrifications: Uneven Development and Displacement,* Bristol: Policy Press: 265–284.

Kirkness, P. (2014), 'The *cités* strike back: Restive responses to territorial taint in the French banlieues', *Environment and Planning A,* 46(6): 1281–1296.

Lees, L., Slater, T. and Wyly, E. (2008), *Gentrification,* London: Routledge.

Lees, L., Shin, H. and López-Morales, E. (2015), 'Conclusion: Global gentrifications', in Lees, L., Shin, H. and López-Morales, E. (eds), *Global Gentrifications: Uneven Development and Displacement,* Bristol: Policy Press: 441–452.

Lis, L. (2009), '31 barracos são derrubados no Varjão' *Correio Braziliense,* 3 September. Available online at: http://www.correioBraziliense.com.br/ (accessed 1 November 2015).

Lúcio, M. d. L. (2007), *Nova Periferiazação Urbana. Políticas Públicas com financiamento internacional e o impacto nos direitos sociais,* Doctoral Thesis, University of Brasília.

Lúcio, M. d. L. (2009), 'O lugar da juventude na Vila Varjão: Política pública de intervenção urbana integrada e implicações educacionais', *Revista Brasileira de Gestão Urbana,* 1(1): 43–60.

Medeiros, C. d. S. (2004), *Viver e resistir: Luta por moradia. Vila Varjão 1961–1988,* Master's Thesis, University of Brasília.

Paviani, A. (1996), *Brasilia: Moradia e Exclusão,* Brasília: Editora Universidade de Brasília.

Paviani, A. (2010a), *Brasília: A Metrópole em Crise: Ensaios sobre Urbanização,* Brasília: Editora Universidade de Brasília.

Paviani, A. (2010b), *Brasilia. Ideologia e Realidade, Espaço Urbano Em Questão,* Brasília: Editora Universidade de Brasília.

Rodrigues, A. O. and Rodrigues, L. (2014), 'Estigmatização territorial na Cidade de Montes Claros – MG', *IV Congresso em Desenvolvimento Social. Mobilidades e Desenvolvimentos.* Available online at: http://www.congressods.com.br (accessed 1 November 2015).

Sakizlioglu, N. B. and Uitermark, J. (2014), 'The symbolic politics of gentrification: The restructuring of stigmatized neighborhoods in Amsterdam and Istanbul', *Environment and Planning A,* 46: 1369–1385.

Sibrian, N. (2015), 'Medios de comunicación, violencia delictiva y estigma territorial en Venezuela', *Anagramas,* 14(26): 95–114.

Slater, T. (2006), 'The eviction of critical perspectives from gentrification research', *International Journal of Urban and Regional Research,* 30(4): 737–757.

Slater, T. and Anderson, N. (2012), 'The reputational ghetto: Territorial stigmatisation in St Paul's, Bristol', *Transactions of the Institute of British Geographers,* 37(4): 530–546.

Smith, N. (2002), 'New Globalism, New Urbanism: Gentrification as global urban strategy', *Antipode*, 34(3): 427–450.
Steinbrink, M., Ehebrecht, D., Haferburg, C. and Deffner, V. (2015), ‚Strategische Interventionen und sozialräumliche Effekte in Rio de Janeiro', *Sub\urban: Zeitschrift für kritische Stadtforschung*, 3(1): 45–74
Tadini, N. (2013), *Chasing the Real: The Gentrification of Rio de Janeiro's Favelas*, Master's Thesis, Roskilde University.
Wacquant, L. (2007), 'Territorial stigmatisation in the age of advanced marginality', *Thesis Eleven*, 91: 66–77.
Wacquant, L. (2008a), 'Ghettos and anti-ghettos: An anatomy of the new urban poverty', *Thesis Eleven*, 94: 113–118.
Wacquant, L. (2008b), 'The militarization of urban marginality: Lessons from the Brazilian metropolis', *International Political Sociology*, 2(1): 56–74.
Wacquant, L., Slater, T. and Pereira, V. B. (2014), 'Territorial stigmatisation in action', *Environment and Planning A*, 46(6): 1–11.
Zibechi, R. (2012), *Territories in Resistance: A Cartography of Latin American Social Movements*, Oakland, CA: AK Press.

11 Researching territorial stigma with social housing tenants

Tenant-led digital media production about people and place

Dallas Rogers, Michael Darcy and Kathy Arthurson

Introduction

This chapter considers how social housing tenants produce digital countercultural products to talk about, represent and analyse people and place relationships, and particularly how they used these products to explore territorialised representations of poverty and class. We draw on three examples from the *Residents' Voices – Advantage, Disadvantage, Community and Place* project (hereafter Residents' Voices): (1) digital story telling disseminated through a website; (2) tenant-driven media analysis of the popular Australian television parody '*Housos*'; and (3) a short dramatic film written and directed by social housing tenants. Each example uses digital media production to represent, and perhaps even challenge, territorial stigma, but represents social housing tenants and their neighbourhoods in different ways. The aim is to expose the methodological challenges within each digital cultural production process in relation to representations of territorial stigma. Bourdieu (1986) has shown how social order is inscribed through 'cultural products'. These products include education, language and the media. Cultural products work through framing and reworking alliances over culture both symbolically and materially. This leads to an unconscious sense of acceptance of social differences and one's place in society both in a social/cultural and geographical/spatial sense. In other words, through these cultural products meanings are attached to certain practices, places and events and these meanings are internalised even by those who themselves are being culturally defined.

Goffman's (1986) seminal work on stigma grouped the concept into the three categories of abominations of the body, blemishes of individual character, and tribal stigma (race, nation, and religion). From Goffman's perspective stigma arises through negative labelling and stereotyping of people who are depicted as possessing discrediting attributes, which leads to a 'spoiled identity'. Wacquant (2007) argues that a key omission in Goffman's (1986) thesis is a link to 'blemish of place' or a discredited neighbourhood reputation, which leads to what he terms 'territorial stigma'. From this perspective analogous to the situation of tribal stigma, territorial stigma can project a virtual social identity on families and individuals living in particular neighbourhoods and thus deprive them of acceptance from others. Place and person become intertwined in negative representations although these may well conflict with tenants' own realities. Consequently, 'blemish of place' can add an additional layer of disadvantage to any existing stigma that is associated with people's poverty, culture, or ethnic origins. In this way, others, often outsiders, construct the community identity, and therefore stigma is associated not just with individual persons but also with the geographical spaces in which they live.

The consequences of 'territorial stigma' include, but are certainly not limited to: discrimination by employers on the basis of postcode, address, or other spatial markers (Bradbury and Chalmers, 2003; Ziersch and Arthurson, 2005); changes to the nature and quality of service provision (Hastings, 2009; Pawson *et al.*, 2015); the disposal of social housing to the private market so as to disperse stigmatised neighbourhoods resulting in reduction of social housing stock (Darcy and Rogers, 2015; Rogers and Darcy, 2014); and impacts on residents' health and well-being and mental health in particular (Dufty(-Jones), 2009; Kelaher *et al.*, 2010). In Australia, Warr (2005) draws particular attention to the role of television and other media whose 'negative . . . attention amplifies and cements the quotidian prejudices that are experienced by people living in 'discredited' neighbourhoods'. Warr (2005) concludes that, while global economic forces and government policy intervention are important mediators of territorial stigma, the 'unwarranted and unsympathetic attitudes and actions of outsiders . . . are key contributors to the difficulties of those living in stigmatised neighbourhoods' (p. 19).

Indeed, digital media is a key medium through which distinctions of class and territorial stigma are shaped, imposed and reproduced. Television and other digital media are easily accessible through 24-hour Internet, so its realm is pervasive. In Australia, as elsewhere (Arthurson *et al.*, 2014; Devereux *et al.*, 2011; Hastings and Dean, 2003; Warr, 2005) scholars have shown that:

> The media has played an active role in supporting and embellishing patho-
> logical depictions of social housing estates as sites of disorder and crime,
> drawing on explanations that cite individual agency and behaviour as the
> problems.
>
> (Arthurson, 2012, p. 101)

Stressed urban communities are frequently sought out by the media to set 'night-marish portrayals of urban life' that may serve or extend negative stereotypes. In the end it matters little if these localities in fact are, or are not, run down and dangerous places, and their populations comprised essentially of minorities and poor people, 'the prejudicial belief that they are suffices to set off socially noxious consequences' (Wacquant, 2008, p. 239).

It is a mistake, however, to view 'the media' as solely responsible for the rep-resentations of people and place they produce. For example, the television shows that are sometimes referred to as 'poverty porn' (Jensen, 2014), which include alarmingly exaggerated and often territorialised portrayals of poverty, are the product of complex cultural production processes that draw together different narratives about people and place. Warr (2016) identifies at least three symbiotic components to this complex cultural production process: poverty news, poverty stories and poverty research (i.e. research data about poverty). Social researchers and the organisations that fund their research are not separate from the media's cultural production of stories or news about people and place. Rather, the research process and research data *itself* are cultural products. For example, the latest release of unemployment statistics for a particular low-income suburb might contribute to the production of news stories about poor places and people, as well as other, at times more problematic, televised dramatic portrayals. It is often the case that researchers who are working with social housing communities are required to recount and construct these familiar narratives about low-income people and places to secure research funding. These funding narratives require researchers to focus on deficits, and researchers – much like journalists and television producers – regularly cite demographic features, such as high levels of unemployment and incarceration, or low average incomes for a suburb, as evidence of people and place poverty and, therefore, as a rationale for research funding. Thus, methodologies that create the type of data, which contributes to territorial stigma, might be more likely to secure grants and other funding success than those deployed in attempt to challenge the construction of categories through which territorial stigma narratives are produced. The latter, we argue, should be a key concern for housing and urban scholars. Slater's (2014, p. 955) scholarship is, therefore, refreshingly critical, arguing this type of research provides 'the evidence base' that appeases funding bodies while 'buffering politicians and their audiences from viable alternatives and inoculating them against the critique of autonomous scholarship'.

Research from Australia, in addition to international studies, explores the disjunct that often exists between media representations of social housing estates and the lived experience of tenants (Hastings, 2004; Arthurson *et al.*, 2014; Wacquant, 2007). Lapeyronnie (2008), for instance, identified the tension between internal self-perceptions of the French *banlieue* experience and external images. Similarly, the often-cited suite of research that represents estates, for example, by way of unemployment, income status, school retention and crime rates, may be poorly aligned with the reported lived experience of tenants. Residents of stigmatised places bemoan the fact that researchers, housing authorities and the media, and particularly news and current affairs programs, stigmatise their neighbourhoods and occupants, often without even having visited the area or knowing the people (Lapeyronnie, 2008). An alternative standpoint is that mainstream media and academia can be recruited to challenge negative perceptions of estates (Jacobs *et al.*, 2011; Hastings and Dean, 2003). In a time of sensationalist and xenophobic media discourse, and under highly rationalist research funding schemas that position 'objective' quantitative research data as more valid than 'subjective' qualitative research data, this can be hard to achieve in practice. Recording the subjective experiences of tenants is important, because acceptance of the negative stereotype invalidates the legitimacy of any claims upon place making by social tenants in identified 'poor' urban areas. These negative stereotypes can be, and often are deployed to legitimise redevelopment of such areas through forced relocation and disposal of public assets (Darcy and Rogers, 2015). Current urban studies debates offer limited engagement with micro-scale analyses of the 'creative destruction' of estates, or, alternatively, with the 'creative potential' of local communities and social tenants themselves who are typically viewed as either passive beneficiaries or victims of redevelopment. Researchers need to pay more attention to the cultural production and intersection of the narratives about social housing tenants and estates, across a broader range of discursive modalities.

Residents' Voices project and collaborative research with tenants

Faced with threats of the demolition and redevelopment of their dwellings, dispersal of tenants and communities (Darcy and Rogers, 2015), and with persistent stigmatisation and demonisation in mainstream media (Arthurson *et al.*, 2014; Jacobs *et al.*, 2011; Wacquant, 2008), some of the tenants of social housing that we have been working with as part of Residents' Voices in Australia have used video and other digital media to create alternative cultural products. Residents' Voices was a four-year project that was funded by the Australian Research Council, St Vincent de Paul Society, Western Sydney University and Loyola University Chicago. The aim of Residents' Voices was to collaborate with tenants to challenge conventional outsider approaches to understanding place and disadvantage by facilitating the emergence and validation of situated knowledge and 'insider' theorising about this relationship. Residents' Voices was broadly guided by action research and digital media production (see Rogers *et al.*, 2012).

Although participatory research has become increasingly common, particularly following the crisis of classic anthropology with its emphasis on outsider knowledge production, the actual role of participants in studies varies greatly. Biggs (1989, cited in Rowe, 2006) suggests four levels of participation: 'contractual' (researchers contract for services from local people); 'consultative' (local people are asked for their opinions or advice about the research); 'collaborative' (researchers and local people work together on a study that is designed, initiated and managed by institutional researchers); and 'collegiate'. In our *collegiate participatory research*, the academics and tenants worked together as colleagues for mutual learning and to develop a system for independent research among local people. Collegiate participation presents the greatest challenge to the university sector which views itself as the holder of expert knowledge. Nevertheless, Residents' Voices aimed for the collegiate approach and the full determination and active involvement in management of the knowledge production process by the lay tenant researchers. Significantly, beyond the local collegiate approach, Residents' Voices was concerned with the sharing of experiences, reflections and understanding, and most particularly the learning and new knowledge produced at the *intersections of local knowledges*. Consequently, Residents' Voices was designed around cross-cultural knowledge exchange and production. This approach draws on the work of

Gumucio-Dagron (2008, p. 5) who argues that the process of knowledge creation is 'dynamic', and that 'individuals and communities start with the knowledge they already have, and put it in dialogue with the information they receive from other sources.' Such a process rests on the fact that cultures are not closed and insular, 'but living bodies of knowledge and experience that are constantly undergoing evolution and social transformation' (Gumucio-Dagron, 2008, p. 5)

The remainder of this chapter is organised into three 'Acts' to discuss these cultural production processes within Residents' Voices, and it makes use of the metaphor of the storytelling in a play as a way of organising the empirical cases. Each act describes a project undertaken under the banner of Residents' Voices, and analyses the response to stereotyping and stigma. Acts 1 and 2 focus on a digital storytelling project (Act 1) and the 'Housos' research study (Act 2), which were components of Residents' Voices. Act 3 covers a project that tangentially emerged from Residents' Voices, the 'Lost in the Woods' film project conducted by the Woodville Community Centre with tenants in Western Sydney. We show how Residents' Voices provided inspiration and conceptual guidance for this project without having any formal or practical role in the making of the film. This guidance took the form of digital storytelling project planning documents for tenants on the Residents' Voices website. Importantly, the Lost in the Woods project was independently conceived of and managed by tenants who were supported by two non-government organisations. In each of these examples, social housing tenants, researchers and non-government organisations speak back to popular negative stereotypes. Yet, for tenants, in each case their main purpose was not primarily to influence public perceptions or the policy agenda, but rather *to reclaim and reinforce their own identity and connection to place.* By comparison, a key finding from Residents' Voices is the observation that when we as researchers were prepared to step back from controlling the research process, tenants and local community organisations were not only willing to initiate their own projects, *they produced more complete and effective counter-cultural products when freed from academic framing and constraints.* Therefore, we conclude with some conceptual reflections about the methodological challenges we experienced or recorded while researching territorial stigma with tenants as co-researchers in Residents' Voices.

Act 1: Digital storytelling project

Residents' Voices was designed to create opportunities for social housing tenants to develop and express their own knowledge and understanding of the links between place and disadvantage in their own terms (Darcy and Gwyther, 2012). The methodology for Residents' Voices placed tenants at the centre of the research process, and included encouraging tenants to frame the research questions, undertake empirical research tasks, analyse any 'data', and publish and report the data in formats deemed appropriate by tenants. The project created a space for tenant-led projects to develop and one such project, the *Housos* study, is outlined later in this section. At the outset, to initiate the formal start of Residents' Voices we organised a suite of smaller collective storytelling projects with tenants who were already engaged with the project. One of these smaller projects was the Residents' Voices digital storytelling project.

Digital storytelling is a narrative-driven form of digital media production that allows people to share aspects of their life by making a short film, audio or photographic essay. While digital storytelling is a relatively new visual methodology within academia, it has a much longer history in the media and the corporate spheres (Lovejoy and Steele, 2004, p. 72). With the addition of increasingly affordable high quality digital media tools, such as photography equipment, video cameras and voice recorders, and online publication tools, digital storytelling is a performative practice that can be undertaken by almost anyone. As a digital media practice, digital storytelling is very diverse and might produce short radio documentaries, photographic essays, participant-directed autobiographical films or stop motion animation stories.

For Residents' Voices we commissioned Information and Cultural Exchange (ICE, http://ice.org.au/), a community arts organisation in Western Sydney, to run digital storytelling workshops as a capacity building process for both university researchers and tenant co-researchers. Lundby (2008) argues that increasing access to the Internet, low cost or free software and the rise of social media have allowed some marginalised groups to redeploy 'the age-old practices of storytelling' (p. 1) to self-represent their own social experience. Therefore, we wanted the tenants to not only create a digital story through the workshops, but more importantly, to acquire the skills and knowledge to create and teach others how to create additional digital stories in the future. This workshop involved five tenants and three university researchers (one of whom was also a resident as outlined below). We conducted two technical sessions in a film studio space and four content development sessions in a computer room at the local library near the tenants' homes. In the first two sessions the digital storytelling facilitators conducted classes on 'talking about personal stories', 'storyboarding for narrative development', 'using digital recording equipment', and 'using digital editing software'. In the four content development sessions, the tenants and researchers drafted their own personal narrative, and recorded it on a voice recorder. They also collected a suite of photographs to match their audio narrative. With help from the digital storytelling facilitators, the tenants and researchers produced their digital story by building a photographic essay over the top of their oral narrative using movie-editing software. The tenants'[1] stories covered topics including living in social housing with a mental illness, criminal activity and violence, interactions with law enforcement agencies, experiencing and addressing both personal and geographical stigma and living in social housing with family members with complex needs.

Guillemin and Drew (2010, p. 175) describe the academic digital storytelling process as follows, 'participants are asked by the researcher to produce photographs, video, drawings and other types of visual images as research data'. However, we set out to challenge the relationship between the researcher and the researched. In our first digital storytelling project we found that the production and consumption of the digital cultural products was more dynamic and contested than Guillemin and Drew suggest. Participant-generated visual methodologies are not bound by the constraints of positivist empirical research frameworks, which define clear roles for the researcher and the researched. Indeed, one of the (PhD candidate) researchers from Residents' Voices participating in this workshop was at the time a social housing tenant living in the neighbourhood alongside the other participants. Thus our digital storytelling project involved multiple stakeholders with complex identities that very clearly called into question the researcher/researched dichotomy. At the request of tenants in our project we exposed the 'researchers' to the same digital storytelling process as the tenants, and the 'researchers' created digital stories alongside the tenants about their experiences with researching and creating 'data' about social housing. The researchers' stories covered ethical questions about conducting research on social housing with tenants as co-researchers. Our project became a collective process of knowledge creation whereby all the participants became autobiographical researchers.

Much of the literature on digital storytelling is focused on the way the participants have sought to deploy their stories to *talk back to*, or to *talk up to*, audiences of power (see the edited volume by Lundby, 2008). Others show how the storytellers have used their narratives tobb pursue 'transformations' of the social, cultural or political 'context in which it operates' (Lundby, 2008, p. 10). Residents' Voices had a different target audience in mind, and a cultural production process that involved more nuanced forms of political action. The storytellers had complex motivations for using digital storytelling and when freed from the constraints of external mediators, including researchers, the stories that they choose to tell challenged our assumptions about their political motives for being involved in the project. In many cases, the people making digital stories do so with the aim of sharing their story with a particular audience (Lovejoy and Steele, 2004). In Residents' Voices, the digital storytelling project participants were offered the chance to share their stories with other tenants, in locations around Australia and in the United States through publication on a website. Some tenants decided to share their stories, others did not. The communicative aim was focused on horizontal connections between tenants rather than, at this stage of the project, of speaking directly to the powerful. However, participants did not all share the same understanding of the purpose or impact of their cultural products, and two tenants in particular wanted to get their hands dirty with more 'academic' social research.

Act 2: Housos project

The term 'houso' has long been in common use amongst Australian social housing tenants, signifying identification with a common community experience. The term doubtless has wider currency and forms part of the stigmatising language used by non-housos, and it took on an unequivocally pejorative tone in the 2011 television show *Housos*. This show is a satirical parody about the daily life of tenants in the fictitious 'Sunnyvale' social housing estate (Arthurson *et al.*, 2014). In the Australian context, the use of the term housos as the title of the program immediately identifies a subject that is associated with very specific and well-defined urban spatial localities, evoking well-rehearsed and exaggerated stereotypes and popular perceptions concerning a jobless underclass. Even before the show was aired, the group of social housing tenants we had been working with approached Residents' Voices to collaborate on a research project focussed on tenants' perceptions of, and reactions to, the television show. The digital storytelling had built a collaborative relationship between the tenants and researchers to build upon.

Housos is a highly embellished representation of Australian social housing estates as lawless zones where people act outside of the law and common norms of society. In this depiction 'housos' is a proxy for an 'underclass' that is explicitly spatialised through clearly recognisable signifiers that identify tenants of specific urban spaces. The depictions of the social housing tenants draw on overdrawn but common caricatures and stereotypes. Characters such as Dazza, Shazza and Franky are portrayed as feckless individuals, who shun work, survive on welfare benefits, indulge in substance abuse, routinely commit crimes and cause generalised disorder. Highly dysfunctional families and relationships surround them. On the spectrum of Australian television programming *Housos* pushes the boundaries of mainstream televised comedy and attracts a relatively small, but devoted, audience following. Three key subject–concept relationships dominate the construction of the houso in the show. These are the housos' relationships to employment, criminal activity, and drug use (Arthurson *et al.*, 2014). *Housos* joins a growing list of television programmes whose central themes rework conventional concepts of class distinction, such as another highly popular Australian comedy programme *Kath and Kim* (Davis, 2008). Other programmes in this genre that attempt to portray a 'postindustrial underclass' include the UK-produced *Shameless* (Creeber, 2009), or 'documentaries' such as *Benefits Street* (based in Birmingham, UK) and *The Scheme* (based in Kilmarnock, Scotland). Recent UK work on cultural representations of class, drawing on Bourdieu (1986), has identified the dominant contemporary depictions of the working class in the media as based on ridicule, disgust (Lawler, 2005), and mockery (Raisborough and Adams, 2008) claiming that disgust is winning out (Lawler, 2005: 443). These representations of the working classes are used as part of the processes of maintaining middle-class distinction, authority, and security (Wacquant, 2007; 2008). Promotional materials provided by the producers of *Housos* include a satirical 'dictionary definition' of a houso that points towards similar class distinctions and derogatory representations of social housing tenants:

> *houso* [how-zo], Informal: Often Disparaging. noun: 1. an uneducated person who lives in social housing. 2. a bigot or reactionary, especially from the urban working class. Adjective: 3. also, Housoish, narrow, prejudiced, or reactionary: a Houso attitude.
>
> (Superchoc Productions, 2011, p. 1)

As noted above, the media is a key medium through which distinctions of class and territorial stigma are shaped, imposed, and reproduced. Thus, the Housos research commenced from acknowledging the importance of directly involving those being stigmatised and experiencing territorial stigma of place in all phases of the study design, implementation, and analysis. As social scientists who do not belong to this stigmatised group we recognised that the lived experiences of those being stigmatised may well reflect very different personal perspectives to our own. Indeed, the research was thought up and developed by the tenants themselves. The questions and methods for the Housos study emerged after the tenants asked us to organise a screening of the first episode of *Housos* at an inner-city social housing estate for an audience consisting of social housing tenants and community workers from across the greater Sydney metropolitan area. This was followed by hosting discussions with a panel of experts comprised of social housing tenants, including those who had raised concerns about the show. Audience responses to the programme varied on a continuum, with some 'enjoying the show' and others expressing the viewpoint that the stereotypes drawn on in the programme would 'reinforce the stigma attached to social housing' (Arthurson *et al.*, 2014). The discussion and question and answer (Q&A) session that followed resulted in a group of tenants developing a set of research questions to further investigate this issue. The themes of the questions encompassed: the role and focus of satire in society; the wider public's conceptualisations of social housing estates; stigmatisation of residents of estates by the media; narrow and prejudiced understandings of social housing; and the dangers of 'glamorised' portrayals of disadvantage in the media.

These questions were then taken up in the tenant-led research project conducted over the nine-week first season of *Housos*. Two tenants who had been working closely with Residents' Voices, Ross Smith[2] from Central Sydney and Peter Butler from Western Sydney, joined us as tenant-researchers and recruited tenants from their local area to participate in their study. Residents' Voices provided institutional and research assistance during recruitment and throughout the project. The tenant–academic research team then recruited tenants from Adelaide (South Australia) and also non-tenant viewers of the show to participate in the research. Each week the 19 participants were sent an episode of *Housos* on DVD with a set of research questions. Participants watched each episode in their own time and responded to each week's questions by writing or recording an audio or video diary. The audio and video diaries were often recorded on a mobile phone while the written diaries were sent by email.

The tenant-academic research team wanted the analytical framework to enable social housing tenants access to specific tools, including media and research resources. This was, at first, in order to contest derogatory and stigmatised narratives of social housing estates that so often go unmediated and unchallenged, especially by tenants (Hastings, 2004; Jacobs *et al.*, 2011; with some notable exceptions, e.g., Darcy and Rogers, 2015). Second, we wanted to enable participants to identify any counter narratives that emerged. The tenant members of the research team accordingly informed the selection of a theoretical framework for the analysis. Peter Butler, one of the tenant-researchers managing the study, remarked in the Q&A session at the end of the screening on the inner-city estate about *Housos* that:

> It strikes me that the programme is a bit like a mirror. And it depends who's holding the mirror and which direction it's pointing towards . . . . I think all this show does, it reflects back a lot of the stereotypes that the public already has about people who live in social housing and it provides a convenient, sort of, stereotype or image up there on the screen, to help the public dump all their negative perceptions on these characters.
>
> (Arthurson *et al.*, 2014, p. 1339)

The academic researchers (the three authors) on the team felt that it was important not to impose an analysis onto tenants. Thus the research team decided to hold a final focus group to conclude the study whereby the two tenant-researchers, a tenant community worker, and tenant participant from the study, reviewed and interpreted participants' contributions including tenant and non-tenant diaries. Some tenant researchers worried that *Housos* might provide a symbolic vehicle that will organise representations of Australian social housing tenants' experience well into the future, so they wanted the voices of tenants' to be heard in response to the show. In our presentation of findings we provided extensive quotes as representative of some of the key themes that emerged from the qualitative material – especially from the focus group – in order to convey these voices as directly as possible (see research findings in Arthurson *et al.*, 2014).

At the conclusion of the focus group, we had undertaken a thematic analysis of the tenant and non-tenant diaries and developed a set of broad tenant-driven research findings. The two tenant-researchers felt that it was important to disseminate the research in both academic and general media publications. In the first instance, they asked Residents' Voices to join them in writing up the research for presentation at an academic conference. Soon after, we wrote up the study collaboratively and presented it at the Australasian Housing Researchers Conference (Rogers *et al.*, 2012). Residents' Voices funded the travel and conference costs of tenants. One of the tenants also produced a number of publications for tenant newsletters and industry journals from this study. Additionally, the Residents' Voices team prepared an academic article for publication to meet our funding and research institution requirements.

Act 3: 'Lost in the Woods' film project

Our third example of resident cultural production has it roots in the first Residents' Voices digital storytelling project, but the multi-directional transfer of knowledge in this case was by no means direct. While Residents' Voices was drawing the digital storytelling project to a close, the residents and a non-government organization on the Villawood East social housing estate, 25 km west of central Sydney, made plans for their own 'Residents' Voices Project'. Although the group asked for approval to use the name on the final product, this project had no formal institutional relationship to Residents' Voices. It provides revealing insights about the emancipatory power of participant-driven visual methodologies, especially when these practices are freed from the disciplinary constraints of academia and the discursive constrains of talking back to the powerful with policy discourse.

As noted above, Residents' Voices asked Information and Cultural Exchange to run the first digital storytelling workshop as a capacity building project for the academic and resident researchers. The Residents' Voices team also included an academic filmmaker from Western Sydney University, who helped to run the initial content development workshops. Throughout the workshop the university researchers took detailed notes about the (1) structure of the workshop and (2) the workshop content. The university researchers wrote this information up in easy-to-read fact sheets and placed these fact sheets on the Residents' Voices website, largely as a resource for the tenants who were completing the workshop. The assumption of the Residents' Voices team during the initial digital storytelling workshops was that the tenants might come back to and use these resources in the future, if they decided to create further stories or to train other tenants in digital storytelling techniques. At this stage of Residents' Voices, the tenant-researcher team that had been involved in the Housos study had moved on to other projects – some of which involved using digital technologies to create transnational communication networks with tenants in Chicago.

Meanwhile, the residents of one particular street in the Villawood East estate took up the fact sheets and digital storytelling documents in a different way. After watching the Residents' Voices digital stories and reading the fact sheets, these residents and community workers from the Woodville Community Centre contacted Information and Cultural Exchange (ICE) to run a similar project in Villawood. Woodville Community Services funded the project. In this project, residents took a leading role in all aspects of the production process, and importantly, this included deciding to produce a film in a genre that was entirely different to the stories that were produced in the first Residents' Voices digital storytelling project. Without the formal involvement of Residents' Voices, the tenant filmmakers were free to produce their own cultural products on their own terms, which, furthermore, would not be constrained by the politics of academic research or housing policy. Surprisingly for the Residents' Voices team, although we would not find out about the project until it was well advanced, the Villawood tenants were about to help us meet one of the central aims of Residents' Voices – which was *to create opportunities for social housing residents to develop and express their own knowledge and understanding of the links between place and disadvantage in their own terms.*

As suggested in our planning documents, Woodville Community Services contacted ICE about running the film-making project. ICE employed film director Vanna Seang and a creative producer and dramaturge Nicholas Lathouris to work with the residents through a digital storytelling process that was longer and more refined than our own. The residents worked with these professionals to brainstorm their individual stories, in much the same way as we had done. They then further refined and developed these individual stories into a set of collective and mostly fictionalised stories about life on their estate. This process involved many sessions where they collectively workshopped the storylines and wrote or improvised the script. After they had a working storyboard and script the project moved into the next stage of training. The residents completed training in locational film and sound recording, film directing and film production and acting. Working alongside ICE, the residents acted in, directed, shot and produced three films. They released these films under the collective title *Lost in the Woods*, with the centrepiece being a naturalistic fictional drama set in the Villawood estate. The films were showcased as part of 2014 Indi Gems emerging filmmaker festival held in Western Sydney.[3]

In mid 2014, a final Residents' Voices workshop was organised to draw together all of the projects and learning undertaken under the 'Residents' Voices Banner'. The highlight of the workshop was hearing about the *Lost in the Woods* project from the residents who were involved in the project. We also heard from the staff of Woodville Community Services who praised the film project and the hard work of the tenants. The tenant film-makers talked passionately about how their fictional film was purposely scripted to cover topics that reflect the trope of discourses about social housing, such as domestic violence, community violence and drug dealing. However, the film moves well beyond the commonly deployed narratives about social housing to present lived experiences in a new light. The filmmakers also address issues of asylum seeker settlement and detention in Western Sydney, which are largely absent from mainstream media. The film portrays a far more complex socio-cultural landscape than the one portrayed in *Housos*, which has to be navigated by tenants in Villawood on a daily basis. Tenants are negotiating cultural and class differences that many of Sydney's residents are not exposed to, such as welcoming new refugee populations into their neighbourhoods, and indeed, into Australian society more generally. They challenge the boundary between fiction and experience, with one of the filmmaker residents stating in the film *The Making of Lost in the Woods,* 'Some people could look at it as fiction, but for some people it could touch home.'

Conclusion: methodological challenges for researching territorial stigma

Residents' Voices sought to explore the use of social media and new communication technologies to develop innovative methods through which knowledge might be collaboratively developed, critiqued and distributed. Residents' Voices built on the emerging potential of visualisation as a process of knowledge disclosure. The production process itself is as important as the visual outcome in understanding the knowledge quotient – where the process of designing the visual representation – as successive iterations of proposals and responses – reveals new insights into the situation. Residents' Voices drew on epistemological and methodological traditions within social science, including the sociology of knowledge and more specifically Insider/Outsider epistemology and the social construction of reality (Berger and Luckmann, 1966; Rogers *et al.*, 2013); and community-based collaborative research methodology (Rowe, 2006) to provide support for the practice of knowledge production through the collaboration of lay and expert researchers as a counterpoint to knowledge produced by institutional experts alone. Martin (1996, p. 5) argues that 'the dominant group of experts in any field is usually closely linked to other power structures, typically government, industry, or professional bodies'. The trans-national spread of certain ideologically based theories and practices in regard to the social problem of poverty concentration within areas of social housing and subsequent 'solutions' suggests a close relationship between housing theorists and policy makers that leaves little space for alternate forms of knowledge, particularly from those most affected by the issues themselves (Allen, 2008). This challenge is one that the Insider perspective tries to resolve. Insider doctrine developed out of the long-standing problem in the sociology of knowledge relating to the differences in access to certain types of knowledge based on socio-economic position and the claim that at certain times particular groups have *privileged* (even *monopolistic*) access to particular kinds of knowledge (Merton, 1972; Biesta, 2007).

In more recent times Insiderism has broadened its frame, going beyond the academy and institutionalised knowledge production to incorporate lay researchers as Insiders (see for instance Biesta, 2007). Allen (2009) lays the foundation for lay people as Insider researchers arguing that social science has been so successful at defining and defending its position as a producer of 'superior' knowledge that it has developed an elite, symbiotic relationship with policy makers seeking 'solutions' to 'policy problems' (e.g. evidenced based policy). Rather than this being a benign enterprise, he argues, 'the imposition of a social scientific way of understanding the social world violates, at the most basic level, the understanding that ordinary people have of the social world' (Allen, 2009, p. 109). Indeed, it was the task of Residents' Voices to ensure that any conflicting perspectives between institutional researchers and those who actually experience living in social housing is made more visible. We deployed participatory research as a methodological tool to expose the conflicts and tensions, and in an attempt to produce countercultural digital media products.

Sociologists have been analysing and exposing the role played by the media in shaping and reproducing territorialised notions of class and disadvantage for well over a decade (Devereux *et al.*, 2011; Blokland, 2008; Warr, 2005; Palmer *et al.*, 2004). It is clear 'the media' is a key technology through which embellished depictions of class and disadvantage are mediated, and these familiar narratives often focus on both interpersonal and neighbourhood level disorder and crime. Therefore, the three resident-led media projects described above did not set out to collect 'data' about locational poverty that simply signifies the effects of territorial stigma. Rather, the Residents' Voices methodology sought to generate data that would shed light on how stigmatising categories and narratives are generated and reproduced. We set out to facilitate the framing of new questions and the reframing of old questions based on collaboration with local tenants in the design and implementation of Residents' Voices.[4] Certainly the cultural analysis and counter-cultural products produced by tenants in these projects reconceptualised the media methods and tools that they could access and use to *talk back* to housing managers or policy makers. But perhaps more importantly, Residents' Voices provided a space for tenants to *talk back* to other housing research methodologies and knowledge systems. Central to this approach to knowledge creation was the reconceptualisation of the social actors who design and participate in these media projects. The digital media products that are produced by the resident groups have the power to directly challenge conventional approaches to understanding place stigma and disadvantage.

On a number of occasions, during these three projects, participants made it clear that they needed to talk to other tenants first, and did not feel comfortable or free to discuss their experiences or ideas with housing managers or researchers – especially while redevelopment and relocation is proceeding (e.g. some tenants did not share their digital stories). A central concern is that tenants in areas targeted for redevelopment have severely limited choice, or voice, in key debates and decisions affecting their living environments, and furthermore, that conventional policy-driven research on neighbourhood social conditions has effectively devalorised the situated knowledge of social housing tenants, compounding their relative powerlessness. These projects aimed to create a space where tenants are able to express, exchange and theorise about the impact of the places they live their lives, to validate their own knowledge, and to use it in ways which best suit their interests. As Wacquant (2007) shows, the 'social exclusion' of social tenants extends well beyond the individual tenant and the housing management arena. Tenants have long been excluded from the research processes that define the 'problems' with disadvantaged people and places. They have been excluded from producing counter-narratives about these people and places, and they are excluded from the policy discussions about how solutions should be framed and implemented.

Notes

1 For example, see stories by Anita, https://www.youtube.com/watch?v=Ve2dXKWHyk4 and Peter, https://www.youtube.com/watch?v=-lCSLmswhrA).
2 The tenants gave us permission, indeed encouraged us to cite their names.
3 See trailer at https://www.youtube.com/watch?v=rPjSqYroYDg
4 After completing the Residents' Voices digital storytelling training one of the authors went on to study radio documentary at the Australian Film, Television and Radio School. He then co-found the SoundMinds Radio project (www.soundminds.com.au) as a research communication project, which broadcasts a weekly national radio show. It is funded by the Community Broadcasting Foundation of Australia and broadcast the Deborah Warr interview about poverty porn cited above. In many ways, SoundMinds Radio is Act 4 of this chapter.

References

Allen, C. (2008), 'Gentrification "research" and the academic nobility: A different class?', *International Journal of Urban and Regional Research*, 32: 180–185.
Allen, C. (2009), 'Clarity, coherence and social theory: comments on housing studies between romantic and baroque complexity', *Housing Theory and Society*, 26: 108–114.

Arthurson, K. (2012), 'Social mix, reputation and stigma: Exploring residents' perspectives of neighbourhood effects', in van Ham, M., Manley, D., Bailey, N., *et al.* (eds) *Neighbourhood Effects Research: New Perspectives*, London, and Berlin: Springer.

Arthurson, K., Darcy, M. and Rogers, D. (2014), 'Televised territorial stigma: How social housing tenants experience the fictional media representation of estates in Australia', *Environment and Planning A*, 46: 1334–1350.

Berger, P. and Luckmann, T. (1966), *The Social Construction of Reality: A Treatise in the Sociology of Knowledge,* London: Penguin.

Biesta, G. (2007), 'Towards the knowledge democracy? Knowledge production and the civic role of the university', *Studies in Philosophy and Education*, 26: 467–479.

Biggs, S. (1989) 'Resource-poor farmer participation in research: A synthesis of experiences from nine national agricultural research systems.' OFCOR Comparative Study Paper 3. The Hague: International Service for National Agricultural Research

Blokland, T. (2008), '"You got to remember you live in public housing": Place-making in an American housing project', *Housing, Theory and Society*, 25: 31–46.

Bourdieu, P. (1986), 'The forms of capital.' In J. Richardson (ed.) *Handbook of Theory and Research for the Sociology of Education*, New York: Greenwood: 241–258.

Bradbury, B. and Chalmers, J. (2003), *Housing, Location and Employment*, Sydney: Australian Housing and Urban Research Institute.

Creeber, G. (2009), 'The truth is out there! Not!: *Shameless* and the moral structures of contemporary social realism', *New Review of Film and Television Studies*, 7: 421–439.

Darcy, M. and Gwyther, G. (2012), 'Recasting research on "neighbourhood effects": A collaborative, participatory, trans-national approach', in van Ham, M., Manley, D., Bailey, N. et al. (eds) *Neighbourhood Effects Research, New Perspectives*, London, and Berlin: Springer.

Darcy, M. and Rogers, D. (2015), 'Place, political culture and post-Green Ban resistance: Public housing in Millers Point, Sydney', *Cities* iFirst.

Devereux, E., Haynes, A. and Power, M. (2011), 'At the edge: Media constructions of a stigmatised Irish housing estate', *Journal of the Built Environment*, 26: 123–142.

Dufty(-Jones) R. (2009), '"At Least I Don't Live in Vegemite Valley": Racism and rural public housing spaces'. *Australian Geographer*, 40: 429–449.

Goffman, E. (1986), *Stigma: Notes on the Management of Spoiled Identity*, New York: Touchstone.

Guillemin, M. and Drew, S. (2010), 'Questions of process in participant-generated visual methodologies', *Visual Studies*, 25: 175–188.

Gumucio-Dagron A. (2008), 'Six degrees and butterflies: Communication, citizentship and change. In Fowler, A. and Biekart, K. (eds) *Civic Driven Change: Citizen's Imagination in Action*, The Hague: Institute of Social Studies, Essay 4.

Hastings, A. (2004), 'Stigma and social housing estates: Beyond pathological explanations', *Journal of Housing and the Built Environment*, 19: 233–254.

Hastings, A. (2009), 'Poor neighbourhoods and poor services: Evidence on the "rationing" of environmental service provision to deprived neighbourhoods', *Urban Studies*, 46: 2907–2927.

Hastings, A. and Dean, J. (2003), 'Challenging images: Tackling stigma through estate regeneration', *Policy & Politics*, 31: 171–184.

Jacobs, K., Arthurson, K., Cica, N., et al. (2011), 'The stigmatisation of social housing: Findings from a panel investigation', Melbourne: Australian Housing and Urban Research Institute.

Jensen, T. (2014), 'Welfare Commonsense, Poverty Porn and Doxosophy', *Sociological Research Online*, 19: 1–10.

Kelaher, M., Warr, D., Feldma, P., et al. (2010), 'Living in "Birdsville": Exploring the impact of neighbourhood stigma on health', *Health and Place*, 16: 381–388.

Lapeyronnie, D. (2008), *Ghetto urbain. Ségrégation, violence, pauvreté en France aujourd'hui*, Paris: Éditions Robert Laffont.

Lawler, S. (2005), Disgusted subjects: The making of middle class identities, *Sociological Review*, 53: 429–446.

Lovejoy, T. and Steele, N. (2004), 'Engaging our audience through photo stories', *Visual Anthropology Review*, 20: 70–81.

Lundby, K. (2008), *Digital Storytelling, Mediating Stories: Self Representation in New Media*, New York: Peter Lang Publishing.

Martin, B. (1996), 'Introduction: Experts and establishments', in Brian, M. (ed) *Confronting the Experts*, Albany, NY: State University of New York Press, 1–12.

Merton, R. K. (1972), Insiders and outsiders: A chapter in the sociology of knowledge, *The American Journal of Sociology*, 78: 9–47.

Palmer, C., Ziersch, A., Arthurson, K., *et al.* (2004), 'Challenging the stigma of public housing: Preliminary findings from a qualitative study in South Australia', *Urban Policy and Research*, 22: 411–426.

Pawson, H., Milligan, V., Liu, E., et al. (2015), *Assessing Management Costs and Tenant Outcomes in Social Housing: Recommended Methods and Future Directions*, Melbourne: Australian Housing and Urban Research Institute.

Raisborough, J. and Adams, J. (2008), 'Mockery and morality in popular cultural representations of the white, working class', *Sociology Research Online*, 13: 1–10.

Rogers, D. and Darcy, M. (2014), 'Global city aspirations, graduated citizenship and public housing: Analysing the consumer citizenships of neoliberalism', *Urban, Planning and Transport Research*, 2(1): 1–17.

Rogers, D., Arthurson, K. and Darcy, M. (2013), 'Disadvantaged citizens as co-researchers in media analysis: Action research utilising mobile phone and video diaries', in SAGE Publications (ed) *SAGE Research Methods Cases*, London: SAGE Publications.

Rogers, D., Darcy, M., Butler, P., et al. (2012), 'Satirical mediation: Stigma, social satire and the representation of public housing tenants and estates in Housos', *6th Australasian Housing Researchers' Conference*, Adelaide.

Rowe, A. (2006), 'The Effect of Involvement in Participatory Research on Parent Researchers in a Sure Start Programme', *Health and Social Care in the Community*, 14: 465–473.

Slater, T. (2014), 'The myth of "Broken Britain": Welfare reform and the production of ignorance', *Antipode*, 46: 948–969.

Superchoc Productions (2011), *FrankyFalzoni.com: An ethnic icon of "working" class Australia*. Available online at: http://www.frankyfalzoni.com/housos/.

Wacquant, L. (2007), Territorial stigmatisation in the age of advanced marginality', *Thesis Eleven*, 91(1): 66–77.

Wacquant, L. (2008), *Urban Outcasts: A Comparative Sociology of Advanced Marginality*, Cambridge: Polity.

Warr, D. (2005), 'Social networks in a discredited neighbourhood', *Journal of Sociology*, 41: 285–308.

Warr, D. (2016), 'Poverty Porn: How journalists, audiences and researchers produce stigma', SoundMinds Radio interview with Deb Warr, broadcast 26 January 2016 on Bay FM. Producer Dallas Rogers. Available online at: http://www.soundminds.com.au/poverty-porn-how-journalists-audiences-and-researchers-produce-stigma/

Ziersch, A. and Arthurson, K. (2005), 'Social networks in public and community housing: The impact on employment outcomes', *Urban Policy and Research*, 23: 429–445.

12 'You have got to represent your ends'

Youth territoriality in London

Adefemi Adekunle

Introduction

This chapter looks at youth belonging, safety and territoriality in various parts of Islington, in London. As such it hopes to add nuance and substance to debates over the 'functional disconnection of dispossessed neighbourhoods from the national and global economies' (Wacquant, 2008, p. 67), allowing us a new perspective on previously abstracted policy debates (Rose *et al.*, 2013). This chapter emerged from observations of how young people (aged between 13 and 21) acted and moved around their neighbourhood. Through participation in a youth work intervention project, this group of engaged young people was asked through a number of reiterative methods how they used and thought about their specific area (see Adekunle, 2013).

The research question itself seemed simple and originates in my own personal experience. As a volunteer youth worker, I tried to understand the motivation of young people who at times – point blank – refused to go into neighbouring areas and certain parts of London that resembled their own. There was a curious mixture of fear mixed with a desire to express bravado and explore these other areas. I witnessed young people eager to explore certain areas in order to '*rep*' or '*represent*' their neighbourhood, even though they were aware of the possibility of being '*rushed*' or (physically) challenged. They were also eager to 'rush' unfamiliar faces who would enter 'their' neighbourhood, despite the fact that, if they looked hard enough, they would realise that there were potential webs of familiarity connecting them to these visitors: these 'intruders' could well have been cousins of someone from the area; they might find that they had gone to the same primary school as this person; or they might realise, quite simply, that this was a 'friend of a friend'. What appeared to be a form of territorial defence seemed all the more surprising when one considers how much London resembles a series of inter-linked villages: it is remarkably easy to find connections amongst young people living in areas neighbouring each other, whether these were established in school, through family or friends. My personal experience and my awareness of the literature prompted me to consider a number of questions. First, do young Londoners from areas with negative representations actually absorb the stigmatising imagery about the place in which they reside? If, as I argue here, they somehow manage to avoid internalising the stigmatic imagery about the specific neighbourhood in which they live, the case study that I develop asks whether they then conceive of nearby neighbourhoods 'out there' – which are also stigmatised – as threatening. After all, these other neighbourhoods are located beyond what the youths consider to be the familiarity of spaces that they understand and that they are able to appropriate. Does this change 'who they are', or merely how 'others' see them? And finally, how precisely does the mechanism of 'blemishing' or 'denigrating', described by Wacquant (2007) function, if at all? Whilst defensive 'rushing' or the physical challenge of outsiders hinted at a 'code of the street' (Anderson, 2000; Brookman, *et al.*, 2011), or new dimensions of 'badness' (Gunter, 2008) or performative edgework (Lyng, 1990), there was a different dimension to this. There was even a hierarchical aspect to this social construction based on age and maturity, since 'Olders' ('older' young people) consciously differentiated themselves from 'Youngers'. I call the development of this particular constellation of spatial stigmatisation and appropriation 'youth territoriality' and this chapter will be focused around why it exists and on its significance for young people.

The point of the chapter is not to affirm that territoriality has no significance outside an audience of young people. Young people, as a category of the population, are often absent from accounts in urban studies, and when they are present, they are cast as victims or inactive recipients of those interventions that try to reshape and change the image of particular neighbourhoods (Lees, 2008). Consequently, my focus on lived experiences is meant to act as a counterpoint to top-down expositions of urban policy professionals, civil servants and local elected officials that rarely see children and young people as fully capable urban agents (for an exception see Butcher and Dickens, 2015). Indeed, territoriality also acts as a corrective to traditional policy accounts that uncritically perceive social mix as an unalloyed benefit (Sarkissian, 1976; Cole and Goodchilde, 2000) rather than an idea 'imported from elsewhere or imposed by another level of government' (Damaris *et al.*, 2012, p. 446). By focusing on this mobile, often unacknowledged but active portion of the population, certain powerful insights can be brought together about the dynamics of urban stigmatisation and their movements as a response to this.

In order to understand questions about why young people move, or refuse to move, around their own city, various things must be acknowledged. This chapter can only provide a partial perspective. It does not go into detail about how local government and community safety infrastructure interact and responds to youth territoriality. The focus is deliberately concentrated on the voice of young people, which give the dynamics of territorial stigmatisation a particular timbre. Who, when and how space is appropriated by young people does not always follow the contours that some of the literature might suggest (see Wacquant *et al.*, 2014). Where and how young people feel comfortable represents a confirmation and a counterpoint to the traditional narrative of internalised territorial stigmatisation, effectively translated by inhabitants into an 'exit strategy' (Wacquant, 2008).

As the next section will outline, this does mean that the work sets itself within a certain political context – specifically, the chapter aims to counter the binaries that are created between constructions of young people as either 'angels' or 'demons' by 'youth geographers' (see, in particular, Valentine, 1996, 2003). Additionally, I do not focus on the motivation behind acts of violence or crime here, since the emphasis is very much on the everyday experience of young people and on their mundane lived experiences. Howard Davies, a former head of Children Services in Wales, stated that research on young people is 'too preoccupied with studying the spectacular, deviant and bizarre. This makes for interesting reading, but distorts the ways in which we understand young people' (cited in Robb, 2007, p. 123). This chapter is an attempt to move away from research as a tool to uncover the sensational, and as such, my emphasis is on *'resisters'* (those who have never offended) and *'desisters'* (those who had offended but have now ceased)[1]. Instead of the common depiction of the supposed 'lost youths' of London streets, both the groups that I portray here represent young people who stay *well away* from trouble, even as they know very well where it is likely to happen. For these groups, although becoming the victim of a crime is likely, the ability to evade it is based on a keen understanding of space appropriation and territoriality.

The empirical focus of this chapter provides an example of the benefits that an approach which is predicated on research *with* young people can have, as opposed to research conducted *on* young people. The research is based entirely on a participatory methodology (Cairns, 2001; Kellet, 2004) and it stems from an analysis of peer-group led surveys, which were followed up by a series of focus groups in which the results of the surveys were debated. The young people themselves formulated the research questions, conceptualised the problems, managed the research project and analysed the findings. The second stage of the research that I present here is based upon a presentation of how this group spoke about its experiences. We will also see how contemporary youth research is shaped not only by changes in the experience of youth, but also by the conceptual resources deployed to predict and explain it (Furlong *et al.*, 2011). As I show here, the research themes were determined by the youths and they orbited around how these young people view their place and neighbourhood, their age and class and discussions of gender.

'Youth Geography' in research focus

It is important to acknowledge that young people and teenagers are very active public beings (Childress, 2000). Youth researchers have already described how, in public spaces, there is an unconscious and inadvertent timetable. In skate-board parks, for instance, researchers in various contexts have noted how truants and older kids (both boys and girls) used these spaces in the early morning and afternoon, schoolchildren in the late afternoon and older teenagers and even adults in the evening, creating a temporal social ecology in the same territory (see McGloin and Collins, 2015; L'Aoustet and Griffet, 2001). The various overlapping temporalities – day of the week; time of day; season; traffic pattern – all hinted at the different socio-temporal rhythms that run through young lives.

As I have already stated, this chapter concentrates on young people's experiences and its focus is on their everyday lives, in places such as the home, their schools, playgrounds, the neighbourhood streets and so on. It adds to the various, increasingly sophisticated, portrayals of young people that describe how they use different social spaces and on the identities that they deploy (Shildrick *et al.*, 20009). Anoop Nayak reminds us that:

> [Young people often have] knowledge of places where dangerous driving, accidents or car theft were likely to occur. Similarly, their knowledge of drugs (who the dealers were, in what places they operated and who were their respective clients) was equally sensitive. They appeared able to identify 'hot spots' and had developed a complex mental map of 'safe' and 'risky' zones within their neighbourhood. This enabled them to develop an elaborate local micro-geography through which to navigate their communities.
>
> (Nayak, 2003, p. 305)

The premise is that young people who are not perpetrators of violence (the 'resisters' and 'desisters') have to confront and navigate around the hostility of other teenage groups who occupy local areas where they at times hang out. In one apposite study, Matthews and colleagues explain how

> 'hassle' from other, often older 'kids' and fear of assault among the girls and fear of attack and fear of fights among the boys, kept these teenagers to tightly defined areas, where they felt 'safe' and free to do what they wanted.
>
> (Matthews *et al.*, 1998, p. 196)

My ambition here is to go further than outlining contesting micro-geographies by looking at the ways in which these young people differentiate themselves from each other and how they resist, affirm or follow processes of spatial identification and stigmatisation. Much Like Matthews *et al.* (ibid.), my aim is to show how neighbourhoods becomes a site of identity struggle forming a component within the rich tapestry of shared interests, behaviours and circumstances of street life. The still nascent literature on youth and urban studies suggests that young people have a different qualitative knowledge and experience of 'place' that could potentially be the site of valuable insights into urban life (Skelton, 2013). The challenge is to find out the best way in which to harvest and interpret this knowledge and to do this whilst also giving due weight to those who demand that we consider young people as valid social agents and competent social actors in their own right (Travlou *et al.*, 2008; Skelton and Gough, 2013).

A definition of youth territoriality

'Youth territoriality' is a concept that lends agency and prominence to young people on their own terms, perhaps as an exaggerated form of 'place-belonging', cited by some as the reason for conflict amongst and between young people

(Bannister, 2011; Childress, 2004). And yet, defining territoriality and situating it within the myriad ways of conceiving space is a formidable task. In effect, the idea that territory is essentially 'humanly differentiated geographical space' (Wolch and Dear, 1989, p. 1) seems too wide to reduce into an easy research question.

The literature tends to describe a wide range of theoretical propositions that links various types of occupancy with some degree of control (Brower, 1980; Altman, 1975). Hall states that 'the act of laying claim and defending territory is called territoriality' (Hall, 1959, p. 187) and Shils sees territory as 'a meaningful aspect of social life, whereby individuals define their scope of their obligations the identity of themselves and others' (Shils, 1975, p. 26). Foucault's description of territory as '[t]he area controlled by a certain power' (Foucault, 1980, p. 68) has been extended immeasurably to provoke a rich and evolving political-geographical take on territoriality as 'a strategy which uses bounded space in the exercise of power and influence' (Johnson, 1996, p. 871; see also Sack, 1986; Kärrholm, 2007). Of course, there are also the dense metaphysical conceptualisations of 'territorialisation', 're-territorialisation' and 'territorial assemblages' that Deleuze coined (see Legg, 2011). At its most basic however, all agree that territoriality is a means by which *X* can affect, influence, or control *Y* (Wolch and Dear, 1989) on a scale that encompasses ranges from interpersonal distances to the spatial arrangements of cities and regions, as well as the flows of people, goods and ideas (Sack, 1993).

Territoriality is significant as 'an important organizer of activity on [various] levels: community, small group and individual' (Edney, 1976: 42) whilst its constituent, 'territory', is 'a meaningful aspect of social life, whereby individuals define the scope of their obligations and their identity and that of others' (Shils, 1975: 26; see also Kärrholm, 2007). Spatial patterns such as territoriality make the world knowable and familiar through the creation of everyday rhythms. It connects the past and the present and provides a relatively secure basis to the future (Skey, 2011). Moreover:

> The temporal structure of our environment . . . adds a strong touch of predictability to the world around us, thus enhancing our cognitive well-being.
>
> (Zerubavel, 1985: 12)

Indeed, within the literature, the closest description to the account described by the young participants involved in my research was Bernard Poche's definition of territoriality as the '*spatial extension of the material world elements on which a group defines itself*' (1986, p. 2, own translation). His stress on the relationship of space to territory also provides the basis of my understanding of the term, since it underlines the fact that territoriality does not describe a geography but a topology. It is not concerned with where things are but rather, where they are in relation to others. Accordingly, it is governed not by geometric distances but by a more symbolic frame of reference. Childress described 'territoriality' as practised by young people in small town America as a mode of communication, serving to convey information, about the location of individuals dispersed in space. By contrast. . . [the adult mode of] tenure is a mode of appropriation, which persons exert claims over resources dispersed in space.

(Ingold, 1987, p. 133; cited in Childress, 2004, p. 195)

In other words, because teenagers cannot control private space, they must appropriate public spaces until ownership rights are conferred upon them (Childress, 2000). Their use of space is predicated on occupation and stopping others from occupying it. In the latest in-depth work on the subject within Britain, Bannister *et al.* (2008) have stressed how only *some* groups of young people engage in such behaviours and that when they do, they tend to hold a close spatial relationship to one another. At this stage, a division must also be made between both territory and the practices that are associated with it (territoriality), that create those spatial 'constellations of relations and meaning' (Pickles, 1985). Indeed, 'territoriality' is and remains a geographical concept. It refers to both a quality of space and yet also to space itself (Hills, 2006). The use of the word is intended to show how:

Place [. . .] often becomes the locus of exclusionary practices. People connect a place with a particular identity and proceed to defend it against the threatening outside with its different identities.

(Cresswell, 2009, p. 176; cited in Tomeney, 2013, p. 301)

The version of territoriality that this chapter will follow somewhat draws upon each of these incarnations. It will attempt to use the voice of young people to explain and articulate the dynamics of youth territoriality: an 'etic' definition that must be contextualised within local circumstances. In order to fully appreciate this, an understanding of the local environment is necessary.

The neighbourhood

Islington is both a district inside inner London and a London Borough. For ease of understanding, rather than referring to the geographical area, 'Islington' will denote the area covered under the remit of the Islington Local Authority. In geographical terms, it is comprised of 17 wards, though my research was focused on a fraction of these (see map below). At the time of the research, Islington was ranked 65th out of 533 of the most deprived parliamentary constituencies according to the government data. And yet, it also had some of the most desirable properties in London[2] coupled with a sizeable concentration of social housing, creating a *de facto* class divide which helped to define how, where and why young people congregated in certain spaces. Indeed, a related research project (Adekunle, 2013) uncovered how class and youth interacted to affect which young people were seen in public space and in which areas of Islington (the so-called CAPABLE project: see Mackett *et al.*, 2007). Research with the police revealed that organised crime had a presence in the neighbourhood but that there was no appreciable 'gang' activity related to territory of the drugs trade (Adekunle, 2013), particularly when compared to neighbouring boroughs. Nonetheless, the area was the site of a rich night-time economy. The inference to make here is that the area had a rich street-level ecology that defined when, where and how young people used public space; when they felt comfortable and made themselves visible in it.

The study originates in a professional curiosity about the topic – specifically from a 'detached youth work' perspective. Detached youth work is something slightly different to the traditional provision: while the vast bulk of youth work took the form of provision via clubs and units, this research arose out of the simultaneous efforts of young people themselves – and involved their approaching or being approached by me, as a street worker (Smith, 2005) for us to both find out more about these neighbourhoods. I had been working as a volunteer detached Youth Worker on and off for the ten years prior to embarking on the research project. As a result, I have spent a great deal of time trying to work out how to investigate where young people like hanging out and how and why they interact with each other so easily in certain contexts but not in others. This has meant developing a working understanding of the social importance for young people of the malls, cafés, playgrounds, sport pitches, private and social housing estates – in good and in bad weather – over extended periods of time. Accordingly, my fieldwork was allied to all the benefits (and prejudices) of extensive experience, and from a pragmatic appreciation of the difficulties faced in trying to effectively engage young people.

Study protocol

I have attempted to generate an understanding of territoriality from a participatory point of view. In practice, this meant bringing young people on board as co-participants. I recruited a small cohort of 11 engaged local young people that became a sounding-board for my initial discussion groups. They were the main driving force behind questionnaire construction and survey implementation – a process that required all aspects of their local knowledge and insights.

From their number, I selected two young people (aged 17 and 19) who wanted to be trainee youth workers. I personally trained them in social research techniques and we were given the resources through the youth work organisation that I was working with in order to fully implement the survey. The youths used their familiarity with Islington to identify the circumstances that would ensure that the survey was as representative as possible. As the survey focused on young people's everyday routines, it was primarily conducted in schools, and those other areas that the participants often frequented. For ethical reasons I was always in the background for this stage of the research. My conscious decision to give the youth workers time and space to engage with young people led to them to perform research in ways that I could not have imagined alone. Harriet and Rowena (not their real names) were aware of rhythms, temporalities and similarities based on things as mundane as the color of the school ties within a school. This provided catalysts for conversations that, as an 'outsider', I would not have considered.

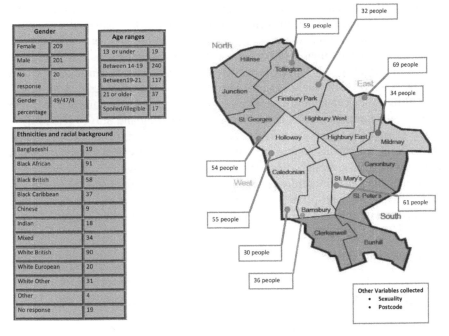

Gender	
Female	209
Male	201
No response	20
Gender percentage	49/47/4

Age ranges	
13 or under	19
Between 14-19	240
Between19-21	117
21 or older	37
Spoiled/illegible	17

Ethnicities and racial background	
Bangladeshi	19
Black African	91
Black British	58
Black Caribbean	37
Chinese	9
Indian	18
Mixed	34
White British	90
White European	20
White Other	31
Other	4
No response	19

Figure 12.1 Survey area details and map N=430.

Ultimately, their situated knowledges, techniques and questions generated very important data which, as a researcher, I would not have attempted to uncover.

I then attempted to corroborate the findings with a series of focus groups comprised of those who were part of the original survey. Based on this foundation, I looked at how and if these findings actively constituted and reflected a 'reality' that young people would recognise. Since youth territoriality was an issue so obviously shaped *by* young people, it seemed reasonable for it to be researched *by* them as well.

The survey responses: crime and safety

Safety

'How safe do you feel?' This was the first question of the survey and it was a first step to understand the extent of territoriality in a somewhat oblique manner. If the experience of crime and victimhood were underlying dynamics behind territoriality, the survey questions were meant to capture it. As a Youth Worker, I had become used to hearing spectacular accounts of violence and I was interested in learning to what degree these were true. Nevertheless, responses to this first question showed that only 10 per cent of responders thought they were not safe in Islington. The majority of young people responding to the questionnaire (or survey) described themselves as safe (48 per cent) and the remainder as feeling very safe (35 per cent). My research diary notes at the time record the fact that there was a significant majority of young people who answered some variation of '*I've always felt safe as I have always lived here.*' There seemed to be a 'neighbourhood dogma' – to borrow a coinage by Karen Evans (1997, cited in McGrellis, 2004, p. 10) – that seemed to equate safety with familiarity, which was itself linked to the length of time that each respondent had resided within the area. Still, to look outside the confines of the study, the only comparable youth survey – the Longitudinal Study of Young People in England Survey (LSYPE)[3] – asked exactly the same question and received a 70 per cent 'not safe' response suggesting that young people in Islington did generally feel far safer than a nominal national average. The apparent fact that feelings of safety are lower nationally than in Islington is a crucial finding. It forms the basis for youth territoriality and an appropriation of space that worked in tandem with those areas (stigmatised or not) that young people occupied as a matter of routine.

Crime and victimhood

Box 12.1   Selection of questions within the Community Survey

1   **How safe do you feel?**
2   **What is your main crime and safety concern?**
3   **How threatened are you by your main crime and safety concern?**
4   **Have you been a victim of crime within the past 12 months?**
5   **Is there an estate/area that you have crime/safety concerns about?**

The second question – 'Have you been a victim of crime in the last 12 months?' – was intended to elicit answers that would deepen the responses obtained to the question about safety (see Table 12.1). The results suggested that only a minority of respondents had some direct experience of crime and/or violence. However, this was not enough to interpret this as a reason behind a territorial mind-set. As is often the case when surveys are being conducted, the answers that are given generate more questions. In this case, participants often asked 'Of which specific crime were you a victim?' Rather than shutting down potentially fruitful avenues of inquiry, this was left open and definitions of crime were left solely to the respondent.

Area and crime: is there an estate/area that you have crime/safety concerns about?

Along with and following from the initial answers, questions about crime and victimhood could have provided be clearest evidence of territoriality. High figures would hint at the clear picture of a 'dangerous other', located outside one's

Table 12.1 Number and percentage of young people who could name an area/estate they had crime/safety concerns about

a)

	No	*Yes*	*No response*	*Grand Total*
Percentage	85	14	1	100

b)

No	*Yes*	*No response*		*Grand Total*
347	62	17		426
81	15	4		Percentage

c)

Amount	*Yes*	*No*	*No Response*	*Grand Total*
Number	62	347	17	426
Percentage	15	81	4	100

neighbourhood, and one that could be named and located. For this reason, the question was one of the few to open up potential responses that might actively point out which parts of Islington the young respondents were concerned about. Linking fear of crime with an identifiable locale would be an important step in formulating a simple measurable output. The low figure in Table 12.2 suggested that this output did not exist. It should be stated that a mere 14 people were specific about identifying an 'unsafe place' and that their responses added up to ten different neighbourhoods that were considered in some way threatening. It is vital to note that, while respondents did identify some neighbourhoods or estates with criminality and associated those areas with threatening behaviours, not one associated this with their own residential area. It is also interesting that the neighbourhoods that were linked to safety concerns appeared to be often far removed from the neighbourhood in which the respondents themselves lived.

Analysis

The survey seemed to argue against the idea of crime and violence as a correlating factor behind territoriality. And yet, the young people I spoke to were adamant that there were areas where they felt anxious. I was able to pinpoint these areas using the survey results, which were based upon amassing information from different areas at different times. Consequently, pinpointing 'unsafe' locales was thus relatively easy.

The most commonly described 'unsafe space' was Finsbury Park – a tube station within the London Borough, which was too large and disparate an area for any easily identifiable group of young people to appropriate as their 'own'. To provide some context, a large amount of school and college pupils – more than 50 per cent according to one police estimate (personal correspondence) – do not live in Islington and come from neighbouring areas, variously Hackney, Haringey and Camden. Consequently, they have up to 45-minute journeys to go home. Accordingly, there are a number of schools in Islington – 9 secondary, 54 primaries – reflecting its status as one of the youngest boroughs in London (source: ONS). Within each secondary school there are, on average, 1,000 people. Consequently, there are a range of transport nodes and corridors where young people congregate. Finsbury Park is a transport intersection between Hackney, Archway and Islington and within it there are various parks and spaces where one can expect to see several young people congregate in the afternoon. Despite the routine, banal nature of the school or college journey, the presence of friends and the dense public transport infrastructure nodes (railway, tube station and bus depot) some respondents described a feeling of disquiet in Finsbury. Here some local context is needed. At the time of the research, the bus from Wood Green to Trafalgar Square had until recently been a 'bendy bus', meaning that it was possible to travel from the deprived locale of Wood Green to central London via Finsbury Park for free if one kept an eye out for bus inspectors. At the same time, the Finsbury Park tube station's idiosyncratic design meant that there were no tube barriers for reasons of fire safety. It is perhaps the only station in London from which one can walk onto a tube without paying a penny or seeing a conductor. In short, the lack of surveillance opened up the possibility of free travel between two deprived and stigmatised areas, which seemed to have some correlation to the possibility of territorial conflict between young residents.

The data suggested other territorial markers. It soon became clear that after a day at school or college, cheap food became a priority, perhaps as an opportunity to meet other young people socially. Accordingly, McDonalds, a Kentucky Fried Chicken and a kebab shop on Finsbury Park Road can become very busy. The large groups of young people and the fact that there are some conflicts between some schools which have traditionally had longstanding 'beefs' with others explained why passing through the area made some apprehensive. The survey reported secondhand accounts of fights arranged after school and stories of the detours that young people had to make on their way home from their educational institution – an anxiety assuaged only by walking in large groups, which paradoxically made them a more visible target and increased anxiety in others. In the survey, Finsbury Park is an anomaly as it was one of the few areas where young people *actively* asked for more visible police presence. This confirms that there was a general sense of unease amongst young people travelling through the area – a view that the focus groups described and confirmed.

The implication was that the areas that generated the most anxiety were transient and typified by a high footfall as well as anonymity. This also seemed to corroborate Childress' view of territoriality: physical presence, regularity of attendance and thoroughness of occupancy were communicative ways of showing territoriality. Respondents explained that they 'just know' which places to avoid while stating that, conflicts in these areas were to be expected (Childress, 2004; see also Ingold, 1987). In another real sense the anxiety measured also embodied what Garland (1996, p. 461) has called '*criminologies of the other*', that is the association of danger with 'the threatening outcast, the fearsome stranger, the excluded and the embittered'. This was made extremely clear and was confirmed when these findings were tested against the backdrop of a focus group.

The focus groups

In order to get a richer sense of its intricacies and to triangulate the survey's outcomes, I presented a short summary to two focus groups composed of survey respondents. The first focus group participants were recruited from each of the six survey areas whilst the second were mainly recruited from a nearby home-less hostel for young people. Members of both focus groups were, however, respondents in the survey. The focus group participants were generally the same age as the largest tranche of survey respondents (aged 17–21), both groups were recruited from a more self-consciously 'street literate' (Cahill, 2000) cohort and they included an equal number of young men and women (four men and four women in each focus group).

What both groups shared was that neither saw anything surprising about the idea of territoriality. They all suggested that it 'was worse if you were young', reinforcing the notion of the transitional or intergenerational aspect of territoriality. Though very few spoke of their mobility being personally constrained, it was clear that victimisation, crime, violence, harassment and fear had some role to play in their experience of the city, or at least, these were presented as reasons for anxiety. Within the first group, this was made clear when one respondent stated that:

I know guys who have been stabbed for £10 of drugs.

(19-year-old, medium term resident, male)

It was the intergenerational perspective that generated the most discussion. Though all in the focus groups agreed that territoriality was to some extent a learned behaviour, Kintrea *et al.*'s (2010) definition of an intergenerational adherence to historical boundaries had limited applications here. Rather, the consensus formed around the idea that it was 'younger's fault' [sic]. 'Something' had changed. One respondent believed that territoriality was inevitable, since:

There is nothing else you can do. You have to change a whole generation. Basically, it's a line that everyone is following. You go to kids and say 'don't sit in the street' unless you give them a PS3 [Playstation 3]. Even if they have a PS3, they are always thinking of taking from another person: having two.

(19-year-old woman, long-term Islington resident)

Others expressed variations of what one young man described:

> This may sound narrow-minded but young people nowadays are ignorant. It's not even the drugs, it's the chavs [. . .]. They can't stand up for what they are trying to say. It won't sit well with their friends. Back in the day if my brothers had a fight, with someone younger, they would go down. Take off their shirts and have a fist fight. Nowadays it's like a gun fight, or taking a knife, and is how silly situations escalate into killing someone.
>
> (20-year-old man, short-term resident)

In a surprising inversion, it appeared that older teenagers were just as afraid of their younger peers, though this was often explained through other markers of difference, such as class ('it's the chavs'). Others attributed territoriality to a complicated nexus of circumstances.

> You have to wonder why kids are hanging around at certain times. Where are the parents? What is going on? Who want to go to college? How do you survive? If your parents got other kids to feed, you know, you are not going to college knowing you could work, but there's no work so kids are hanging around the street and getting bored, so they find something to occupy their time.
>
> (20-year-old man, short-term resident in Mildmay)

There were a number of references to social structural factors, such as the absence of work. Nevertheless, others blamed their anxiety squarely on the individual.

> Actually, I think instead of worrying about how to change the world, first of all, you have to change yourself, to influence you, and the choices you make for the people around you. You can show people the way, but if people are used to their life being a certain way, if they are used to violence, then that is just it.
>
> (20-year-old woman, long-term resident)

Leaving aside this last partially dissenting opinion, all the respondents united around a form of 'othering' that was generational. The challenge was to see what was behind this: either a novel method of creating 'criminologies of the other' based on age or a sense that territoriality was something that one 'grew' out of once one 'got on' and had dealt with the messy process of becoming an adult.

Crime

It is unsurprising that crime emerged as a topic of discussion given the places in which the survey was conducted. The most obvious research finding was that violence was never truly random. One exchange exemplified this:

Did you hear about the 15-year-old who got stabbed in Victoria?

> (19-year-old male from Cannonbury)

[I knew someone who knew] this guy who got stabbed. You feel bad. It's mad.

> (20-year-old male, long-term resident)

What kind of person did he have to be to be involved? I think that is how I think of it. He could have been the nicest person on earth but at the same time he could have already shot someone and you wouldn't know. You hear it and think 'oh well'. You never know. That's what I mean. It could be karma coming back, and it doesn't make me feel any different now. Maybe a little more careful.

> (19-year-old male)

This was indicative of a number of processes: first, the way that it was presented pointed towards ways of learning about events on the street – 'I knew the guy' – through channels other than the media.[4] Second, and more significantly, it provides further evidence of the view that crime – even serious violent crime – was never seen as random or, even in this instance, totally undeserved, cementing the respondents' status as either 'resisters' or 'desisters'. Nevertheless, despite this view, the idea that the police could 'do more' was unanimous, even though respondents left open where this was necessary. This must however be balanced by the fact that the police and other groups in charge of security, such as security guards, were not regarded as a benign or benevolent presence:

> I don't like it sometimes when people look at me and I go into a shop and security people follow me everywhere.
>
> (18-year-old male, short-term resident)

The entire group quickly reached a consensus and agreed that young people in Islington were over-policed, yet paradoxically under-protected.

> When you look at crime related to young people they are not really seeing the young person's view. We need to understand where young people are coming from. They are talking to parents, policemen, councillors.
>
> (20-year-old girl, long-term resident)

Gender

Gender was invoked in both focus groups without any prompting. One contribution in particular focused strongly on certain gendered street representations. The young women in both groups had a clearly different take on what kept them safe. Given the relatively small spectrum of views expressed, a representative view would tell us how:

[I]t's just about knowing what is right and wrong, being streetwise. I can see what you are like from your walk. Yeah. If I see someone, I might cross the street, if that keeps me from getting mugged, being dead, then yeah. If someone is walking behind me with his hoody covering his face, I will cross the road. I'll hold onto my phone so I can make a call. I'm a girl. I've got priorities. Girls are more. . .Usually when I go clubbing, I make sure that I leave with someone. Walk with friends. We all wait together for everyone to get the bus. Even when we went out the other day, the only reason I'm going out because I know some-one is coming home with me. I'm not silly.

(20-year-old female)

The conflation of 'being streetwise' (a supposition that seemed to focus on the agency of the participant) seemed at odds with what else was said – although, the rest of group did react in support of this statement. The respondents' statement according to which she 'can see what you are like from your walk', establishes the fact that she feels she has the skill to negotiate dangerous situations because of a particular awareness which she links to her gendered vulnerability ('I am a girl'). Despite the fact that the young men found this statement problematic and admitted to feeling just as vulnerable as the girls, all the young women tended to agree with the above statement, stating that for girls 'things were different'. This provides a gendered interpretation of the idea that was posited above and relating to the ways in which violence is never perceived as random, yet is felt to be 'different for girls'. Still, the idea that a participant could just say 'I am girl' with this apparently signifying something so obvious that it required no further elaboration was intriguing to me. It implied an implicit knowledge that changed the angle of engagement for resisters and desisters. Nevertheless, it did provide some justification for why an interpretation of territorial violence might very well be gendered. Her insistence on coming home with a group of girls would appear to mean that her mobility was, on an individual level constrained. Paradoxically, the implication was that a *group* of girls might very well have a mobility that a group of boys might not: the boys tended to believe that they were not welcome in new/unfamiliar contexts. Young women described a situation in which it was apparent that their gendered vulnerability could be tactically reversed as a strength considering that they were not seen as threats by others.

Neighbourhoods

It was the conversations about the area and crime that generated the richest data. The males in both focus groups viewed territoriality as based upon a clear and real risk, attuned and diminished through a hard-won appreciation of how to keep safe, aligned with a continual focus on the 'word on the street'. Overall, for these young men, territoriality signified a reliance on their own 'street literacy' and their ability to navigate within a neighbourhood that they had fully appropriated.

If people see something, for example, from my own experience, there was a road where a young boy was stabbed and my friends avoided that road, that road for a number of reasons. Someone had got stabbed, different factors, you think about when you hear someone got stabbed. Some of the stabbings don't even make the paper. Some people get paranoid and think that they can't go anywhere.

(20-year-old male)

This incarnation of territoriality has built within it a certain appreciation of 'street practices' and of how to read them which were sometimes hyperbolically exaggerated.

Certain areas, some people [say] I'm not going to go to Brixton. I think it is not even the area. It's just the people in the area. Even in the posh areas there are shootings and stabbings. It's not just the area itself.

(20-year-old male)

This demonstrates the ways in which male participants linked violent crime to an area and to social class ('even in the posh areas') and saw nothing intrinsically threatening about any one of the neighbourhoods in particular ('It's not just the area itself'). In fact, they seemed well aware of the injustice and arbitrariness of area stigmatisation to the extent were they were ready to challenge each other about the role of myths in their own negotiation of unfamiliar neighbouring locales.

If you avoid an area, you make that area bad. That's why a lot of people don't go to Hackney. But if every person did not care every time something happened, then trust me, there would be nothing there. I've lived in Brixton. There is nothing there. It's just what you hear. No one ever spoke to me. Why avoid a place just because you hear of something happening?

(20-year-old man, long-term Barnesbury resident)

In this account of delegated agency ('If you avoid an area, you make that area bad'), the respondent undermines the prevailing street mythos in stressing how a level of maturity diminishes the power of youth territoriality: if only, like him, no one cared, there wouldn't be any issues. If, like him, people stopped avoiding stigmatised places and started to ignore their reputations, then perhaps we could start to envisage an end to issues linked to negative territoriality. His account featured an idiosyncratic mixture of bravado and (over)confidence but it was a salutary point that members of both focus groups managed to cohere around.

Conclusion

Ultimately, the research described here acts as a riposte to those social theorists that have downplayed the importance of place (Calhoun, 1991; Giddens, 1984) or ignored calls for research that focuses specifically on everyday youth issues such as transportation and the spatial contexts of im/mobilities (Skelton and Gough, 2013). It confirms that 'the continuing importance of the home area can be perhaps in part explained by specific locational and transport deficits and by the stigmatising attitudes of outsiders. . . towards residents' (Pickering *et al.*, 2012, p. 950).

To go back to the definition of youth territoriality described above, the literature traditionally offers as explanation a nexus of symbolic, social and physical space in which stigmatisation leads to a blemish of place in which taint is spatialised and then transmitted. Residents in this situation, conventionally cope by utilising a spectrum of techniques ranging from submission to defiance (Wacquant *et al.*, 2014). It appeared that for the young participants, the process of appropriation of stigmatised areas was not so clear since both facets of the research did not generate any named geographical areas. With the exception of Finsbury Park, the survey did not identify a single neighbourhood that more than 5 per cent of respondents felt was unsafe. Instead, it led to 10 different names being highlighted, an outcome that has real theoretical significance since a belief in causal neighbourhood effects is now the dominant paradigm among policy elites, mainstream urban scholars, journalists and think-tank researchers. It did not appear that these young people could readily identify a 'bad area' despite their proximity to crime and the street. They instead held a more multi-variate understanding of stigma and area: -an understanding that urban studies should perhaps duplicate. Moreover, to extend this point, this work is also presented as a counter-blast to the tendency of the 'neighbourhood effects' to become not just more abstracted but spatially deconcentrated. Indeed, as Tom Slater has been eloquent in pointing out:

> The genre of 'neighbourhood effects' stems from an understanding of society that adheres to one overarching assumption, that 'where you live affects your life chances'. It is seductively simple, and on the surface, very convincing . . . The striking simplicity and inherent 'fait accompli' of this line of thinking in a complex world has led to the emergence of analytic hegemony in urban studies: neighbourhoods matter and shape the life of their residents (and their young residents most acutely), and therefore, urban policies must be geared towards poor neighbourhoods, seen as incubators of social dysfunction.
>
> (Slater, 2013, p. 368)

By complicating that link between space and stigma, young people in the neighbourhoods that I studied reconfigured the negative imagery of their neighbouring locales. Territoriality should be seen as a moment in transition to adulthood, something that was transcended by some unforeseen milestone on the way to maturity. Indeed, in this analysis, space and spatial exploration could very well be signifiers of maturity and conceptual resources deployed to predict and explain the unknown (see also Furlong *et al.*, 2011). As one focus group participants said with more than a superficial awareness of nuance, attacks or crime took place in a 'bad area' for a number of reasons that transcended easy explanations of agency or structure:

You never know [what's going to happen]. That's what I mean. It could be karma coming back, and it doesn't make me feel any different now. Maybe a little more careful.

(19-year-old male)

Navigating those 'bad areas' that were perceived to involve a high risk of attack meant that respondents thought they had to resort to using a range of strategies in order to keep themselves safe. This is a process which was described in ways that seemed surprisingly easy and straightforward, at least for most participants. One respondent explained that, even in an area with clearly identified troublemakers, territoriality was just another obstacle to navigate around:

I've got used to knowing who is where and who is doing what, right now I say hello so it's not like I'm ignoring them, I don't know them, so if anything happens to me, at least I know there might be some people to help me, but I'm never going to get involved, or try and be involved in what they are doing.

(Long-term male resident (1))

Indeed, for others, it meant a sense of place attachment that generated its own comfortable inertia since:

Everyone knows where they are. There is nothing we can't find in East London.

(Long-term male resident (2))

It should be noted that this did not automatically translate into territoriality, as various participants were at pains to stress.

I don't feel that just because I'm from East London I'm the only person who can walk around East London. I'm not going to be territorial and say get out, I like to mix around a lot of types of people.

(Long-term male resident (1))

In total, it should be noted that based on the participants' status as 'resisters' or 'desisters', crime was described as something that was fixed in a number of points, usually outside one's neighbourhood, and that respondents had to navigate through or glide around. It appeared that a sense of danger was important for identity formation: a spatial test of my respondents' capacity to understand and overcome the unfamiliar and potentially dangerous city on the route to becoming adult. The complexity of this process illustrates the dynamics of territorial stigmatisation by showing how it is not a static condition or a neutral process. As Wacquant *et al.* (2014, p. 1271) have asserted it is 'a consequential and injurious form of action through collective representation fastened on space' to which it appears that the young people who are represented in this chapter have fashioned specific, locally attuned responses. I have endeavoured to show that when young people have a platform to voice their opinions, they demonstrate a nuanced understanding of why they and their peers do what they do. Insights like these can only run counter to those facile policy proposals to exogenously dilute spatial concentrations of long-term poverty by increasing so-called 'social mix'. In the face of this, new ecologies are created that defy or reframe efforts to rebrand areas. Potentially much could be learnt by looking at the subtleties of youth territoriality to undermine the 'invention of the notion of "underclass area" as a precinct of concentrated pathology' (ibid, p. 1272) or of 'Broken Britain' (see Slater, 2014), particularly if we see the lived experience of young people as valuable in and of itself (Skelton, 2013). It must be worth acknowledging that a way out of the typical (youth and urban) policy debate impasse is to enable young people to do the talking and policy makers to listen to them.

Notes

1 These terms are borrowed from Murray (2009).
2 According to the website home.co.uk, average prices in September 2012 were over £650,000 with detached housing selling for over a million pounds sterling.
3 Collected by the social research companies BMRB (Social Research); NOP World and MORI, the survey had seven reiterations and over 15,770 participants.
4 The event itself was reported in the Guardian "Victoria station stabbing: 20 arrested over knife killing of teenager", Glenn McMahon, *The Guardian*, Friday 26 March 2010.

References

Adekunle, A. B. (2013) '"You have to represent your endz": Youth territoriality in London.' Unpublished PhD dissertation, University College London.

Altman, I. (1975), *The Environment and Social Behavior: Privacy, Territoriality, Crowding and Personal Space*, Monterey, CA: Brooks-Cole.

L'Aoustet, O. and Griffet, J. (2001), 'The experience of teenagers at Marseilles' skate park: Emergence and evaluation of an urban sports site', *Cities*, 18(6): 413–418.

Anderson, E. (2000), *Code of the Street: Decency, Violence, and the Moral Life of the Inner City*, New York: W. W. Norton & Company.

Bannister, J., Pickering, J., Reid, M. and Suzuki, N. (2008), *Young People and Territoriality in British Cities*, York: Joseph Rowntree Foundation.

Biggs, S. (1989), 'Resource-poor farmer participation in research: A synthesis of experiences from nine national agricultural research systems.' OFCOR Comparative Study Paper 3. The Hague: International Service for National Agricultural Research.

Bourdieu, P. (1986), 'The forms of capital.' In J. Richardson (ed.) *Handbook of Theory and Research for the Sociology of Education*, New York: Greenwood, pp. 241–258.

Brookman, F., Bennett, T., Hochstetler, A., and Copes, H. (2011), 'The "code of the street" and the generation of street violence in the UK', *European Journal of Criminology* 8(1): 17–31.

Brower, S. N. (1980), 'Territory in urban settings', in Altman, I., Rapaport, A. and Wohlwill, J. F. (eds), *Human Behavior and Environment*, New York: Plenum Press, pp. 179–207.

Butcher, M. and Dickens, L. (2015), *Creating Hackney as Home*. Available online at: http://www.hackneyashome.co.uk/sites/default/files/documents/CHAsH_2015_FinalReport_webversion.pdf (accessed 26 January 2016).

Cahill, C. (2000), 'Street literacy: Urban teenagers' strategies for negotiating their neighbourhood', *Journal of Youth Studies*, 3(3): 251–277.

Cairns, L. (2001), 'Investing in children: Learning how to promote the rights of all children', *Children & Society*, 15(5): 347–360.

Calhoun, C. (1991), 'Indirect relationships and imagined communities: Large-scale social integration and the transformation of everyday life', in Bourdieu, P. and Coleman, J. S., (eds), *Social Theory for a Changing Society*, Boulder, CA: Westview Press,

Childress, H. (2000), *Landscapes of Betrayal, Landscapes of Joy: Curtisville in the Lives of its Teenagers*. Albany, NY: SUNY Press.

Childress, H. (2004), 'Teenagers, territory and the appropriation of space', *Childhood*, 11(2): 195–205.

Cole, I. and Goodchild, B. (2000) 'Social mix and the "balanced community" in British housing policy – a tale of two epochs', *GeoJournal* 51(4): 351–360.

Cresswell, T. (2009) 'Place.' In Kitchin, R. and Thrift, N. (eds) *International Encyclopaedia of Human Geography*. Oxford: Elsevier, pp. 169–177.

Edney, J. J. (1976), 'Human territories: Comment on functional properties', *Environment and Behavior*, 8(1): 31–47.

Evans, K. (1997) '"It's alright 'round here if you're a local": Community in the inner city', in P. Hoggett (ed.) *Contested Communities: Experiences, Struggles, Policies*. Cambridge: Polity Press.

Foucault, M. (1980), *Power/Knowledge: Selected Interviews and Other Writings, 1972–1977*, New York: Pantheon Publishing.

Furlong, A. (1992), *Growing up in a Classless Society? School to Work Transitions*, Edinburgh: Edinburgh University Press.

Furlong, A., Woodman, D. and Wyn, J. (2011), 'Changing times, changing perspectives: Reconciling "transition" and "cultural" perspectives on youth and young adulthood.' *Journal of Sociology* 47(4): 355–370.

Giddens, A. (1984), *The Constitution of Society: Outline of the Theory of Structuration*, California: University of California Press.

Gunter, A. (2008), 'Growing up bad: Black youth, "road" culture and badness in an East London neighbourhood', *Crime, media, culture* 4(3): 349–366.

Hall, E. T. (1959), *The Silent Language*, New York: Anchor Press.

Ingold, T. (1987), *The Appropriation of Nature: Essays on Human Ecology and Social Relations*, Manchester: Manchester University Press.

Johnston, R. (1996), 'Territoriality'. In A. J. Kuper (ed.) *The Social Science Encyclopaedia*, London: Routledge.

Kärrholm, M. (2007), 'The materiality of territorial production a conceptual discussion of territoriality, materiality, and the everyday life of public space', *Space and Culture*, 10(4): 437–453.

Kellett, M. (2004), '"Just teach us the skills please, we'll do the rest": Empowering ten-year-olds as active researchers', *Children & Society*, 18(5): 329–343.

Kintrea, K., Bannister, J. and Pickering, J. (2010), 'Territoriality and disadvantage among young people: An exploratory study of six British neighbourhoods', *Journal of Housing and the Built Environment*, 25(4): 447–465.

Kintrea, K., Bannister, J. and Pickering, J. (2011), '"It's just an area: everybody represents it": Exploring young people's territorial behaviour in British cities.' In Goldson, B. (ed.) *Youth in Crisis? 'Gangs', Territoriality and Violence*, London: Routledge, pp. 55–71.

Lees, L. (2008), '"Gentrification and social mixing: Towards an inclusive urban renaissance?" *Urban Studies* 45(12): 2449–2470.

Legg, S. (2011), 'Assemblage/apparatus: Using Deleuze and Foucault', *Area*, 43(2): 128–133.

Lyng, S. (1990), 'Edgework: A social psychological analysis of voluntary risk taking', *American Journal of Sociology*, 95(4): 851–886.

Mackett, R., Brown, B., Gong, Y., Kitazawa, K. and Paskins, J. (2007). 'Children's independent movement in the local environment. Built Environment.' *Final report*, London: University College London.

Matthews, H., Limb, M. and Percy-Smith, B. (1998), 'Changing worlds: The microgeographies of young teenagers', *Tijdschrift voor Economische en Sociale Geografie*, 89(2): 193–202.

McGrellis, S. (2004), 'Pushing the boundaries in Northern Ireland: Young people, violence and sectarianism', London: London South Bank University, 2004.

McGloin, J. M. and Collins, M. E. (2015), 'Micro-level processes of the gang', in Decker, S. and Pyrooz, D. C. (eds), *The Handbook of Gangs*, 276–293, New Jersey: Wiley-Blackwell.

Murray, C. (2009), 'Typologies of young resisters and desisters', *Youth Justice*, 9(2): 115–129.

Pickering, J., Kintrea, K. and Bannister, J. (2012), 'Invisible walls and visible youth territoriality among young people in British cities', *Urban Studies*, 49(5): 945–960.

Poche, B. (1986), '"Localité" et subdivisions spatiales du social: pour une définition culturelle', *Espaces et Sociétés*, 48–49: 225–238.

Nayak, A. (2003), *Race, Place and Globalization: Youth Cultures in a Changing World*, New York: Berg Publishers.

ONS (Office of National Statistics) (2011), Regional Trends, No. 43, Edition. Permalink: http://webarchive.nationalarchives.gov.uk/20160105160709/http://www.ons.gov.uk/ons/rel/regional-trends/regional-trends/no--43--2011-edition/index.html (accessed 19 January 2017).

Robb, M. (ed.) (2007), *Youth in Context: Frameworks, Settings and Encounters*, London: Sage.

Rowe A. (2006), 'The effect of involvement in participatory research on parent researchers in a sure start programme.' *Health and Social Care in the Community*, 14(1): 465–473.

Rose, D., Germain, A., Bacqué, M-H., Bridge, G., Fijalkow, Y., and Slater, T. (2013), '"Social Mix" and neighbourhood revitalization in a transatlantic perspective: Comparing local policy discourses and expectations in Paris (France), Bristol (UK) and Montréal (Canada)." *International Journal of Urban and Regional Research*, 37(2): 430–450.

Sack, R. (1986), *Human Territoriality, Its Theory and History*, Cambridge: Cambridge University Press.

Sack, R. D. (1993), 'The power of place and space', *Geographical Review*, 83(3): 326–329.

Sarkissian, W. (1976), 'The idea of social mix in town planning: An historical review', *Urban Studies* 13(3): 231–246.

Shildrick, T., Blackman, S. and MacDonald, R. (2009). 'Young people, class and place', *Journal of Youth Studies*, 12(5): 457–465.

Shils, E. (1975), *Periphery: Essays in Macrosociology*, Chicago: University of Chicago Press.

Slater, T. (2013), 'Your life chances affect where you live: A critique of the "cottage industry of neighbourhood effects research", *International Journal of Urban and Regional Research*, 37(2): 367–387.

Skelton, T. (2013), 'Young people's urban im/mobilities: Relationality and identity formation', *Urban Studies*, 50(3): 467–483.

Skelton, T. and Gough, K. V. (2013), 'Introduction: Young people's im/mobile urban geographies', *Urban Studies* 50(3): 455–466.

Skey, M. (2011), *National Belonging and Everyday Life*, Basingstoke: Palgrave Macmillan.

Smith, M. K. (2005), 'Detached, street-based and project work with young people', in *The Encyclopaedia of Informal Education*, Available online at: http://infed.org/mobi/detached-street-based-and-project-work-with-young-people/ (accessed 12 August 2015).

Tomaney, J. (2013) 'Parochialism – a defence', *Progress in Human Geography*, 37(5): 658–672.

Travlou, P., Owens, P. E., Thompson, C. W. and Maxwell, L. (2008), 'Place mapping with teenagers: Locating their territories and documenting their experience of the public realm', *Children's Geographies*, 6(3): 309–326.

Valentine, G. (1996), 'Children should be seen and not heard: The production and transgression of adults' public space', *Urban Geography*, 17(3): 205–220.

Valentine, G. (2003), 'Boundary crossings: Transitions from childhood to adulthood', *Children's Geographies*, 1(1): 37–52.

Wacquant, L. (2008), 'Ordering Insecurity', *Radical Philosophy Review,* 11(1): 1–19.

Wacquant, L., Slater, T. and Pereira, V. (2014) 'Territorial stigmatization in action', *Environment and Planning A*, 46(6): 1270–1280.

Wolch, J. and Dear, M. (eds) (1989), *The Power of Geography: How Territory Shapes Social Life*, Boston, MA: Routledge.

Zerubavel, E. (1985), *Hidden Rhythms: Schedules and Calendars in Social Life*, California: University of California Press.

13 Call it by its proper name! Territory-ism and territorial stigmatisation as a dynamic model

The case of Old Naledi

Klaus Geiselhart

Abstract

Emerging debates point to possible shortcomings in relation to Loïc Wacquant's elaboration of the concept of territorial stigmatisation. One of the most frequently raised issues relates to the assumption that residents of neighbourhoods or cities with negative reputations necessarily internalise the stigmatising discourses. Indeed, in a number of his writings, Wacquant formulates the thesis according to which certain neighbourhoods become vilified in increasingly polarising cities. This, he argues, contributes to the marginalisation of the residents of these areas. The process operates in much the same way that stigmatisation does in Erving Goffman's now famous account, whereby it 'spoils the identity' of those who are forced to bear it. However, in his work Wacquant adopts Goffman's theory of stigmatisation without reference to the critiques that it has generated. This chapter highlights the ways in which this initial shortcoming has resulted in a theory that is defamatory, in that it ignores many highly relevant processes of identity building, moving beyond theoretical depictions of inescapable marginality and suffering. I argue that an integrative perspective on territorial stigmatisation, and respective discrimination (territory-ism), is more adequate in order to avoid the vilification of certain groups or the population of a number of neighbourhoods. On top of this, the chapter distinguishes category-based processes from individuating processes of stigmatisation and discrimination, which is particularly helpful in order to analyse the local dynamics of segregation or integration.

Introduction

In a valuable contribution on the topic, Tom Slater (2015) summarises the different branches of literature on territorial stigmatisation. He states that some works tend to focus on the production of territorial stigma through an analysis of the symbolic power of negative images that are reproduced by the media, politicians or other public actors. By focusing on processes of investment and gentrification, these writings point the finger at a number of private interests (e.g. investors, politicians) that use negative images about neighbourhoods in order to justify lucrative upgrading schemes. Yet another series of works has concentrated on the activation of territorial stigma and assesses how penal policies are put into action.

These have focussed on the ways in which residents of vilified places take on the negative images of their neighbourhood. As a consequence of this and of other experiences associated to stigma, they are said to feel ashamed. The last series of works that Slater identifies (2015, p. 11) contest Wacquant's assumption that residents cannot escape the internalisation of negative images and prejudices. These are works that concentrate on residents who contest territorial stigmatisation, and 'present a more positive picture of collective defiance and defence of residents in response to the denigration of their communities' (ibid., p. 12). Summarising that territorial stigmatisation is 'not a "neighbourhood effect", but a gaze trained on the neighbourhood', Slater (2015, p. 12) uses the word 'stigmatisation' to denote the perpetuation of existing prejudices or in the sense whereby it is something unloaded onto something or someone – something or someone is then considered through a discrediting image. In this chapter, I argue that when we use the word 'stigmatisation', we often do so when we really mean 'discrimination'. When discrediting images are repeatedly reproduced in the media, by politicians, or through the language of administrations and state institutions, this has been referred to as structural discrimination or structural violence, by Galtung (1969), one of the key thinkers on the topic. We need to think about whether it is more adequate to call it discrimination when someone deliberately produces discrediting prejudices about a certain group of people. In my research on the academic usage of both 'stigmatisation' and 'discrimination', I have found that the ways in which these terms are discussed in the social sciences has tended to confuse the two, resulting in an overall vagueness.

When the words 'stigma' and 'discrimination' are referred to in the same publication, they are often used interchangeably and these scholarly works are thus not discriminative enough.[1] 'Being discriminative' can also mean 'being capable of making fine distinctions', and indeed the term derives from the Latin origin '*discriminare*', which simply means 'to distinguish'. Discrimination can be understood as the act of distinguishing individuals on the basis of the perception of attributes that they bear, and then on establishing what the consequences might be. My point here is that we as scholars should be more discriminative when it comes to distinguishing between discrimination and stigmatisation.

Interestingly, there is a strong tendency to associate specific issues with *either* stigma *or* discrimination (Geiselhart 2009, p. 52). Social exclusion on the grounds of affiliation with a race or religion, for instance, is usually explained in terms of discrimination. But what exactly is the essential difference between those attributes of people that are mostly described as having suffered from discrimination and those that are regarded as stigmatised? It is not easy to identify an essential difference between belonging to a certain race, sex, or religion, and the property of having a life-threatening illness, a mental disorder, or that of being poor. The latter traits are mostly associated to stigmata. The commonsense understanding of the criteria for differentiation between the two categories is that race or gender, for instance, are socially understood not to constitute a meaningful difference. Structural disadvantages, or the unequal treatment of persons that differ in these categories would be called racism or sexism. On the other hand, it is assumed people who might be disabled suffer from their special attributes and that this hampers their ability to participate in society on an equal footing – in that sense, they are not denied equal rights. When we think of stigmatisation, it is assumed that the character of the attribute separates or disadvantages these people. Society is not blamed for the existence of a stigma; rather, a stigma exists per se, but society can become more integrative and benevolent. Such commonsense understandings facilitate Goffman's (1990 [1963]) catchy definition of stigma as a deviation from the norm and, as a consequence, the ubiquitous use of his concept of stigma – as well as Wacquant's (2007) concept of territorial stigmatisation, which plays a central role in this book.

What I try to show is that the conceptualisation of stigma that Goffman develops, and which is followed through by Wacquant, has a tendency to reify stigmas as given, inescapable discrediting marks. Indeed, the understanding presents marginalised people as victims and victims are not usually considered to be in control of the situations that they find themselves in. The status of a victim automatically brings to mind images of helplessness and subjection. Goffman and Wacquant's conceptualisation does not truly engage with the agency of those who have a stigma. In this way, although they would perceive themselves as advocates of the marginalised, they somehow deny these groups and individuals the possibility of being truly respected. If one assumes – as these authors do – that marginalised or stigmatised groups cannot escape the roles that the majority of society imposes upon them (Goffman 1990 [1963]), this leads to a further assumption in the case of Wacquant: populations of certain deprived neighbourhoods will automatically feel ashamed about their geographical origin. This disregards the potential for people to cope with lives under difficult circumstances, and to develop self-esteem, pride and a sense of dignity. It also ignores the potential for these same people to feel a sense of belonging to the places where they reside.

I suggest that discrimination and stigmatisation are intertwined social processes that trigger each other. Those who are discriminated against can feel ashamed and might begin to blame themselves for the reasons that have led to their discrimination: this is stigmatisation. Stigma-associated behaviours, such as self-exclusion or resistance, can foster discrediting prejudices and thus lead to discrimination. However, this is not always the whole picture. The case of Old Naledi, which is a deprived and often vilified township in Botswana's capital city, Gaborone, will be introduced later in this chapter to show that there are multiple ways of dealing with place-attached stigma. There, and in many marginalised townships, emancipatory movements argue for their right to reside in stigmatised places and to organise and design these spaces themselves. Expressions of popular culture often represent deprived townships as places where one feels a sense of home as well as pride. It is essential to look at these politics of place which include both: discriminating actions against certain neighbourhoods and their inhabitants, which I refer to as *territory-ism* and discuss below; and second, territorial stigmatisation, a process during which people are confronted with prejudices about their neighbourhood or town of origin, but it is a two-way process that implies the possibility that they can learn to cope and counter such notions.

I will first show how Wacquant and Goffman are brothers in spirit and explain how their use of words bears consequences in contemporary debates. I shall then develop an integrative concept of territory-ism and territorial stigmatisation before applying this to Old Naledi.

From Goffman to Wacquant: reductionist accounts of stigma

Goffman's view on stigma has been frequently contested, in large part because of its assumption of a normative order (Weiss and Ramakrishna, 2001; Kusow, 2004). The notion of stigma that he develops has also been described as too narrow, too focused on the individual level, and it has been critiqued for the fact that it is unable to help design interventions that might reduce stigmatisation. As such, it is descriptive rather than prescriptive (Sayce, 1998; Parker and Aggleton, 2002; Jewkes, 2006).

However, it is not one of Wacquant's main purposes to critically examine the term stigma, and as such he borrows the concept directly from Goffman:

> It is remarkable that Erving Goffman (1963) does not mention place of residence as one of the 'disabilities' that can 'disqualify the individual' and deprive him or her from 'full acceptance by others'. Yet territorial infamy displays properties analogous to those of bodily, moral, and tribal stigmata, and it poses dilemmas of information management, identity, and social relations quite similar to these, even as it also sports distinctive properties of its own. Of the three main types of stigma catalogued by Goffman (1963: 4–5) – 'abominations of the body', 'blemishes of individual character' and marks of 'race, nation and religion' – it is to the third that territorial stigma is akin, since, like the latter, it 'can be transmitted through lineages and equally contaminate all members of a family'.
>
> (Wacquant 2007, p. 67)

Undoubtedly, residing in a disreputable neighbourhood – or township – can have serious consequences on the sense of self of those living in these places. Along with the resulting self-blame, this can further result in the mistreatment and discrimination of these very residents. However, Wacquant falls into the traps that emerge from within Erving Goffman's theory.

In *Stigma*, Goffman (1990 [1963]) describes the grievances of people who are socially segregated. The term 'stigma' is 'used to refer to an attribute that is deeply discrediting' (Goffman 1990 [1963], 13). He analyses stigma from the 'common sense' point of view, whereby stigma is seen as an attribute that is contrary to existing norms. He lays the foundation for an understanding of how salient attributes may lead to processes of social exclusion. His theory thus gives us an insight into how the so-called 'stigmatised' exclude themselves and anticipate rejection. According to Goffman, stigma emerges in 'mixed contacts', whereby the bearer of a discrediting attribute meets someone who does not bear that sign. The interaction within a 'mixed contact' will be clouded by the inability to handle the deviance. The bearers of a stigma necessarily experience that they cannot truly blend in as 'normals'. Consequently, the stigmatised person will continuously anticipate rejection in further social situations. Ultimately, the bearers of a stigma will have to accept that the only chance of living within a community will be to take on the role that the 'normals' provide for people 'like them'.

At this point, it is important to note that Goffman does not consider that norms might be flexible. Accordingly, he only lists three possible reactions that the 'stigmatised' might have in response to their 'stigma' (Goffman 1990 [1963], pp. 154f and 19ff). The first of these possibilities is that a 'stigmatised' person can attempt to 'correct' his or her stigma. If the direct elimination of the respective attribute is not possible, the person can try to indirectly correct his or her identity. For instance, this could happen if a stigmatised person attempted to gain respect in realms that are ordinarily restricted to people without the attributes that they bear – for example when a physically handicapped person participates in sports competitions with non-handicapped people. Second, Goffman states that the 'stigmatised' can attempt to avoid 'mixed contacts' entirely. This can be accomplished by avoiding appearances in places where one would assume such contacts to be possible. It can also be performed through 'covering' and 'passing'. If the stigma is not visible, the 'stigmatised' person can in effect bluff and act like other people (covering). Similarly, if the person cannot hide the stigma, they can attempt to learn to behave as if they did not bear the respective discrediting characteristic (passing). The example does not visibly do away with the stigma but the adjusted behaviour diminishes the counterpart's awareness of it.

Goffman describes the third way that one can handle a stigma in the following way:

> Finally, the person with a shameful differentness can break with what is called reality, and obstinately attempt to employ an unconventional interpretation of the character of his social identity. The stigmatized individual is likely to use his stigma for 'secondary gains', as an excuse for ill success that has come his way for other reasons.
>
> (Goffman 1990 [1963], p. 21)

It is tempting to read this passage and to find it compelling and astute, yet it can also be read as deeply problematic. For instance, it can lead us to think of social movements in a tainted light. It reads as though we should assume that all activists who stand up for their rights as marginalised groups must in fact be shameful about their difference. In affirming slogans such as 'Black is beautiful', or the famous feminist slogan 'We can do it', the bearers of 'stigmas' project their difference as something that is positive and a trait that one should be proud of. However, if we take Goffman's description literally, the implication is that people using such slogans do so in order to gain some publicity, as a secondary gain and as a compensation for their 'unfortunate' shortcomings. In this description, the inadequacy of Goffman's theorisation and approach becomes apparent. An over-reliance on Goffman's work pushes us to conceive of these activists, or of anyone with a stigma, as being ultimately 'spoiled' in their identities, eliminating the credibility of certain types of declarations, such as those I have listed above declarations, or if they act in ways that would otherwise suggest that they do not feel 'discredited'. Goffman creates a vilifying vision of the way in which people try to cope with discrimination. This notion is not only defamatory but it ultimately contests some form of human dignity, in that it denies certain people the option of feeling proud of themselves (both individually, and collectively).

With this in mind, it is somehow curious that those groups that are marginalised often make use of Goffman and his stigma theory in order to underline the situations in which they find themselves (e.g. Visser and Mhone, 2002 [2001]). This is most likely linked to a common perception of Goffman's work where he is often perceived as an advocate of those who are marginalised. Indeed, he certainly provides insights into the situations and the hardships of socially excluded people. However, there is something fundamentally wrong about the fact that, for him, in the end it is the stigma, not society that produces inequalities.

Wacquant's contributions to sociology work in similar ways. He never refers to the possibility that inhabitants of deprived places might develop attachments to their local environment despite the vilification of their neighbourhood. Of course, residents certainly experience hardship and rejection because of where they live, but there are plenty of examples of situations in which such residents manage their lives and cope with the pressures that are imposed upon them (Kirkness, 2014; Slater and Anderson, 2012). The limitation of Wacquant's conceptualisation boils down to a lack of focus on the ways in which poor people experience their neighbourhoods as their homes, or the social networks that are often established through families and friendships. If we limit ourselves to reading Wacquant, it is not possible to understand how such residents might be proud of themselves and of the place in which they live – probably as a result of having successfully mastered unfortunate circumstances. The mere possibility that they might, at the very least, feel ambivalent about the neighbourhood is also not construed as a possibility. With this in mind, it is clear that a more discriminative concept of stigmatisation is necessary.

An integrative concept of territory-ism and territorial stigmatisation

Initially, the surprising finding that HIV and AIDS-related stigmatisation was higher in the urban deprived township of Old Naledi when compared to rural villages was the foundation for developing the framework that I present here. It is commonly assumed that urban dwellers tend to be more open with regard to innovations and I had also observed that Old Naledi had much better access to effective health services than rural areas. So, I was intrigued by the fact that urban dwellers did not cope better with the stigma. It seemed natural to assume that the negative image of the area contributes to the negative self-images of people living with HIV and AIDS. However, my research revealed that it was less the negative image of the urban neighbourhood of Old Naledi that enhanced stigmatisation, rather it was the social cohesion that could be found in the rural communities which effectively reduced stigmatisation (Geiselhart 2009; for similar findings see Wutich *et al.*, 2014).

The concept of the interplay of stigmatisation and discrimination that I present below draws from a number of academic writings about stigma and discrimination. Some of the most recent work on stigma no longer sets itself the objective of revealing the internal processes of stigmatisation. Instead, it has tended to concentrate on causes and effects that are mostly derived from applied sciences (Nyblade *et al.*, 2003; Kidd and Clay, 2003; Bond and Mathur, 2003; Ogden and Nyblade, 2006; Duvvury *et al.*, 2006). These works are very helpful for intervention programmes. For instance, some develop tools to measure stigma (USAID, 2006); others aim at identifying the public health implications of stigma (Link, 2001); and others are especially useful because of the tools for intervention that they design (Kidd and Clay, 2003). It is also of note that these recent publications contribute to our understanding of stigma as more complex than it is described in the past. For instance, new important conceptualisations of stigma, such as the distinction between 'identity stigma' and 'treatment stigma' (Stuber and Schlesinger, 2006), indicate that there is a self-related and a community-related aspect of stigma.

Along with Sayce (1998) and Jewkes (2006), I would argue that the common usage of the term stigma is deeply problematic in that this notion does not do enough to understand the social dynamics of social differentiation, such as segregation and exclusion, as well as the complete ostracism or mistreatment of those who bear 'abnormal' attributes. Thus I argue that, in addition to stigma, we need the notion of discrimination. It needs to be applied because feeling stigmatised is simply not possible without the experience of having suffered discrimination. Each and every stigma finds its respective form of discrimination, whether it is racism, classism, sexism and so on (Lott and Maluso, 1995). Each [discrimin]-ism finds its own characteristic patterns of related actions, its special form of enactment. Labelling, stereotyping and discrimination are important factors of the interplay between discrimination and stigma (see Link, 2001).

The basic framework, as it is outlined here (see Figure 13.1), derives stigma and discrimination from interpretive processes of the people involved.[2] Stigma and discrimination are like the two sides of a coin and should always be considered together. Stigmatisation refers to the perspective of those who bear a salient attribute, and discrimination to that of those who become aware of a person bearing a salient attribute. It is the latter who might either regard the salient attribute as discrediting or out of the 'ordinary'. Processes of stigma and discrimination are interlinked on an interpersonal level where people experience being discriminated against and might thus also experience stigma in the form of self-blaming, for instance. Following from that, stigmatisation and discrimination can influence and enhance each other.

One main distinction needs to be made between category-based and individuating processes (see Figure 13.1). In category-based processes, people employ interpretations that are made on the basis of pre-existing stereotypes and social identities; these are reflections of the mechanisms that propel the vicious circle

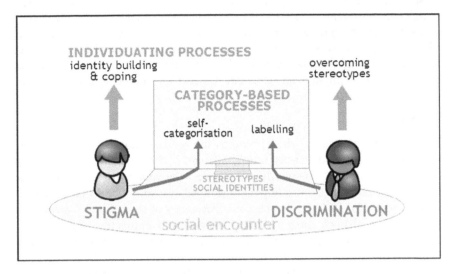

Figure 13.1 A basic framework of stigma and discrimination.

Source: Klaus Geiselhart.

of self-categorisation and labelling. However, the individuating processes point to ways in which one can exit the circle. Once a person becomes known as an acquaintance, other attributes besides the stigma come into play. The person is no longer viewed solely in terms of the prejudices that are attributed to the stigma but for other characteristics as well. When the social counterpart of a person who bears a stigma is made aware of the individuality of that person, the person stops being considered a 'typical' member of the defamed group. The social counterpart can thus overcome existing prejudices. On the other hand, the bearers of a stigma can commit themselves to a process of identity building, coping and emancipation. In this way, they might help non-bearers to learn more about a certain attribute and underline the falsities buried within any set of given prejudices. On an interactional basis, individuals might overcome stereotypes. The individuating processes are thus the key to social change in the field of stigmatisation and discrimination. It is a change that is induced either from the 'bottom up', or through activists or advocates aiming to change societal conditions they they experience as unbearable, divisive or intolerable. People who counter stigmatisation and discrimination try to change conditions by reflecting on the way they themselves and others interpret and deal with stigmata.

To illustrate this short depiction of a dynamic concept of the interplay (Wechselwirkungen) between stigmatisation and discrimination, I will discuss the example of Old Naledi, where I have been conducting research since 2002. A main finding of my research was that where people know each other through regular social encounters, prejudices and stereotypes lose the power that they have to determine how a person is looked upon. This insight gives a hint of how to reveal the specific form of discrimination against people from a certain geographical space. I call this *territory-ism*. This functions to assess if and how people experience stigmatisation due to their spatial origin or where they live. Furthermore, it is necessary to look at the dynamic of the social structure and social interactions within a neighbourhood and to ask what situations beyond the borders of the neighbourhood might make its dwellers to feel they are being discriminated.

Old Naledi

Old Naledi is an urban neighbourhood in the southern outskirts of Gaborone. According to the official census, it officially had 21,693 inhabitants in 2001 and 19,079 in 2011. Yet it is very difficult to assess the veracity of these estimates, and one can find figures indicating that the population might be 43,000[3], 46,001[4] or even more than 50,002.[5] The obvious inaccuracy of the official figures is an indication that urban planning has partly lost control over this township.

Old Naledi is exceptional in that it was fully accepted as a squatter settlement by the government in the early 1960s, during the initial phase of the construction of Gaborone. The workers who settled there had been expected to leave Gaborone at the end of the construction work, but as is often the case in such situations, they stayed put (a situation that is reminiscent of what occurred in the Varjão, see Haferburg and Huchzermeyer's chapter, this edited volume). In the course of the growth of the city and as urbanisation progressed, an increasing number of people settled in Gaborone, and as a result of a general lack of available housing, a great many migrants settled in Old Naledi. Eventually, in 1975, the government recognised the neighbourhood as an official residential one, and ceased to consider it an industrial zone. Whereas there had been strong restrictions on residential plots in Gaborone, stipulating that plots in residential areas should be at least 400m², this law only applied to a minority of the compounds in Old Naledi where plot size initially varied greatly. The SHHA[6] altered the standards in order to help Old Naledi to develop and ultimately, the minimum standard was reduced to 200m² as long as plots had access to a road and that a pit latrine was set up. Whereas the SHHA also imposed a standard that allowed constructions for one household per plot only, this was not the case in Old Naledia, where several 'upgrading programmes' were implemented. As a result, the neighbourhood has retained its unlegislated character and Old Naledi is still one of the first places where rural-urban migrants find shelter when they head for Gaborone, safe in the knowledge that it will be easier for them to find accommodation than in the rest of the city (Krüger, 1994; Gwebu, 2003).

Today, Old Naledi sports many structures that do not meet official standards. The typical building is set up on a rectangular ground plan with a corrugated iron roof. These housing units have several doors, each leading into rooms of approximately 9m² and accessible directly from the compound. Each room is typically rented out to families with up to five or six people, and with an average of 27 people living on a plot, the structures – along with the entire township – are extremely overcrowded, causing social, environmental and health problems (Gwebu, 2003). No matter the construction type, plot owners know that tenants can be found in Old Naledia, even for the simplest of wooden shacks. Ultimately, Old Naledi's 'illegal' character and the fact that no one knows how many residents live there prevent a sustainable upgrading of the area. During an interview, an employee of the Gaborone City Council town planning unit pointed to this issue and expressed the fact that an upgrading of the very dense housing area would require the relocation of inhabitants. But the council realises that this is a problem, since there are no accurate figures relating to how many people live in these homes.

Informal occupation and small-scale self-employment are the major sources of monetary income, but many dwellers face wretched living conditions. The level of education of residents is low, which minimises the opportunities for formal employment (Gwebu, 2003), and as Gaborone's population is continually rising, competition is also increasing. The expansion of Gaborone's peripheral residential suburbs is a clear indicator of the dynamism that Gaborone is experiencing. The downside of this urban vitality can be witnessed in Old Naledi.

Old Naledi – homestead or unfamiliar place?

In Gaborone and beyond, Old Naledi is publicly known as a place in which outsiders would not venture into at night. For Wacquant, Old Naledi could be described as one of 'those urban hellholes in which violence, vice, and dereliction are the order of things' (Wacquant, 2007: 67). It is also true for Old Naledi that 'discourses of vilification proliferate and agglomerate [. . .] 'from below', in the ordinary interactions of daily life, as well as 'from above', in the journalistic, political and bureaucratic (and even scientific) fields' (Wacquant, 2007: 67).

In reality, there is in Old Naledi a local social structure that originates in the township's historical development. After it was recognised as a residential area in 1975, the neighbourhood has been partitioned into compounds and these were allocated to the residents who had been living there legally. Today, the owners of these plots have a privileged role in the neighbourhood as they can rely on a steady income generated through renting out rooms. On top of this, since they are often the direct descendants of the initial residents, they can also rely on social contacts amongst the other long-established families and from the fact that they grew up in the neighbourhood. There is a strong sense of social cohesion amongst these residents due to personal relations and deep social networks.[7] The people of Old Naledi organise social support initiatives such as HIV/AIDS self-help groups or leisure activities such as sport teams which proudly represent Old Naledi in regional matches and competitions. There are plenty of other examples, in other world regions, where social links are 'important in facilitating a sense of safety and belonging, even in estates which to outsiders could well be regarded as "rough" or dangerous places' (Watt, 2006: 786). This is a vitally important point, and Old Naledi does not stand alone in this respect.

However, it is true to say that not all Old Naledi inhabitants are part of these networks. Many urban-rural migrants remain unknown to established communities. Due to the overcrowding and the tendency of residents to hide illegal structures through the use of fences and walls, the traditional venue of Botswana's social living, originally taking place in open compounds, was relocated to other places hidden from view, often inside the houses. As a consequence, the degree of anonymity rose, especially for those who had no chance to socialise and meet with others. Migrants experienced difficult financial situations in which they had to rely on occasional poor-paying jobs, but also because they had to pay rents to their Old Naledi landlords. In part because of the anonymity in which they evolved, these residents have mostly remained relatively secretive about their HIV infection if they carried it. This was in large part a consequence of hoping that they would not end up being one of those people who contribute to the prejudices against Old Naledi. But how can these findings be transferred to the issue of place-attached stigmatisation and discrimination? In order to assess the dynamics of Old Naledi's stigmatisation and discrimination it is necessary to look at incidences that influence the image of the neighbourhood and events or efforts that either enhance social cohesion or diminish it. In the following section, my own observations from regular visits to the neighbourhood over the past years, and information from talks with residents will be supplemented with an analysis of articles about Old Naledi in the main Botswana daily newspaper *Mmegi*.

Old Naledi's politics of place

Since the early 1980s, the opposition party, Botswana National Front (BNF), has garnered massive support in Old Naledi. People there have long felt neglected by the ruling Botswana Democratic Party (BDP), which they perceive as representatives of the privileged classes. 'In the 1990s, BNF support swelled so much that BDP rallies there often turned violent as the people of the ghetto threatened to remove ruling party activists from their neighbourhood' (*Mmegi*, 2009b). In 2009 the BDP won the elections as usual. However, *unusually* a large number of Old Naledi's residents had voted for the BDP. The elected president, Ian Khama, then addressed the township during a massive victory rally. For some residents, this was an affront and a sign of the loss of Old Naledi's identity as a township, which was once regarded as the very heart of the BNF political party (*Mmegi*, 2009b). Some residents have always muttered that 'the BDP-run government won't avail funds to develop Naledi until they vote for BDP' (*Mmegi*, 2004).

The latest upgrading of Old Naledi started in 2007 and was intended to improve the sewer and water network, storm water drainage and street lighting. Some roads were also built, replacing the gravelly pathways in the township. As a result of the infrastructural work in the neighbourhood, 50 plot owners had to be removed. According to official accounts, the residents all agreed on compensation before being relocated; however, forced evictions were enacted in a number of cases (*Mmegi*, 2008b). Because of this, the project was often subject to complaints. Ultimately, work did not proceed smoothly which affected the lives of number of residents and imposed several constraints. For instance, when the work on the drainage system came to a standstill for some months, the trenches were left unattended leaving some inhabitants without access to their homes. Accustomed to making do, inhabitants resorted to assembling home-made bridges, which then blocked the drainage system and caused floods during the rainy season. Several homes then had their water taps disconnected and thus people had to queue at public stand pipes. Originally, the project was supposed to end in 2010, but it was not before 2012 that the facelift was officially completed (*Mmegi*, 2012). However, in 2013 it was still not possible to shut down the public water standpipes because so many residents still relied on them. The quality of the work, especially the tarring of the roads, was frequently criticised. Officials had to confess this: 'The project is a disaster and there is nothing to celebrate because the works are substandard, and there is need for resurfacing of some of the roads' (*Mmegi*, 2011b). The improvement of the neighbourhood overall as a result of the project is questionable.

It is all but impossible to state for certain that residents were attended to so poorly as a consequence of discrimination of officials aimed towards Old Naledi and its residents, yet for many this conclusion was an obvious one to reach. Regardless of whether the area suffered from deliberate discrimination, a clear case of structural territory-ism affected the neighbourhood. Leaving a house for several months without proper direct access would have been unimaginable in more well-off areas.

Old Naledi was also subject to other initiatives aimed at improving its image, and reducing stigma and discrimination. For instance, a market-place was built as a centre of art and gardening. The colourful structure of rock and brick was intended as a piece of art but was also suitable as a meeting place. As a direct result of the existence of the market, people planted flowers and amateur artists created sculptures all over Old Naledi. 'The new Old Naledi is symbolic of a paradise regained [. . .] where newly invented gardens are speaking volumes of the residents of this location's determination to rid themselves of the "shabbiest, dirtiest location in town" stigma.' (*Mmegi*, 2009a). In 2011, the Dulux Botswana Company trained '20 unemployed youths from Old Naledi' in order 'to give this iconic settlement a bit of colour' (*Mmegi*, 2011a). This novel initiative was part of the 'Let's Colour the World' project, where deprived neighbourhoods worldwide became the subjects of mass painting projects destined to infuse colour in places that were conceived of as otherwise bleak. In 2013, '[*Arts for Change*], a new youth-based initiative in Gaborone, had a vision to not only destigmatize street art but utilize it to uplift the community and inspire the youth' (Modak, 2013). Graffiti artists sprayed tuck shops and hosted workshops for youths all over the neighbourhood.

Members of a new generation have chosen the township in which they grew up as their homes, feeling the place is a great place to live in. A young graphic designer with whom I talked explained that:

I have lived in both poor neighbourhoods such as here in Naledi, and very wealthy ones like when I was studying outside. Naledi remains close to my heart. I would not live anywhere else. Naledi has a unique character about it that no other place has.

(*Mmegi*, 2004)

A picture in the newspaper *Mmegi* displays Tebogo Sebogo – who had become an affluent lawyer – as a role model for outsiders to demonstrate the potentials of the township, as well as for insiders because he runs a football team and sponsors matches: 'Sebego is proud of his Old Naledi roots and feels it is the township that made him what he is' (*Mmegi*, 2008a). In opposition to its defamatory image, which is still prevalent, in some media reports Old Naledi is described as 'iconic' and is presented as a vivid, creative and progressive place.

All these initiatives vary in the long-term effects that they will have had on the townships' image and on the given structure and problems of the township. Notwithstanding, in Old Naledi it would appear that residents of the deprived township 'enjoy a strong sense of community; they have access to dense networks of friendship and support, local amenities and convenience, and services and agencies that suit their needs' (August, 2014, p. 1317). This is not to say that Old Naledi is not facing real problems deriving from poverty, overpopulation, and insecurity and criminality. Such factors obviously influence the development of discrediting images that represent residents of certain neighbourhoods. However, when one takes into account the opinions and life experiences of insiders, a more multifaceted picture emerges than when only the perspective of outsiders, officials or media reports is taken into account. Currently, the future of Old Naledi looks favourable. Due to the immense growth of Gaborone, the bordering villages of Tlokweng and Mogoditshane are expereriencing intense overcrowding and a growth in crime, probably more uncontrollable than Old Naledi. Furthermore, the fact that Old Naledi is situated within the city boundaries entitles its residents to certain provisions from the city council – even if services are relatively basic, and some would say, poor. As space in town gets increasingly scarce it is just a matter of time before plots in Old Naledi will see a rise in value. Notwithstanding the potentially difficult consequences for some residents, this will change the dynamics surrounding the township's image.

Assessing dynamics of territory-ism and territorial stigmatisation

A discriminative use of the terms stigmatisation and discrimination can help us to analyse the social dynamics of a neighbourhood's reputation. On the one hand, the focus should lie on analysing discrimination, and we should ask which actors discriminate against a place. This might be done with positive intentions in order to organise help or to implement welfare, but we need to recognise that such efforts can easily end up reproducing negative images. It can also be done viciously in order to serve interests, for example if investors deliberately produce negative images because of their profits (Slater, 2015). The latter certainly deserves the name territory-ism. On the other hand, it is necessary to analyse whether people from inside a residential area take on and internalise prejudices and stereotypes and how they participate in combating them. The latter is the perspective on stigma.

In this respect, all sorts of actors need to be taken into consideration. They may be individuals, community groups or local authorities. But beyond the local level

Table 13.1 Measures to reduce or enhance stereotypes.

Measure	Inclusive	Rejective
Influencing bonds between people	Supporting acquaintance	Enhancing anonymity
Creating present-time orientation	Festival	Rioting
Commanding	Prohibiting excessively/ aggressively discriminative behaviour	Segregation law
Giving examples	Role-model strategy	Scapegoat strategy
Inducing expectations	Benefit from approximation	Benefit from segregation
Emphasising norms	Overall norms, e.g. humanity	In-group norms, e.g. loyalty
Contradictions	Display	Gloss over, disguise

also national and international levels need to be taken into account. Political will, metaphors, cultural values, international aid, economic interests, various service providers and civic society can interfere in local conditions. It is particularly interesting to know whether an actor supports the assumption that a negative image applies to the whole of the neighbourhood or whether efforts are intended to make clear that a neighbourhood always hosts a multifaceted picture of different individuals and that it is worth to look on their individual capacities and capabilities. This is the difference between category-based processes of stigmatisation and discrimination or individuating assessments of bearers of a stigma. In this respect it is also important whether a deprived neighbourhood is unquestioned equated with other structurally similar townships without looking at the uniqueness of that particular neighbourhood.

Table 13.1 shows several measures that can be taken to either reduce or enhance category-based thinking. The table is compiled from the works of Fiske and Neuberg (1989), Bar-Tal (1989) and Zimbardo (2004). Each measure can be taken to either enforce or alleviate socially divisive practices and can thus be used either in an integrative or a rejective way.

- Social cohesion in a neighbourhood is not independent of influences from outside. Infrastructural developments that cut through important places or that result in the development of incongruous structures may destroy the basis of sound neighbourhood bonds. At its most extreme, the relocation of people is a possibility. Social institutions are dependent on financial support and thus rely on political will. As such, political influences from inside and outside can also disrupt a neighbourhood's solidarity. If there is pressure to implement substantial change through urban 'redevelopment' and planning, it is possible that efforts will be taken to destroy social cohesion within the township because less resistance will be expected from neighbourhoods in which anonymity reigns.
- The degree of people's social accountability can be reduced or enhanced by inducing intense emotions. This can be observed whenever a mass

movement is organised and its members join together. In such moments, the readiness to follow suggested ideas is high, and worries about self-evaluation are reduced. While festivals can be utilised to reinforce a sense of community, rioting against certain groups, people or situations is also possible. The latter can be used for furthering the negative reputation of inhabitants of an area.

- Authorities, or actors who claim to be authorities, can try to command certain modes of behaviour. Such pressures restrain the choices that people have, and they can then obey the rule or instruction, or end up bearing the consequences. An obvious example of massive structural spatial discrimination was Apartheid which enforced state racism by imposing spatial divisions under threat of (often extreme) punishment. Such commanding structures can also be installed informally by gangs or criminal organisations.

- Individual cases of an event or a person originating from a specific neighbourhood can be referred to and made public in order to confirm or refute existing prejudices and stereotypes. For example, incidences of crime or violence might be publicly asserted as typical examples of the ways in which people from a township or neighbourhood behave. Such scapegoat strategies are meant to reinforce stereotypical thinking. On the other hand, a role model strategy would point to events or people who were successful in some way, or creative individuals originating from a township in order to promote identification. This can also be done in order to show outsiders that stereotypical thinking is an over-generalisation, and aims at integration.

- It can be argued that acknowledging the potential of a township's residents will ultimately benefit the wider population, e.g. the whole city. Otherwise, it might be asserted that the township will threaten the wider community for the foreseeable future, leading to increased segregation and securitisation. Raising such expectations will result in integrative or rejective behaviour.

- Moral norms can be put into place. Calls for mutual respect and the appreciation of others are one possibility, as well as calls to promote the integration of out-groups (emphasis is placed on 'humanity' instead of difference). On the other end of the spectrum, the loyalty developed through in-group relations can be utilised to reinforce discrimination against out-group members.

- There might be incompatibilities between a general assumption or belief and what the acceptance of a stereotype would imply. For example, regarding a township's inhabitants as useless might conflict with the fact that many of them work in low-wage jobs, e.g. in factories or as house servants and are thus important for non-township residents to maintain a certain lifestyle or a degree of economic wealth. Such contradictions can either be displayed, in order to make people aware of their conflicting assumptions, or they can be glossed over and disguised in order to support rejective behaviour.

Ultimately, being aware of such measures helps to analyse the effects and dynamics of place-attached images as well as helping to separate actors who contribute to integration from those who foster exclusion.

Conclusion

Analysing the forces that compete in defining neighbourhood images reveals the dynamics of space-based stigmatisation and discrimination. With discrimination it is difficult to draw the line between good and bad, because naming real problems might help to allocate aid and assistance but it also might reproduce discrediting images and thus contribute to stigma and unequal treatment. Supplying food might be effective in mitigating the worst hunger but some could see it as displaying the indigence of the neighbourhood. Intensified presence of police might help to reduce crime but might also be taken as a visible sign that the place is being driven by violence. It is therefore essential to try to understand how such actions are framed, namely either with an attitude of solidarity or one that casts the beneficiaries as a social burden. However, there are incidences that can clearly be called territory-ism. Is a township deliberately discriminated against for the interests of outsiders? Does a city council deny or restrict its services to a neighbourhood? Do employers or administrations systematically deny employment or services because of where an applicant comes from?

If prejudices lead to inhabitants of a certain neighbourhood to be disadvantaged, then these people have to deal or cope with this experience, which is always challenging. I have argued that it is defamatory to assume that all inhabitants will internalise the stigma. One should always consider the possibility that residents might not regard the negative image as something that applies to them. As a result, there are a number of questions that need to be asked: are they successful in coping with the stigma, e.g. through the establishment of a counter image? Do a township's inhabitants find ways to take organised action against the serious difficulties that they face? Have they developed a strong sense of community, notwithstanding the grievances that they face on a daily basis (perhaps, even as a result of this)? Do the residents make any efforts to alter the public image of the township?

The dynamics of stigmatisation and discrimination can take on a variety of forms. Distinguishing discrimination from stigmatisation helps to separate socially inclusive actions from those that alienate social inequalities. Only an advanced notion of stigmatisation, which does not deny the possibility of successful coping mechanisms, helps to see residents of deprived townships in their capacities and ultimately respects their dignity. Regarding the politics of place in this way is essential to understand how a township might develop in the future and whether actions should be taken in order to diminish social divisions.

Notes

1 For an extensive reflection, see Geiselhart (2009, p. 52).
2 For a more detailed explanation of this concept, see Geiselhart (2009, p. 144).
3 Mmegi (2007) cites a 2001 study conducted by Haas Consult as evidence for the overpopulation of Old Naledi.
4 This figure is given by the "Gaborone City Development Plan, 1997–2021", 1997, quoted in Gwebu (2003, p. 417).

5 Estimate given verbally by an employee of the Gaborone City Council town planning unit during the fieldwork in 2007.
6 SHHA (Self-Help Housing Agency) is an organisation that has supported the expansion of home ownership to low-income households since the early 1990s.
7 For an illustrative report see the four-part essays in *Mmegi*, 2004.

References

August, M. (2014), 'Challenging the rhetoric of stigmatization: The benefits of concentrated poverty in Toronto's Regent Park', *Environment and Planning A*, 46(6): 1317–1333.
Bar-Tal, Y. (1989), 'Can leaders change followers' stereotypes?', in Bar-Tal, D. (ed.) *Stereotyping and Prejudice. Changing Conceptions*, Berlin: Springer-Verlag: 225–242.
Bond, V. and Mathur, S. (2003), 'Stigma associated with opportunistic infections and HIV and AIDS in Zambia.' Available online at: http://www.icrw.org/docs/stigmaoi.pdf (accessed 16 August 2006).
Duvvury, N., et al. (2006), 'HIV & AIDS – Stigma and violence reduction intervention manual.' Available online at: http://www.icrw.org/docs/2006_SVRI-Manual.pdf (accessed 16 August 2006).
Fiske, S. T. and Neuberg, S. L. (1989), 'Category-based and individuating processes as a function of information and motivation: Evidence from our laboratory', in Bar-Tal, D. (ed.) *Stereotyping and Prejudice. Changing Conceptions*, Berlin: Springer-Verlag: 83–103.
Galtung, J. (1969), 'Violence, peace, and peace research', *Journal of Peace Research*, (6): 167–191.
Geiselhart, K. (2009): 'Stigma and discrimination. Social encounters, identity and space.' Avaialable online at: http://nbn-resolving.de/urn:nbn:de:0168-ssoar-290930 (accessed 7 April 2016)
Goffman, E. (1990 [1963]), *Stigma. Notes on the Management of Spoiled Identity*, London: Penguin.
Gwebu, T. D. (2003), 'Environmental problems among low income urban residents: An empirical analysis of Old Naledi-Gaborone, Botswana', *Habitat International*, 27(3): 407–427.
Jewkes, R. (2006), 'Beyond stigma: Social responses to HIV in South Africa', *The Lancet*, 368(9534): 430–431.
Kidd, R. and Clay, S. (2003), 'Understanding and challenging HIV Stigma. Toolkit for action.' Available online at: http://www.aed.org/ToolsandPublications/upload/StigmaToolkit.pdf (accessed 9 February 2006).
Kirkness, P. (2014), 'The *cités* strike back: Restive responses to territorial taint in the French *banlieues*', *Environment and Planning A, 46(6): 1281–1296.*
Krüger, F. (1994), 'Urbanization and vulnerable urban groups in Gaborone/Botswana', *GeoJournal*, 34(3): 287–293.
Kusow, A. M. (2004), 'Contesting stigma: On Goffman's assumptions of normative order', *Symbolic Interaction*, 27(2): 179–197.
Link, B. G. (2001), 'On stigma and its public health implications', *Background paper of the conference: Stigma and Global Health*. Available online at: http://www.stigma conference.nih.gov (accessed 4 November 2005).
Lott, B. and Maluso, D. (eds) (1995). *The Social Psychology of Interpersonal Discrimination*, London: Guilford Press.

Mmegi (2004), 'In the heart of the hood. Part 1–4' (accessed 25 September 2015). https://www.mmegi.bw/2004/June/Tuesday15/1009491627870.html.

Mmegi (2007), 'Old Naledi set for P100m upgrading', (accessed 25 September 2015). http://www.mmegi.bw/index.php?sid=1&aid=10&dir=2007/july/Monday30.

Mmegi (2008a), 'Tebogo Sebego: Up from poverty', (accessed 25 September 2015). http://www.mmegi.bw/index.php?sid=1&aid=1&dir=2008/January/Monday28.

Mmegi (2008b), 'Old Naledi project stalls', (accessed 25 September 2015). http://www.mmegi.bw/index.php?sid=6&aid=11&dir=2008/August/Friday15.

Mmegi (2009a), 'Old Naledi: From "paradise lost to paradise regained"', (accessed 25 September 2015). http://www.mmegi.bw/index.php?sid=7&aid=28&dir=2009/May/Friday22.

Mmegi (2009b), 'The symbolism of Khama in Old Naledi', (accessed 25 September 2015). www.mmegi.bw/index.php?sid=6&aid=198&dir=2009/October/Friday23.

Mmegi (2011a), '"Iconic" Old Naledi township gets colour' (accessed 25 September 2015). http://www.mmegi.bw/index.php?sid=7&aid=1729&dir=2011/March/Thursday17.

Mmegi (2011b), 'New roads lead to new frustrations', (accessed 25 September 2015). http://www.mmegi.bw/index.php?sid=1&aid=32&dir=2011/August/Friday26.

Mmegi (2012): 'Old Naledi facelift completed, says mayor', (accessed 25 September 2015). http://www.mmegi.bw/index.php?sid=1&aid=263&dir=2012/march/tuesday27.

Modak (2013), 'Arts for change: Colour in Old Naledi', (accessed 25 September 2015). http://fulbright.mtvu.com/smodak/2013/07/31/arts-for-change-color-in-naledi.

Nyblade, L., et al. (2003), 'Disentangling HIV and AIDS stigma in Ethiopia, Tanzania and Zambia' (accessed 15 August 2006). http://www.icrw.org/docs/stigmareport093003.pdf.

Ogden, J. and Nyblade, L. (2006), 'Common at its Core: HIV-Related Stigma. Across Contexts', (accessed 16 August 2006). http://www.icrw.org/docs/2005_report_stigma_synthesis.pdf.

Parker, R. and Aggleton, P. (2002), 'HIV/AIDS-related stigma and discrimination: A conceptual framework and an agenda for action', (accessed 15 August 2006). http://www.popcouncil.org/pdfs/horizons/sdcncptlfrmwrk.pdf.

Sayce, L. (1998), 'Stigma, discrimination and social exclusion: What's in a word?', *Journal of Mental Health*, 7(4): 331–343.

Slater, T. (2015), 'Territorial stigmatization: Symbolic defamation and the contemporary metropolis', in Hanigan, J. and Richards, G. (eds) *The Handbook of New Urban Studies*. Pre-print online version, (accessed 19 April 2016). http://www.geos.ed.ac.uk/homes/tslater/terstig_handbookurbanstudies.pdf.

Slater, T. and Anderson, N. (2012), 'The reputational ghetto: territorial stigmatisation in St Paul's, Bristol', *Transactions of the Institute of British Geographers*, 37(4): 530–546.

Stuber, J. and Schlesinger, M. (2006), 'Sources of stigma for means-tested government programs', *Social Science & Medicine*, 63: 933–945.

USAID (2006), 'Can we measure HIV/AIDS-related stigma and discrimination? Current knowledge about quantifying stigma in developing countries', accessed 16 August 2006. http://www.icrw.org.

Visser, A. and Mhone, H. D. (2002 [2001]), 'Need assessment report: People living with HIV/Aids in Botswana', Gaborone.

Wacquant, L. (2007), 'Territorial stigmatisation in the age of advanced marginality', *Thesis Eleven*, 91: 66–77.

Watt, P. (2006), 'Respectability, roughness and "race": Neigbourhood place images and the making of working-class social distinctions in London', *International Journal of Urban and Regional Research*, 30(4): 776–797.

Weiss, M. G. and Ramakrishna, J. (2001), 'Stigma interventions and research for international health', *Background paper of the Conference: Stigma and Global Health*, accessed 4 November 2005. http://www.stigmaconference.nih.gov.

Wutich, A., et. al. (2014), 'Stigmatized neighborhoods, social bonding, and health', *Medical Anthropology Quarterly*, 28(4): 556–577.

Zimbardo, P. G. (2004), 'A situationist perspective on the psychology of evil: Understanding how good people are transformed into perpetrators', in Miller, A. G. (ed.), *The Social Psychology of Good and Evil*, New York: Guilford Press, pp. 21–50.

14 Territorial stigmatisation, gentrification and class struggle

An interview with Tom Slater

Interviewed by Paul Kirkness and Andreas Tijé-Dra

Ultimately, we would like to use this opportunity to invite you to retrace your thoughts on the matter of territorial stigmatisation. But, as an internationally recognised scholar, some people might be more aware of your reflections on the topic of gentrification. Was there any key personal experience that somehow led to your work on issues of gentrification, displacement and neighbourhood stigmatisation?

My interest in the closely related issues of gentrification, displacement and territorial stigmatisation goes back to my undergraduate days in London from 1995–8. I was a student at the wonderful Department of Geography at Queen Mary, University of London (in those days Queen Mary and Westfield College), and the extraordinary teachers there (especially David M. Smith and Miles Ogborn) helped me to understand the profound changes I was observing not only in east London (the university campus is on Mile End Road), but also where I lived in Tooting (south London). The same week in which I completed my final exams, the landlord of the small Victorian terraced house that I was sharing with two others informed us that he had sold the house for a 'packet', which didn't surprise me at all (you could almost smell the gentrification in Tooting at the time!). The landlord had always been prompt if we encountered any problems with the house, and he did show considerable regret that we had to find somewhere else to live pretty quickly, but the minor upheaval created by this particular incident of displacement did make me wonder how we might have reacted and coped if we had put down roots in the neighbourhood, how we would have found somewhere else to live if we had been elderly, socially isolated, or unemployed with young children, to name but a few vulnerable, stigmatised categories. As it transpired, we could not find anywhere else to live in Tooting, and had to migrate to the more expensive district of Clapham, where we moved into the very first flat we could find: a fairly cramped space with numerous maintenance issues and a far less competent landlord. In the midst of a few petty battles with this individual over the inadequacy of the central heating system and his failure to install a new kitchen properly (or even install it at all, for the first few weeks that we lived there), I recall thinking, 'What if we were more vulnerable?'

But the real game-changer for me was Toronto, where I did my PhD fieldwork (2000–2). Following the suggestion from another graduate student, I immersed myself in the neighbourhood of South Parkdale, one of the few areas of central Toronto yet to experience intense gentrification, and therefore one of the last bastions of affordable housing in the central city area. It quickly proved to be the ideal observatory for a methodical analysis of capital flows, displacement, stigma, and class struggle brought about by neoliberal municipal and provincial policy. Much of the neighbourhood's housing stock consisted of large Victorian residences that had been illegally subdivided into tiny apartments known locally as 'bachelorettes'. These apartments had two rooms, one serving as the living and sleeping space with a kitchen in the corner, the other being a bathroom. Many of the occupants of these bachelorettes were people who had been discharged from two nearby psychiatric hospitals in the 1980s under the rhetoric of 'care in the community', the language used by the Ontario provincial government to justify massive mental health care spending cuts. But these bachelorettes also housed new immigrants to Canada, including many political refugees seeking asylum. For over three decades, and due to the affordability of its housing stock and the existence of social networks amongst those newly arrived immigrants, South Parkdale had been a crucial settlement location for people with nowhere else to go.

At the time that I began my research, gentrification in South Parkdale was just beginning to gather momentum, for three reasons. First, surrounding neighbourhoods had gentrified, and the central city housing market was becoming tighter and tighter, causing middle-class people to look towards areas where affordable and attractive properties could be purchased or rented. Second, the City of Toronto government was attempting to eradicate the significant stigma attached to South Parkdale via a policy of legalising and cleaning up its bachelorette buildings – part of an overt effort to attract middle-class families to live in the district, justified by the language of creating a 'more balanced community profile' in a part of the city famous for having a high percentage of single persons living in small rented dwellings. Third, some landlords were taking advantage of the 1998 weakening of tenant protection by the Provincial (Ontario) government by making life very difficult for their tenants, with the hope of getting them out of their properties and attracting a wealthier class of tenant, or selling the buildings they owned for a tidy profit. Once I had established the causes of gentrification via a varied panoply of research methods (analysis of official statistics, the scrutiny of policy documents, lengthy interviews with community organisers, government officials, middle-class homeowners and artists, and ethnographic observation of various neighbourhood forums), it struck me as impossible to gain a full understanding of the impact of gentrification on the existing residents of South Parkdale without talking to those low-income tenants living at the bottom of the class structure. With help from people who I encountered (officially, a strategy of 'snowball sampling'), I managed to make contact with people for whom the gentrification of their neighbourhood was nothing but stressful, worrying, and upsetting. Most of these particular interviews were quite harrowing, listening to the stories of socially isolated people struggling to make rent at the same time as they were struggling with debilitating mental health problems for which 'care in the community' was a spurious, empty slogan. But one encounter in particular has haunted me ever since, and will all my life.

One afternoon in March 2001 I went to meet a young woman, a Tibetan immigrant who arrived in Toronto with her family in 1998 as political refugees. She invited me into her home, a bachelorette apartment that housed her, her husband and their three young children. The place was remarkably tidy and clean given the overcrowding, but there was a pervasive smell of damp in the entire building, and the alarming cracks in the plaster of the walls suggested major structural faults. I sat at their small dining table whilst she served me a cup of Jasmine tea. I asked her about her landlord, and she responded at length with a terrible litany of problems including harassment, bullying, rent increases, and unresolved maintenance issues. By the end of our conversation – interview seems a rather cardboard word for it – she was in tears as she reflected on her family's situation. If they complained, they faced eviction from the building, and if they didn't complain, they remained tenants trapped with an unethical landlord apparently devoid of compassion, living in sub-standard accommodation and dreading another rent increase letter. As I thanked her for her time and stood up to leave, she showed me the awful letters she had received from her landlord (as if proof were needed), and then asked politely if there was anything I could do to help her and her family. These are the situations that 'social research methods', classes and textbooks cannot prepare you for. This woman had welcomed me, a complete stranger, into her home, and told me all about the miserable predicament that she and her family were in, and I was leaving with what social scientists call 'data' whilst leaving her with tears running down her cheeks from the ordeal of describing her situation. As I wiped my own tears away on the streetcar heading back towards my apartment, I felt that the very least I could do was fulfil the moral obligation of writing about the social injustice that was the gentrification of a neighbourhood in the West End of Toronto. This was, after all, just one example of many sad situations brought about by the fact that there was a lot of money to be made from the changes happening in that neighbourhood. The crucial analytic lesson was that the stigma attached to South Parkdale, far from being a barrier to investment, was actually very convenient indeed for institutions and individuals trying to extract profit from urban space.

You were evidently already involved in determining what the material and social consequences of gentrification are, but could this experience have participated in forging the academic that you are today, and your understanding of what academia is supposed to strive to achieve?

After completing my fieldwork in Toronto and then in New York City, where equally depressing accounts of displacement were conveyed by my interviewees, I arrived back in Britain in September 2002, more or less in the middle of writing up my thesis. That month I headed for a conference in Glasgow entitled 'Upward Neighbourhood Trajectories: Gentrification in a New Century', in order to present my work, meet other scholars working on similar questions, and learn about gentrification in other contexts. Over those two days some presentations were very informative and displayed a clear sense of social justice on the part of the presenters, but my lasting memory of the conference is the sense of genuine surprise and dismay I felt listening to the perspectives offered in many presentations, particularly from scholars based in Britain. In a breathtaking retreat from social critique, there were numerous accounts delivered under the rubric of 'positive gentrification', 'sustainable regeneration', 'middle-class belonging' and 'urban renaissance', to name a few. Some scholars were even talking about 'policy-driven gentrification' as if it was a new development – before going on to advocate such policy! After my own presentation, which I gabbled dreadfully because I was nervous that my critical take on the process would not be appreciated by the audience, I was asked by one delegate, 'You told us about all the negative things you found, but where are the positives?'

I hoped that this conference was to prove an aberration (perhaps a reflection of its slightly boosterish title), but in the years that followed it was impossible to ignore a general trend in the literature on gentrification, matching a trend in the research-funding proposals and manuscripts that I was receiving for the duty of peer review. Many scholars were writing about middle-class gentrifiers as if they were the only 'agents' involved in gentrification, the only voices that mattered. A torrent of publications – outputs from a stream of policy-oriented research projects – had emerged on gentrifiers' everyday lives, dilemmas, desires, all divorced from any structural context but positioned as theoretically salient via appeals to Pierre Bourdieu's masterbook *Distinction* (1984), yet completely missing the political purchase of that book and of Bourdieu's scholarship in general (it was telling that nobody felt it relevant to quote Bourdieu on the not insignificant matter of what he famously called *La Misère du Monde*!). Particularly troublesome was the apparent rightward shift of scholars once at the forefront of analysing class inequality in cities, who were now smoothing over inequality and smooching policy with writings that showed not a glimpse of the analytical and political tradition from which they emerged. Endowed with plausibility by the sheer weight of 'policy relevance', and reinforced by thinly veiled sentiments that effaced the working classes whilst empathising with the middle classes, for me such scholarship truncated and distorted an understanding of the ongoing articulation of class and space in cities. This was happening at the same time as high-profile work emerging from America which argued that concerns about displacement caused by gentrification were overblown, as government housing databases apparently offered little evidence of high rates of 'exit' of low-income people from gentrifying neighbourhoods. In February 2006, the day I received yet another manuscript to review on the dilemmas of the middle classes (in gentrifying London), I felt compelled to intervene, so I penned a critique of the literature entitled 'The eviction of critical perspectives from gentrification research.' Judging by the correspondence that I received (and still receive) from those supporting my argument and those angered by it, it seemed to touch a nerve. I have always felt that scholars have something of a civic duty to intervene when myths and misunderstandings begin to circulate, and also to shine light on issues that are too important to be left without critical analysis and reflection.

Your work on gentrification has led to several oft-quoted articles, but it has also led you to collaborate with people like Elvin Wyly and Loretta Lees in one of the most comprehensive texts on the topic (2008). We would be interested to know more about how working with these scholars influenced your own thought.

In 2003 I took up my first teaching position at the University of Bristol in a policy studies department. As most junior scholars in their first teaching post will tell you, it's hard to get any research done when you are starting out, as you have to write a lot of lectures from scratch (or, at the very least, reshape existing lectures you inherit). Bristol was really tough – I was made to do an enormous amount of teaching, at one point teaching on nine different undergraduate courses! When I was putting together a couple of lectures on gentrification, I noticed very quickly – with some surprise – that there was no book that took a panoramic view of a huge literature and made sense of it all for students. E-mail conversations with Loretta and Elvin followed, and we quickly put our heads together and proposed the *Gentrification* book to Routledge. I think we all enjoyed writing it, and I learned a great deal from the challenges of co-authorship, where we had to work really hard to tie together our sometimes quite contrasting thoughts and approaches into a coherent narrative. From Loretta I learned about class analysis and conceptual questions relating to geographical scale; from Elvin I learned about capital flows, land rent, and the power of quantitative methods propelled by a critical imagination. I hope that book helped (and still helps) a lot of students understand gentrification: why it happens, who decides, and who suffers because of it – and how people have fought back.

You have also had the good fortune to dialogue with people like Peter Marcuse and Neil Smith, who has sadly left us a few years ago and you have movingly written about his work – and his personality – in a recent tribute (2012). You also wrote a strong defence of Marcuse's critical stance in 'Missing Marcuse' (2009). Once again, we can only assume that both these renowned urban scholars have had some influence on your own work. Are they your main inspirations for critical scholarship?

There is no question about that! All of us have certain writers who move us, inspire us, guide us in some way. Peter's intellectual stamina is extraordinary – he's nearly 90 years old and I wish I had his analytical and political energy at my current age! Peter was very helpful to me when I started thinking in much more depth about the displacement question. The papers he wrote in the mid 1980s on the causes and consequences of displacement in New York City are masterclasses in the logic of concept formation and in theoretical/empirical clarity. When all that work came out in America denying displacement via statistics on low-income household mobility, it seemed essential to revisit those papers as they offered an understanding of displacement beyond the human ecological tradition of 'who moves in and who moves out' of neighbourhoods. To help with an analysis of displacement in overheated real estate markets, Peter brings to the table *exclusionary displacement* (when a low-income household is prevented from living where it otherwise may have lived due to gentrification) and *displacement pressure* (when a low-income household voluntarily vacates a home because a neighbourhood is changing so dramatically it no longer offers the vital, affordable and support services it once did). I had found evidence of this in my own fieldwork, and so much scholarship was coming out in all kinds of contexts showing the same dynamics.

What Neil Smith's work means to me and to so many critical social scientists is well known. He left behind a glorious body of politically engaged scholarship that will continue to amaze and inspire future generations of students, scholars, and activists. One thing I could add to the tribute I wrote is that there are lessons to be learned from Neil in terms of how to write. He was a beautiful writer, with an extraordinary talent for articulating complex ideas and processes in accessible, lively, provocative, elegant prose. I despair at the grotesquely opaque, obfuscatory, pretentious writing I sometimes encounter in academic journals, and British male cultural geographers are among the very worst culprits as, (mis)guided by Nigel Thrift, they struggle to wrestle the latest faddish theorist or concept to the ground. Perhaps I am missing something in this work, but I have tried to read some of these contributions and I simply do not understand large chunks of them. Now, I am prepared to admit that this could be because I am in need of some kind of cerebral reboot, and it could be because Thrift is not (to me anyway) a clear writer. But it worries me that these authors cannot see that incoherent, deliberately ambiguous, foggy text is likely to achieve very little politically; on that point, many do not even appear to recognise that some of the theorists whose ideas they are working with (e.g. Deleuze) were Marxists.

You have become something of an academic-activist, stressing, as you often have, that the university should remain a critical institution. Some see tension between these two poles and others even go as far as to claim an incommensurability between these two positions, pointing to the fact that the interests involved in each case somehow diverge. How can one respond to this? Is it possible to reconcile academic analysis as data gathering and theory building with emancipative goals, and did you experience such tensions in your own fieldwork?

Not in my fieldwork – more in my workplaces. I think it was the late great Stuart Hall who said that 'the university is a critical institution or it is nothing'. Wise words that should be etched above all university administration buildings! Since getting my PhD in 2003 I have been stunned by the appalling neoliberal assault on universities in the UK, and watched the decimation of morale and collective endeavor via all sorts of mechanisms imposed by the state, of which the 'Research Excellence Framework' (previously the 'Research Assessment Exercise') is most prominent. Neoliberal managers and supine senior professors, via a toxic blend of power, greed and neurosis, bring great shame upon themselves by showing not the least hint of squeamishness when pushing ahead with audit directives from above. The categories are so preposterous: how can scholarship be judged 'world leading' when the panel members are often evaluating work way outside their area of research specialisation, and when they have not read work published in any other languages? (I wish postcolonial theorists would start pulling apart the REF 'world leading' category!). The pawns and quislings of audit games also miss striking ironies: the latest obsession with research 'impact' purports to be about academics engaging with the world, when in concrete practice it requires untold thousands of person hours – and massive financial resources that could be directed towards research (especially postgraduate research) – engaging exclusively with ourselves. I cannot imagine anything more insular, introverted and irrelevant than telling academics to jockey for position and promotion via the production of 'pathways to impact' strategies that are seldom much different from crystal ball fortune telling. I wonder what would happen if all the bigwigs on the REF panels realised the power of collective social action, came together and stated, 'We are not doing this any more!' HEFCE[1] would have a huge problem!

Are there ways to escape from this quagmire? Of course, but they are enormously difficult to achieve under the working conditions typical of the neoliberal university, where so much of academic staff time is spent locked into a bureaucratic cell that drains people of intellectual creativity and energy. Especially for early career scholars who may feel they can do little to subvert and speak back to power, I am increasingly convinced by radical pedagogy: the craft of nurturing and shaping critical citizens through theory and practice. The most exciting university teachers I have ever seen are the ones who invite undergraduate students to reflect on the particular circumstances of their own lives (e.g. getting into grotesque debt in order to graduate), and then help them radically analyse the brutal coupling of state power with free market logic behind it all (that had been naturalised and normalised prior to their arrival at university). I think this is what Stuart Hall was getting at in the abovementioned quote: higher education can and should be transformative in the sense of challenging students to see the world through critical lenses. Only by understanding why it is that higher education has been so utterly commodified can students reject the awful language of whether they are 'satisfied' or not (on which my colleague Julie Cupples wrote a devastating critique[2]), and perhaps gain the ideas and courage needed to formulate strategies of resistance and revolt. Look at what happened in Chile from 2011 onwards: a geography undergraduate who had learned all about neoliberalism acquired the knowledge and courage to lead a student movement that ultimately led to the abolition of tuition fees.

In your early work on gentrification (2004a, 2004b, 2006), you may not say so explicitly, but territorial stigmatisation is always somehow present. How would you say that these two concepts co-exist and do they reinforce each other in some way?

In numerous contexts there is a very strong relationship between the active defamation of declining working-class areas and their revamping for and by private real estate through public policies calibrated to the wishes and purchasing power of middle-class households. A raft of studies has revealed that state agents fashion and fasten territorial disrepute onto neighbourhoods (or even parts of neighbourhoods) and then invoke that very disrepute to justify gentrification strategies of varying intensities. For example, I have lost count of how many times I have read about stigma being activated and amplified in order to raze large public housing estates anchoring it, in utter disregard for the collective needs of their denizens. I think it's also very important to register the multisided political manipulation of territorial taint, whereby it supplies the symbolic springboard for and practical target of state-driven gentrification while also fostering the censorship of alternative policies of social investment that would frontally tackle poverty and housing disrepair in the chosen area. Once certain places become universally renowned and reviled across class and space as redoubts of self-inflicted and self-perpetuating destitution and depravity – and pictured as vortexes and vectors of social disintegration, fundamentally dissolute and irretrievably disorganised – then it becomes far, far easier for policy officials and developers to justify a wholesale transformation of those places and the expulsion of their residents as socially progressive 'regeneration'. Even that term, 'regeneration', is a stigmatising label, for it suggests that the place to be regenerated is full of degenerate individuals (and the same with 'revitalisation' – what was not vital about people living there before?). In England, the 'sink estate' has been invented by journalists and is repeatedly invoked by politicians (particularly since Tony Blair's 1997 visit to the Aylesbury Estate in South London to form his Social Exclusion Unit) to dramatise and denounce the poverty in disinvested council housing estates. Incidents of deviance or violence in and around these areas are routinely sensationalised and referred back to the allegedly intrinsic sociocultural traits of the residents fit to brand them as outcasts. Such symbolic buckling can quickly turn any neighborhood sporting drab but adequate housing into the spectre of a hostile environment ready to erupt in mayhem any minute. This is very convenient indeed for institutions waiting to profit from housing demolition and a thorough class makeover of urban space.

For those who might not have read it, could you perhaps summarise your main argument in the article entitled 'Your Life Chances Affect Where You Live' (2013)? Would you also like to tell us why you think the concept of territorial stigma provides a more adequate analytical lens when doing research on 'deprived neighbourhoods' than some of the literature that has focused on 'neighbourhood effects'?

In February 2010 I travelled to St Andrews to attend an ESRC-funded seminar on neighbourhood effects, the first of three seminars that led to three edited volumes on the subject. The seminar was entitled 'Neighbourhood Effects: Theory and Evidence' and it featured a truly international cast of speakers, drawing an admirably large and diverse audience consisting of academics from several disciplines, urban planners and policy makers. I arrived with an open mind, not knowing very much about the neighbourhood effects literature, and hoping to be enlightened theoretically, conceptually, methodologically, perhaps politically. But although the event took place in an excellent, collegial spirit, and the levels of scholarly accomplishment and analytic perspicacity were impressive, I left the event shaking my head in sheer frustration, bewildered and bemused by the failure of every speaker to engage in matters of political economy, to acknowledge the structural causes of poverty, to pay much attention to the role of the state and of the institutional arrangements that would seem impossible to ignore in all the discussions of how 'where you live affects your life chances'. It seemed that there was an absolutely fundamental structural question being ignored: *why do people live where they do in cities?* If where any given individual lives affects their life chances as deeply as neighbourhood effects proponents believe, it seems crucial to understand why that individual is living there in the first place. Life chances will of course be very different for residents of very different neighbourhoods, but stating the obvious and 'controlling' for various externalities (especially popular amongst statistically oriented urban sociologists) does not explain why such urban inequality exists. So, if we invert the neighbourhood effects thesis to: *your life chances affect where you live*, then the problem becomes one of understanding life chances via a theory of capital accumulation and class struggle in cities. Such a theory provides an understanding of the injustices inherent in letting the market (buttressed by the state) be the force that determines the cost of housing, and correspondingly, the major determinant of where people live.

I set myself the challenge of wading into that huge literature on neighbourhood effects, and the more I read, the more annoyed I became! Most authors in the genre are very sceptical about neighbourhood effects from the outset, yet they conclude by insisting that we need more (and more and more and more) research to see if they exist (via newer globs of geocoded data). So they keep on researching something they keep finding to be not very important! In much of this work there was a truly breathtaking disregard for, and even ignorance of, well over a century of theoretical advances in respect of how differential life chances are created in cities. There was almost no attention to the role of the state as a major determinant of life chances – in work that often boasts of policy relevance! At times there was near-evangelical belief in the 'concentrated poverty' thesis: the idea that when poor people are clustered together it leads to anything from a lack of positive role models to a school dropout culture, teenage pregnancy, crime, and poor employment outcomes, for example. (It is worth remembering that some of the most aggressive housing policies in the US and the UK that have involved demolishing people's homes in the name of deconcentrating poverty have seen academic input). But something like unequal educational attainment is more helpfully considered not as an offshoot of the way young people of different incomes behave, but as an offshoot of the unequal provision of public goods and unequal treatment by the state of the different areas. The degree of inequality between neighbourhoods with bad schools and good schools is not a property of the neighbourhood, but a property of the school system. So I ended that article calling for attention to the stigmatisation of poor neighbourhoods because I felt that scholarship in the neighbourhood effects genre was deeply implicated in that stigmatisation. Why was nobody researching 'concentrated affluence' in extremely rich neighbourhoods, and correspondingly, recommending policies aimed at dispersing the rich when their concentration may have caused such grievous collective disasters as the 2008 financial crisis?

Since the emergence of 'territorial stigmatisation' as a concept in the social sciences and Geography in particular, how would you characterise its main impacts? Has it changed the ways in which scholars research segregation and exclusion? What, if possible, would you say are the shortcomings of the work that has been done on territorial stigmatisation to date?

Geographers have actually been rather slow to embrace territorial stigmatisation as a concept, probably because they would prefer to start by thinking through what is actually meant by 'territorial', and because it's impossible to be a geographer without considering questions of scale. By contrast, sociologists, who have done the vast majority of work on territorial stigmatisation, prefer to think through what is meant by 'stigmatisation', and tend to use 'territory' as a synonym for whatever scale at which the stigmatisation is happening (street, neighbourhood, housing estate, etc). I think there are many fruitful, cross-disciplinary discussions still to be had – but the problem is that different social science disciplines do tend to stick to their own epistemological pumpkin patches (or comfort zones). It's been exciting for me to have many (ongoing) dialogues with sociologists all over the world about territorial stigmatisation – I've learned a great deal about the sociological craft from scholars like Alfredo Alietti, Virgilio Borges Pereira, Catharina Thörn, Imogen Tyler, and Justus Uitermark. My PhD students, and the many students whose PhDs I have examined, have also been incredibly helpful in introducing me to work that I would not otherwise have time to find and read.

The real giant in this field is of course Loïc Wacquant, who has been a strong supporter and enthusiastic mentor. Wacquant's work is important because it reminds us that failure to heed the role of symbolic structures in the production of inequality and marginality in the city means that neighbourhoods are made into the cause of poverty rather than the expression of underlying problems to be addressed. He also warns that scholars who deploy the trope of the 'ghetto' for rhetorical dramatisation in hopes of inciting progressive policy intervention actually contribute to the further symbolic degradation of dispossessed districts and thus to the very phenomenon that they should be dissecting. Viewed through Wacquant's urban sociological lenses, the state is not a bureaucratic monolith delivering uniform goods, nor an ambulance that comes to the rescue in response to 'market failure', but rather a potent stratifying and classifying agency that continually moulds social and physical space, and particularly the shape, recruitment, structure, and texture of lower-class districts. His conceptualisation of territorial stigmatisation is therefore a crucial point of departure for empirical research. However, there is one aspect of his conceptualisation that is troubling, and requires much more evidence. He argues that territorial stigma is closely tied to, but has become *partially autonomised* from, the stain of poverty, class position and subaltern ethnicity (encompassing national and regional 'minorities', recognised or not, and lower-class foreign migrants). Now, this may have been the case in the two sites (in Paris and Chicago) where Wacquant conducted his famous fieldwork that undergirded the arguments in his book *Urban Outcasts* (2008), but close scrutiny of a mushrooming body of work suggests that the links between the stigma attached to the negative branding of a place and the stigmata of poverty, class, ethnicity, crime etc. are very strong indeed, to the point of being interchangeable. In the numerous contexts that I have read and heard about, it is not the case that the bad name of a place has developed its own effects independently of the bad name of the people who live in it; rather, the two feed off each other to have major implications for residents of defamed districts. To take one example, I think it is impossible to sustain an argument that the 'sink estate' label in England only makes people think of degraded tracts of council housing; it has all sorts of very negative, disturbing associations with the class position of people living there. Furthermore, discourses of abjection, regardless of terminology, are usually classed, racialised and gendered *all at once*. So, I think there is a need for more systematic comparative inquiries into the production of both place stigma and 'people' stigma (if you will), and to trace and dissect links between them in different contexts.

It seems to be that the very idea of 'resistance' – used in the widest possible sense – to territorial stigmatisation takes little or no place in these conceptualisations. Those who would somehow develop the means of coping with this form of stigma are said to be somehow reproducing the territorial taint that is associated with their neighbourhood – through blaming others, trying to leave or actually escaping the neighbourhood, for instance. Do you see broader possibilities of resistance to territorial stigmatisation? Some of the work on territorial stigmatisation stresses the fact that structures of domination result in those living in tainted places somehow internalising the negative representations of the places in which they reside. A number of recent papers have suggested that this may not always be the case – and your article on territorial stigma in St Paul's, Bristol (2012) – is part of this emerging literature. Why are the nuances and ambivalences that these papers highlight perhaps quite important?

Whilst it may very well be the case that coping strategies or protest strategies do reproduce territorial stigma, I think scholarship must analyse these strategies in the detail they deserve. It is very easy for scholars who do not have to live in stigmatised places to make these sorts of conclusions, but there are remarkable activities underway that often achieve a great deal against the odds, whether they be individual acts of defiance or collective, organised claims on a place. These tactics may be considered a minor irrelevance to a sociologist, but a stigmatised existence is a brutal one and it is crucial to uncover how people take a stand, and what has, can, or might be achieved. One of the more depressing outcomes of efforts to fight stigma is that the more positive image of a place that unfolds then paves the way for gentrification, ultimately leading to the displacement of those who fought the stigma. This happens because the underlying socio-political interests constituting urban land markets usually remain unchallenged, due to the magnitude of that task. So the question of land ownership becomes utterly crucial – who is going to profit from the efforts of local people to paint a more positive image of a place? Therein lies another example of the strong links between territorial stigmatisation and gentrification. It would seem that collective efforts (especially) to fight back against social abjection have to build in some kind of booby trap further down the road, a defence mechanism against the speculators and developers waiting to pounce on the inevitable media rhetoric that accompanies a reversal in fortunes for a degraded neighbourhood. Collective ownership and commoning strategies are crucial in this regard, but even these can be hijacked by powerful interests. So, I do see possibilities for resistance (and you can learn about these in published scholarship), but the structural constraints are enormous.

In your understanding, how can it be possible that those living in structurally marginalised neighbourhoods – places that are also highly stigmatised – sometimes develop attachments to the place where they live?

Based on what I have read, and based on what I know from my own work, place attachment in such neighbourhoods is *usually* (not sometimes) very strong indeed. The 'concentrated poverty' that is so despised in scholarship, politics and journalism has considerable forgotten benefits. While a top-down discourse of isolation, disorganisation, and danger has long enshrouded certain areas, if one takes the trouble to look closely and especially to listen to people living in them, in spite of physical deterioration and nefarious repute, they value their dwelling place for anchoring dense webs of friendship and reciprocal support, proximate amenities, and local agencies providing services finely tailored to their needs. Martine August (2014) in Toronto has written brilliantly on this, focusing on what happened to the Regent Park housing project there. There is also a very substantial body of scholarship that demonstrates the loss, even grief, people feel after being involuntarily displaced from such districts. I get very tired of people who review my work and who seem to find place attachment in defamed areas inexplicable, and then offer barbs like, 'The author romanticises the working class', even when there is ample evidence to show very deep roots to place that have developed despite everything life has thrown at them there. I think the trick here is, as Elvin Wyly has pointed out so well, to 'let the data speak', let respondents tell us in their own words what their neighbourhoods mean to them despite the discourses of disgust that swirl around and through them. Usually, what you find is the inestimable importance of social networks, kinship, and meeting points for the urban working class, rooted in place. These are things worth fighting for, and it is not 'romantic' to do so.

When rioting occurs in stigmatised neighbourhoods, whether in the USA, France, Brazil or the UK, does this not somehow deny that these attachments exist for a part of the neighbourhood? Does it not actually tend to confirm the 'neighbourhood effects' literature?

The neighbourhood effects scholars are rarely concerned with how clustering and concentration of people leads to radical action in the form of urban uprising! But if you look at the staggering amount of anti-austerity uprisings across European cities since the financial crisis of 2008, what you tend to find is not that people are smashing up their own neighbourhoods, but targeting symbols of the state. Take the 2011 riots in urban England. These were in large part a reaction to the violation of civic claims to equality from young people with almost no political representation (a product of the collapse and disappearance of traditional workers' parties). It was ultimately because of this violation that state institutions such as the police and the school system – through their routine functioning – generated such massive rage and disappointment and were targeted for attack. This does not, of course, explain the widespread looting of stores that happened, but to a considerable degree it accounts for why so many young people took to the streets. The looting scenes of August 2011 made it all-too-easy for many across the political spectrum to make conclusions in the 'mindless violence' register, and even easier to forget two hugely important precedents for the riots. Anger among youth was still simmering following the significant student protests in late 2010 at the tripling of tuition fees in England. The first of these culminated in the storming of the Conservative Party headquarters in central London, but the two subsequent protests carried less symbolic power due to merciless (and still deeply controversial) 'kettling' by the police. The second precedent was the massive J30 (30 June 2011) public sector strike across Britain, involving in excess of 750,000 public sector workers. The cause of J30 was the raising of the retirement age from 60 to 66, and the replacement of final salary pension schemes with a career-average system. However, J30 also included calls for an end to freezes in remuneration, an end to exhausting working hours, and the reversal of the steady privatisation of almost every imaginable aspect of the sector. If we look closely at the systematic assault on the welfare state in Britain (massive public sector cuts were being mooted by politicians with great enthusiasm in 2011, which goes some way towards explaining the anger of that summer) we can move a step closer towards understanding the shared indignity and dishonour among young people (especially) who feel abandoned and betrayed – and it becomes difficult to avoid the conclusion that austerity is, quite simply, structural violence against the working class.

You now reside in Edinburgh, and you recently published an article along with Hamish Kallin which is based on research that you conducted there (2014). Have you been able to uncover attachments to place taking hold in some of the housing estate neighbourhoods there?

The fieldwork for that study was done by Hamish – I just added some conceptu-alisations and helped with the data analysis. What struck me from the encounters Hamish had was the extent to which the despised and feared neighbourhood of Craigmillar was a place where remarkable things had occurred for decades, where people (particularly young working class mothers) had been crafting commu-nity, solidarity and possibility against the odds – and to international acclaim (via the amazing achievements of the Craigmillar Festival Society, led by Helen Crummy). But none of this mattered to those policy elites and private sector lob-byists with major interests in land and real estate, who felt that too many poor people were clustered together. Craigmillar is an all-too-familiar story of a place that could have been improved for people living there being razed to the ground for people who were not.

Some would say that, no matter how much we stress the attachments that residents have for their otherwise stigmatised neighbourhoods, these are ultimately unlikely to have any true and lasting influence on the ways in which urban policy is modelled. Do you believe that targeting a specific neighbourhood or urban area with some specific spatial policy can have any positive effect? Is there, in your opinion, any chance to continue spatial targeting – in terms of housing or the provision of public services – without stigmatising a territory? Is it simply that these urban policies are imbued by a set of neoliberal discourses?

This is a pressing research question and a difficult one to answer! At first glance, area-based policies appear logical, as if one place does have many more social problems than another place, it seems a no-brainer to focus all the policy resources and energy upon it. But to do this would be to ignore all the structural violence visited upon people living in poverty. To make a genuine difference in people's lives, interventions are required on a massive scale in the form of a genuine living wage or basic income, in the provision of truly affordable, decom-modified housing, in healthcare divorced from business interests, in education that is universally outstanding. These are also area-based policies if you think about it, in that they would have a huge positive impact on the lives of people living in stigmatised neighbourhoods, but they are not seen as such. The usual response is that there simply isn't the money to do all these things, but this is nonsense – in the UK, the amount of money handed out in corporate benefits (via direct aid, subsidies and tax breaks) in the most recent financial year was enough to wipe out at a stroke this year's budget deficit. Sadly, the problem lies in the brain damage done by several decades of neoliberal ideology (bolstered by free-market think tanks and a right-wing stigmatising press producing ignorance) which cannot be erased quickly.

You have recently been travelling a great deal to discuss your research, including Chile. In light of the fact that a lot of the work on territorial stigma-tisation has been conducted in what one might loosely call the 'global North',

how is this addressed in regions that have perhaps received less scholarly attention – at least from scholars in Europe and the United States?

I think if we were ever to write a new edition of *Gentrification*, we would expand our geographical and epistemological lenses as the book (and *The Gentrification Reader* [2010] that followed it) is very Anglophone in scope and the most disturbing gentrification strategies today are in Latin America, the Middle East, South Asia and East Asia – and there is some fantastic scholarship emerging from those regions, which speaks to all sorts of larger theoretical questions about the pertinence of certain concepts and whether they are helpful or not in dissecting urban processes beyond where they were formed. However, I do wish that the debates between postcolonial/provincial and Marxist/political economic urban theorists were not so jumped-up (for want of a better description). It strikes me that both these theoretical traditions are hugely valuable, with a great deal to teach students of social science and the humanities, and what matters most is that scholars learn from each other in their efforts to critique the disturbing transformations roiling cities from London to Mumbai to Buenos Aires to Nairobi. Of course, I am very much working in the latter Marxist tradition, and I know that a recent piece I wrote, 'Planetary Rent Gaps' (2016), has annoyed some postcolonial theorists simply because I made the argument – drawing on available scholarship – that Neil Smith's rent gap theory is very useful in understanding urbanisation in parts of the Global South, and the fact that the theory was developed in the US in the 1970s is not a valid reason to 'unlearn' it in very different contexts. The challenge is to take it seriously, and if it turns out not to be useful, then not to use it. In terms of territorial stigmatisation, I would say the same – take the concept seriously, and it turns out to be useful in, say, Cairo, then use it. If not, then don't. It strikes me as anti-intellectual to write off a whole theory or concept for a whole region just because it isn't useful to one particular analyst working in one particular context! The anti-intellectualism is not one-way either: Marxist urban scholars are sometimes guilty of firing off ridiculous political accusations at those who prefer to look beyond flows of capital, class relations and state structures to understand problems in cities. I find these 'debates', if you can call them that, exhausting, but they are very much in vogue right now. The casualty, as ever, is actual research! As Jamie Peck (2015) has recently highlighted, very few people are actually doing the systematic comparative work that the 'new comparative urbanists' are calling for.

This is, of course, a rather difficult question but do you think that there is some way that territorial stigmatisation can truly be challenged successfully and, if so, how do you suggest this could be done?

One of the most important lessons I have learned in the last 20 years of research and in reading the research of others is that problems *in* the neighbourhood are not problems *of* the neighbourhood. Correspondingly, the tendency for people to recoil in horror at the name of a neighbourhood is indicative of the much wider problem of the inequalities produced by neoliberal capitalist urbanisation. This invites a much closer consideration of the changing parameters of class struggle under these conditions. On this issue, I think Imogen Tyler (2013) and Kirsteen Paton (2014) are writing some of the very best analyses right now. As they have demonstrated, a signature outcome of over three decades of neoliberal urbanism has been the production of a double movement of: (1) class decomposition (the growing class/generational splits among the middle classes, and the uncertain prospects facing the children of middle-class parents – all of which result in heightening disgust directed at people lower in the class structure due to fears of 'falling'); and (2) class recomposition (a growing cross-class consciousness of inequality which emerges through spatial/local struggles against what David Harvey refers to as 'speculative landed developer interests' in cities). But then there is a contradictory repertoire of protest tactics, where claims are made for space, centrality and housing as social *rights*, yet somewhat detached from the language and histories of *class* struggle. I do think that territorial stigmatisation can be challenged successfully, but only if that language and those histories are at the forefront of social movements, and at the centre of academic analyses. Quite simply, territorial stigmatisation is an expression of class inequality visited upon cities, and addressing its production and consequences requires a close consideration of class power, anchored by questions such as: urbanisation for whom, against whom, and who decides? And, who decides who decides?! This edited collection guides us in the right direction, and is a significant step forward.

Notes

1 Higher Education Funding Council for England.
2 See Cupples, 2015.

References

August, M. (2014), 'Challenging the rhetoric of stigmatization: The benefits of concentrated poverty in Toronto's Regent Park', *Environment and Planning A*, 46 (6): 1317–1333.
Bourdieu, P. (1984), *Distinction: A Social Critique of the Judgement of Taste*, Cambridge, MA: Harvard University Press.
Cupples, J. (2015), 'I am not here to satisfy you: The NSS and our institutional knickers.' Available online at: https://juliecupples.wordpress.com/2015/05/16/i-am-not-here-to-satisfy-you-the-nss-and-our-institutional-knickers/ (accessed 1 March 2016).
Kallin, H. and Slater, T. (2014), 'Activating territorial stigma: Gentrifying marginality on Edinburgh's periphery', *Environment and Planning A*, 46(6) pp. 1351–1368.
Lees, L., Slater, S. and Wyly, E. (2008), *Gentrification*, London: Routledge.
Lees, L., Slater, T. and Wyly, E. (2010), *The Gentrification Reader*, London: Routledge.
Paton, K. (2014), *Gentrification: A Working Class Perspective*, Basingstoke: Ashgate.
Peck, J. (2015) 'Cities beyond compare?', *Regional Studies*, 49(1), pp. 160–182.
Slater, T. (2004a), 'Municipally-managed gentrification in South Parkdale, Toronto', *The Canadian Geographer*, 48(3), pp. 303–325.

Slater, T. (2004b), 'North American gentrification? Revanchist and emancipatory perspectives explored', *Environment and Planning A*, 36(7), pp. 1191–1213.

Slater, T. (2006), 'The eviction of critical perspectives from gentrification research', *International Journal of Urban and Regional Research*, 30(4), pp. 737–757.

Slater, T. (2009), 'Missing Marcuse: On gentrification and displacement', *CITY: Analysis of urban trends, culture, theory, policy, action*, 13(2), pp. 292–311.

Slater, T. (2012), 'Rose Street and revolution: A tribute to Neil Smith (1954–2012).' Available online at: http://www.geos.ed.ac.uk/homes/tslater/tributetoNeilSmith.html (accessed 28 February 2016).

Slater, T. (2013), 'Your life chances affect where you live: A critique of the "cottage industry" of neighbourhood effects research', *International Journal of Urban and Regional Research*, 37(2), pp. 367–387.

Slater, T. and Anderson, N. (2012), 'The reputational ghetto: Territorial stigmatisation in St Paul's, Bristol', *Transactions of the Institute of British Geographers*, 37(4), pp. 530–546.

Tyler, I. (2013), *Revolting Subjects: Social Abjection and Resistance in Neoliberal Britain*, London: Zed Books.

Wacquant, L. (2008), *Urban Outcasts: A Comparative Sociology of Advanced Marginality*, Cambridge: Polity Press.

15 Conclusion

Tainted urban spaces at the intersection of urban planning, politics of identity and urban capitalism

Paul Kirkness and Andreas Tijé-Dra

This edited collection exposes a number of urban contexts in which territorial stigma unfolds as a result of negative place reputations. It underlines the wide range of ways in which territorial stigmatisation is constituted and dealt with by different local, political and economic actors. Through various – sometimes rather unconventional – case studies, covering the so-called 'Global South' as well as the 'Global North', the chapters trace the genealogies of tainted places, and simultaneously point to generalisable findings that contribute to the theorisation of the production and contestation of place-based stigma. Departing from debates revolving around the strengths and shortcomings of the analytic concept of territorial stigmatisation (see Wacquant *et al.*, 2014), the aim of the authors has been to update and shed light on topics that are now discussed with increasing frequency. These are threefold: 1) the different mechanisms that produce territorial stigma, as well as the distribution of strategies available to the actors involved in such constellations; 2) these have especially focused on how residents of tainted neighbourhoods often remain in tension between submission and resistance to territorial stigma; and 3) connected questions revolving around possible attachment to stigmatised places.

One of the book's contributions consists in expanding the focus on the symbolic consequences of territorial stigma, with case studies on the 'blemish of place'. While place-bound stigma is most often associated to (large) social housing estates and central or peripheral working-class dominated neighbourhoods within the city fabric (in fact, most of the works included in the book share this focus), several contributions herein shed light on different types of territorial stigmatisation. These have included constructions of stigma that essentialise entire cities while imposing a number of characteristic properties onto them. The properties in question in these examples no longer apply to specific neighbourhoods, but affect the whole population and territory of an urban unit. The word 'sin' in Vegas' popularised nickname (Sin City) has real consequences on the ways in which residents construct their understandings of home and belonging, as was demonstrated in Pascale Nédelec's chapter. Lucas Pohl explains that when the residents of Zwickau were confronted to the fact that their home town had been the home of a neo-Nazi group of terrorists, responsible for the murder of at least ten people, bombings, and bank robberies, they had to come to terms with the fact that their city was now tainted. These hegemonic 'taintings' occur as a result of 'district-wide deficiencies', that carry with them a number of serious consequences for residents, even though these often differ in kind to the consequences that are described in detail by Wacquant *et al.* (2014) and Slater (in press).

Being tainted, or (suddenly) isolated symbolically in spatial terms, turns out to be challenging for urban policy makers. In the examples from this volume, they react in different ways: alternately they try to profit from these deficiencies, or (successfully) develop modes of psychological repression. At best, this might result in the touristic exploitation of a symbolic 'extra-territoriality', but at the cost of 'average' residents, who then resort to strategies that are mostly submissive, including that of individual distancing from 'their' place of residence, both psychologically and geographically. At worst, in Lucas Pohl's example, a 'right-wing territory'-stigma can drive local politics and image policies into a deep symbolic crisis, provoking impulsive actions from the authorities in order to strip the stigma away without actually dealing with the existing problem of local Nazism, suppressing the issue for the sake of the city's neoliberal competitiveness.

A second relevant aspect of the dynamics that produce territorial stigma becomes apparent in the decisive role of urban capitalism that profits from powerful everyday 'hearsay', as well as media depictions of neighbourhoods, and the articulations made at the political level. It shows that speculative and evaluative practices of financial or real estate actors can strongly re-enforce existing urban geographies of stigmatisation by turning negative reputations into aspects that can eventually lead to them generating profits in the future. This can be done, as has been demonstrated in the case of Johannesburg, through the denial or restricting of access to financial products through redlining. This occurs when inhabitants of an area are collectively cut off from the loan sector, which results in furthering their already existing economic exclusion due to territorial taint. What Haferburg and Huchzermeyer show is that when residents possess sufficient social and economic capital, they use it to resist the stigma through a time-consuming and financially costly process in order to prove their credit-worthiness to institutions. This 'skill-intensive' resistance can obviously not be performed by all the inhabitants of an identified neighbourhood, but all the residents might ultimately benefit in the long run. Notwithstanding, redlining finance scoring systems as producers of urban tainting continue to be applied in post-apartheid South Africa, as well as in a number of other countries, where politically they are regarded as a minor concern.

The considerations of real estate actors form the other side of the coin in this context. They also point to the intersection of gentrification, urban development discourses and territorial stigma, that are addressed by a number of chapters in the book. As the British (Crookes and Kallin), French (Kirkness and Tijé-Dra) and Chinese (Zhang) case studies have made clear, and as the interview with Tom Slater has also impressively demonstrated, profiting from an ongoing tainting of place is inherent to the market logic of the urban real estate sector. If it is not successfully interrupted, the vicious circle of territorial stigmatisation leads to the decline of local housing property prices and, eventually, to public social housing property being reintegrated into the property market (neoliberal urbanists would no doubt say that it is thus being 'liberated'). This results in subjecting the mostly working-class neighbourhoods to the forces of the real estate market, which usually paves the way to speculation, demolitions, displacement and renewal programmes, intended to attract middle classes and relocate 'problematic' populations. While these attempts to fill the rent gap remain an inherent feature of urban capitalism, the contributions in this book also bring to light the converging interests of actors from the real estate and public (marketing) sector, as well as urban planners, when confronted with stigmatised neighbourhoods. By fostering territorial taint, municipal actors can legitimate large-scale interventions for the sake of a city's image, which is then said to have become 'enriched' or 'bettered' by gentrified neighbourhoods, urban growth, regeneration or 'social mix'. Through setting up the framework for the activation of these potentials and in order to fill a constructed 'reputational gap' – a concept that is coined by Hamish Kallin in his chapter – the tasks are increasingly delegated to private actors due to municipal regimes of austerity, or simply neoliberal approaches.

This should be viewed critically and it should be understood as a wake-up call for planners in order for them to attach more importance to the views and opinions of those who are stigmatised into consideration instead of those of the market. The case studies clearly underline the negative consequences of such modes of urban planning: rather than being solved through what is described as a necessary stage of modernisation, territorialised problems are merely displaced, as are the populations of the stigmatised neighbourhoods. Even when promises are made stating that for every one home that is demolished, one will be built elsewhere, this is often not the case and promises end up being broken (see Epstein, 2011). This comes as a result of insufficient financial means, and as a further consequence, unreplaceable local social networks are literally torn apart. The strong feelings of attachment of some inhabitants – accounted for by the many everyday or artistic articulations that can be found in the contributions from this book – are ignored. All these partly unintended consequences stand in sharp contrast to the 'participative' prose of the policy measures and of real integrative approaches.

Generally speaking, the fact that territorial stigma must not always be accepted as a depiction of the truth, and thus be internalised, leads us to the third main contribution of the collection: the nexus of place attachment and the possibility of resistant practices. Unlike hegemonic views on territorial stigma, which state that leaving one's stigmatised neighbourhood for good becomes desirable and that massive urbanistic interventions are the only realistic *tabula rasa* solution to existing problems in tainted places, several of the accounts presented in this book have pointed to other possibilities. By tracing cases of submission and resistance in such situations, territorial taint turns out to be Janus-faced, just like the role of its researchers, being far from neutral or impartial.

The people who are actually willing to leave their neighbourhood – and who are often evoked in hegemonic discourses – can be opposed to a significant number of residents, who tie manifold symbolic and material bonds to these places, regardless of the stigma that these endure elsewhere. As confirmed by the fruits of research conducted throughout the globe, dwellers of stigmatised places are at times perfectly capable of balancing the hardships and shortcomings of the situation in which they find themselves through a number of ways. Among other tactics, they can valorise the social networks and the relationships that they benefit from as well as develop further a familiar cosmos of proximity. This also applies to the built environment: whether it has been handed down to residents in the form of social housing; whether it is being squatted; or whether people have a stronger intimate relationship with it (as in the examples taken from Shanghai and Brasília in this book, where residents often built the homes in which they live themselves). The built environment forges, if not pride, at least a certain adherence to people's place of residence, especially when they are confronted to 'outsiders' or 'intruders', resulting in a disposition of resistive practices as the case studies from Botswana (Geiselhart), Brazil (Husseini de Araújo and Batista da Costa), China (Zhang) and Scotland (Kallin) have notably demonstrated.

As in the case of Italian Roma that is outlined by Gaja Maestri, it sometimes seems impossible to escape stigmatisation without paying the high price of concealing and rebuilding one's identity. This long existent, powerful stigmatisation even outweighs the option of an exit from the Roma camp through geographic mobility. But there are evidently several positive outcomes to resistive practices, enabled and empowered through local networks of proximity. These result in those living in stigmatised neighbourhoods being able to form a critical mass, or to develop enough social capital in order to collectively counter the situations in which they find themselves.

One crucial aspect of the contributions turns out to be both the possibility of resistive practices towards the tainting of places as well as the temporality of these actions. The examples illustrate the ways in which, notwithstanding their limited means, sometimes large coalitions of resisters can be forged. They unify through a shared, universalistic, and at times rather unspecific goal of emancipation, which allows them to generate a common identity. In some of the examples from the book, these coalitions have unfortunately proven to be temporary. In the course of time, state officials either were successful in co-opting the resistance of inhabitants politically or through monetary compensation. In some extreme examples, they simply enforced eviction. It also shows that emancipative achievements – whether these are partial or more significant – have resulted in threatening such coalitions, transforming the former universalistic goal into a lesser shared, particularistic one. These dynamic features of social struggles are discussed by Laclau and Mouffe (1985) who describe the antagonistic lines in social struggles that shift over time. In the book, this is most evident in the case of the Varjaõ, in Brasília, and this allows for more sophisticated explanations of the outcomes of political struggle in situations of territorial stigmatisation.

It is important to clarify these attachments and entanglements in order to gain a more complete understanding of the different responses to territorial stigma. This notably entails the researcher's role. On the one hand, it is necessary to reflect upon and to clarify the social scientific vocabulary. For instance, differentiating more rigorously between processes of stigmatisation and discrimination is argued to be a move in the right direction in Geiselhart's chapter. This allows us to better isolate the corresponding aspects of stigmatisation as a whole. Further, ten years after Liamputtong's (2006) useful invitation to research those who are vulnerable by being sensitive to their concerns and their fears, this book is a reminder that we need to keep focused on this objective. Several contributors advocate convincingly for innovative participative approaches to territorial taint, by cooperating with a number of inhabitants, who were themselves implicated in designing and helping to conduct the research. Unlike more conventional forms of analyses, these are less likely to impose the researchers' categories upon the researched. They help to avoid the reproduction of common stereotypes, but they also account for agency and ambivalent attitudes towards territorial stigma. This should not obscure the fact that there will always continue to be a certain distance – as well as subtle power relations – between the researched and the researcher.

It still is vital for the theorisation of the urban, and likewise the constructive evaluation of urban planning measures, to expand our view on territorial stigma by producing more case studies. They help to better understand the nuances of place attachment, negative reputation and to find applicable, affordable, and more respectful responses to it. This edited volume is a collective call for more sensitivity in dealing *with*, but also conducting research *on* those who live in stigmatised neighbourhoods, cities and regions. It is essential to scrutinise our 'approved' representations and practices towards territorial taint. Only then can we begin to imagine that the hierarchies between urban areas and cities that produce highly fragmented, imagined and exoticised communities cease to be of interest. For now, however, they should remain central to our academic and activist-oriented endeavours.

References

Epstein, R. (2011), 'Politiques de la ville: bilan et (absence de) perspectives', *Regards croisés sur l'économie,* 9(1): 203–211.

Liamputtong, P. (2006), *Researching the Vulnerable: A Guide to Sensitive Research Methods*, London: Sage.

Laclau, E. and Mouffe, C. (1985), *Hegemony and Socialist Strategy. Towards a radical democratic politics*, New York: Verso.

Slater, T. (in press), 'Territorial stigmatization: Symbolic defamation and the contemporary metropolis', in Hannigan, J. and Richards, G. (eds), *The New Handbook of Urban Studies*, London: Sage.

Wacquant, L., Slater, T. and Pereira, V. (2014), 'Territorial stigmatization in action', *Environment & Planning A*, 46 (6): 1270–1280.

Index